Progress in IS

Progress in IS encompasses the various areas of Information Systems in theory and practice, presenting cutting-edge advances in the field. It is aimed especially at researchers, doctoral students, and advanced practitioners. The series features both research monographs, edited volumes, and conference proceedings that make substantial contributions to our state of knowledge and handbooks and other edited volumes, in which a team of experts is organized by one or more leading authorities to write individual chapters on various aspects of the topic. Individual volumes in this series are supported by a minimum of two external reviews.

The Series is SCOPUS-indexed.

Ken Huang
Editor

Agentic AI

Theories and Practices

Springer

Editor
Ken Huang
DistributedApps.ai
Fairfax, VA, USA

ISSN 2196-8705 ISSN 2196-8713 (electronic)
Progress in IS
ISBN 978-3-031-90025-9 ISBN 978-3-031-90026-6 (eBook)
https://doi.org/10.1007/978-3-031-90026-6

This Springer imprint is published by the registered company Springer Nature Switzerland AG
The registered company address is: Gewerbestrasse 11, 6330 Cham, Switzerland

To the pioneers who dared to dream of intelligent, autonomous systems, your vision has illuminated the path toward agentic AI, where machines act not only as tools but as collaborative partners reshaping our understanding of intelligence and agency.
To the researchers and practitioners exploring the complexities of agentic AI, your relentless pursuit of knowledge and innovation drives this transformative field forward, bridging theoretical possibilities with real-world applications.
To the organizations and policymakers working to harness the power of agentic AI for societal good, your commitment to creating frameworks that promote fairness, security, and accountability ensures that this technology benefits humanity as a whole.
To my colleagues and collaborators in the AI security communities such as Cloud Security Alliance, NIST, OWASP, SANS, and ISO, your insights, challenges, and support have been instrumental in crafting this book. It is a reflection of the shared quest to demystify agentic AI and make it accessible to all who seek to understand and apply its principles.

Lastly, to the readers of this book—innovators, students, and visionaries—your curiosity and ambition are what will shape the future of agentic AI. May this book serve as a catalyst for your endeavors, inspiring you to explore the profound potential of intelligent agents and their role in a rapidly evolving world.

Foreword

The journey of artificial intelligence has been one of profound transformation, evolving from abstract theoretical constructs to tangible tools that now shape industries, economies, and everyday lives. As an AI SME, a Distinguished Engineer, White House Presidential Innovation Fellow and someone deeply entrenched in the intersection of AI, healthcare, and quantum computing, I have witnessed firsthand the sweeping impact of AI-driven systems and the paradigm shifts they enable. Yet, amidst this progress, there lies a need for clarity, structure, and guidance for understanding and harnessing these transformative tools effectively. This is precisely what makes *Agentic AI: Theories and Practice* a timely and invaluable resource.

In this book, the author provides an excellent exploration of AI agents, their architecture, and their application across a diverse array of domains. Beginning with foundational concepts in Chap. 1, readers are introduced to the genesis and evolution of AI agents, accompanied by an engaging taxonomy and a glimpse into real-world examples that contextualize their practical significance. The meticulous organization of the book ensures that both novice and seasoned AI enthusiasts will find value, whether seeking an introduction to multi-agent coordination or exploring advanced topics such as AI-enabled offensive and defensive security applications.

What sets this work apart is its balanced approach, seamlessly blending technical depth with practical insights. The chapters on AI agent tools, frameworks, and the seven-layer architecture offer a robust foundation for building and deploying AI agents effectively, while discussions on challenges provide a pragmatic view of the real-world obstacles developers and organizations may face. The exploration of multi-agent systems in Chap. 3 showcases the complexities of coordination, communication, and conflict resolution, while also highlighting the immense potential of these systems in solving global-scale problems.

Equally compelling is the book's attention to emerging domains. From AI agents in the agentic economy, where blockchain and tokenization redefine value exchange, to their role in reimagining workflows, offensive security, and healthcare, the breadth of application is truly remarkable. These chapters not only provide actionable insights but also inspire readers to envision possibilities beyond the current technological landscape.

The author's inclusion of sector-specific applications, such as in banking, insurance, and robotics, further emphasizes the universal relevance of AI agents. These industries, often regarded as traditional or slow to adopt innovation, are now poised to be revolutionized by agentic AI. Furthermore, the book's commitment to addressing safety and governance considerations in its final chapters underscores the importance of developing AI systems responsibly. As we march toward a future increasingly driven by autonomous systems, these discussions on ethical deployment and security are not just necessary; they are imperative.

In *Agentic AI: Theories and Practice*, the reader will find a comprehensive guide that bridges theory and practice, ensuring they are equipped to navigate this rapidly evolving field with confidence and purpose. This book is not just a reference but a call to action for innovators, researchers, and policymakers to collaborate and harness the transformative power of agentic AI responsibly and effectively.

Whether you are an AI practitioner, an entrepreneur seeking to integrate AI agents into your business, or a curious observer of this technological revolution, you will find this book to be a profound source of knowledge and inspiration. It is my honor to contribute this foreword and to commend the author for their exceptional effort in delivering a resource that will undoubtedly shape the future of agentic AI.

https://www.linkedin.com/in/versaggi/

Senior Leader in Artificial Intelligence Healthcare
and White House Presidential Innovation Fellow
in the Special AI cohort of '24
Washington, DC, USA

Matthew R. Versaggi

Foreword

The advent of generative AI has marked a profound shift in the way we perceive and interact with technology. Among the many facets of AI, the emergence of agentic AI systems has paved the way for a new paradigm, where autonomous, intelligent agents operate with increasing sophistication across diverse domains. As a mathematician who has explored both the theoretical and interdisciplinary applications of mathematics, I find this field deeply resonant with the evolving interplay between rigorous theory and practical innovation.

My own academic journey began at the University of Science and Technology of China and continued through my doctoral studies at Harvard University. Over the years, I have witnessed firsthand how advancements in mathematical models and computational techniques have laid the groundwork for the transformative technologies we see today. Mathematics has always been at the heart of artificial intelligence, serving as its structural backbone. The algorithms and models underpinning AI agents—be it optimization techniques, probabilistic reasoning, or machine learning—are all born of mathematical inquiry. As Vice-President for Institutional Advancement at the Hong Kong University of Science and Technology (HKUST), I also see the role of institutions as catalysts in bringing such innovations to life.

HKUST, with its emphasis on interdisciplinary research and global collaboration, embodies the spirit required to nurture breakthroughs in agentic AI.

"Agentic AI: Theories and Practices" is an ambitious and timely work that delves into the transformative world of AI agents. This book provides a comprehensive exploration of the field, from its theoretical genesis to its real-world implementations. Each chapter is meticulously structured to guide readers through foundational concepts, practical frameworks, and the implications of agentic AI across industries. For instance, the detailed discussions on multi-agent systems and their coordination, tokenized AI economies, and safety considerations are not only intellectually stimulating but also provide actionable insights for practitioners and policymakers alike.

One of the book's most commendable aspects is its thorough examination of the dual use of AI agents and their integration with generative AI models, illustrating how these technologies can be harnessed for significant progress while also being exploited for malicious purposes. Chapters such as "AI Agents in Offensive Security" and "AI Agent Safety and Security Considerations" emphasize the vigilance required to mitigate risks and prevent misuse. At the same time, sections on healthcare, robotics, and the agentic economy showcase how these technologies can be leveraged to revolutionize industries, solve complex challenges, and improve human life when applied responsibly.

The future of agentic AI is both exhilarating and complex. It calls for collaboration across disciplines, a commitment to ethical governance, and a shared vision for technological progress. Academic institutions, like HKUST, will play an instrumental role in fostering the research and development needed to ensure that AI agents serve humanity's best interests. As we stand on the cusp of this new era, works like this book are indispensable in equipping researchers, practitioners, and visionaries with the tools to navigate this rapidly evolving landscape.

I wholeheartedly recommend "Agentic AI: Theories and Practices" to anyone seeking to deepen their understanding of this fascinating field. Whether you are an academic, a technology professional, or a policymaker, this book provides a wealth of knowledge and a roadmap for engaging with agentic AI. It is an essential contribution to the discourse on how intelligent agents will shape our world.

Vice-President for Institutional Advancement, Yang Wang
Hong Kong University of Science and Technology
Victoria, Hong Kong

Preface

This book began as an exploration into the burgeoning field of AI agents, at a time when the world was only just beginning to grasp the potential of generative models. When I first put pen to paper in the middle of 2023, the frenzy surrounding generative AI agents that we see in early 2025 was but a distant glimmer on the horizon. Now, many believe 2025 will be remembered as the year of the Agent, and for good reasons.

The capabilities of agentic AI are proving to be truly transformative. Their ability to plan, utilize tools, take actions, reflect on their performance, and manage both short and long-term memory has unlocked a new era of intelligent automation. Crucially, multi-agent systems are extending the reach of foundation models by integrating external knowledge, mitigating hallucinations, and enhancing security. These advancements are not merely incremental; they represent a fundamental shift in how we interact with and leverage artificial intelligence.

Even if the capabilities of foundational models remain static at the level of OpenAI's o3 model, which is widely recognized as the state-of-the-art as of early 2025, AI agents offer a powerful pathway to mine the full potential of this intelligence. We are at the cusp of a new scaling law for AI, one that goes beyond the traditional focus on scaling training data, model size, and computational power—a paradigm that is beginning to show diminishing returns. This new scaling law prioritizes test-time compute, process-based reward functions, and improved synthetic data generation algorithms, improved training optimization as demonstrated by DeepSeek, all aimed at fundamentally enhancing model capabilities while reducing cost.

It's important to note that this evolution in foundational models does *not* diminish the importance of AI agents. In fact, it strengthens their role. Agents can dynamically leverage real-time external data sources, providing a crucial link between the abstract knowledge of foundational models and the dynamic complexities of the real world.

This book delves into both the theoretical underpinnings and practical applications of AI agents. We explore agentic tools, frameworks, the emerging multi-agent landscape, and the nascent agentic economy. We also examine how AI agents are poised to revolutionize a range of practical domains, from business workflows to cybersecurity, banking, insurance, healthcare, and robotics.

A critical thread woven throughout the book is the paramount importance of AI agent security and safety. We cannot afford to treat these concerns as afterthoughts. Chapter 12 is dedicated to this vital area, offering a deep dive into the challenges and considerations necessary to ensure the responsible development and deployment of these powerful technologies.

Due to the constraints of space, we were unable to cover every fascinating application of AI agents. For instance, the potential for AI agents to innovate education, or to revolutionize coding as seen in the emergence of tools like Cognition's Devin, Codium's Windsurf, and Cursor, deserve entire volumes in their own right. We have instead chosen to focus on areas where we possess deep expertise, allowing us to offer a more focused and insightful analysis.

The chapters that follow offer a comprehensive journey into the world of AI agents:

- **Chapter 1: The Genesis and Evolution of AI Agents**
- **Chapter 2: AI Agent Tools and Frameworks**
- **Chapter 3: AI Agent Ecosystem—Multi-agent Coordination**
- **Chapter 4: AI Agent Economics**
- **Chapter 5: AI Agents and Business Workflow**
- **Chapter 6: AI Agents in Offensive Security**
- **Chapter 7: AI Agents in Cyber Defense**
- **Chapter 8: AI Agents in Banking**
- **Chapter 9: AI Agents in Insurance**
- **Chapter 10: AI Agents in Healthcare Practices**
- **Chapter 11: AI Agents in Robotics**
- **Chapter 12: AI Agent Safety and Security Considerations**

You need to read this book because it's your guide to understanding and navigating the most significant technological shift since the rise of the internet: the emergence of AI agents. Here's why, in clear, direct terms:

1. **Stay Ahead of the Curve:** AI agents are moving from research labs to real-world applications at an unprecedented pace. This book gives you a crucial head start in understanding this rapidly evolving landscape, allowing you to anticipate changes, adapt your skills, and seize new opportunities.
2. **Make Informed Decisions:** Whether you're a business leader, a developer, an investor, or simply a curious individual, this book equips you with the knowledge to make informed decisions about AI agents. You'll understand their capabilities, limitations, and potential impact, enabling you to navigate the hype and make strategic choices.
3. **Gain a Competitive Edge:** Understanding AI agents will be a key differentiator in the near future. This book provides you with insights into how agents can be leveraged to improve efficiency, drive innovation, and create new value in various industries. This knowledge will give you a competitive edge in your field.

4. **Understand the Risks and Opportunities:** The book doesn't shy away from the potential risks associated with AI agents. It provides a balanced perspective, exploring both the immense opportunities and the ethical and security challenges. This understanding is crucial for responsible development and deployment of this powerful technology.
5. **Grasp the Economic Implications:** AI agents are poised to disrupt entire industries and create a new "agentic economy." This book helps you understand the economic forces at play, allowing you to anticipate shifts in the job market, business models, and value creation.
6. **Unlock the Potential of Foundation Models:** You'll learn how AI agents are the key to unlocking the full potential of even existing foundation models. This understanding will help you to maximize the value of AI investments and drive practical applications.
7. **Become a Part of the Conversation:** AI agents will increasingly shape our world. This book empowers you to become an informed participant in the crucial conversations about their ethical development, societal impact, and future trajectory.
8. **Go Beyond the Headlines:** If you want to understand the substance behind the AI agent hype, not just the sensational headlines, this book is for you. It offers in-depth analysis and insights you won't find anywhere else.

The field of AI agents is advancing at an extraordinary pace, with new innovations continuously reshaping their development and potential impact. In Agentic AI: Theories and Practices, we aim to lay a foundation for innovative thinking and strategic planning in this dynamic domain. However, we do not pretend to have all the answers. Instead, we embrace the complexity of the field by asking many open-ended questions—questions about AI safety and security, the influence of AI agents on economic theories, and their applications across diverse industries. These are challenges that remain unresolved and invite deeper exploration.

Our hope is that this book inspires you to think critically, approach problems creatively, and ask the kinds of questions that drive meaningful progress. As Elon Musk aptly put it, "The question is harder than the answer," referencing Douglas Adams' The Hitchhiker's Guide to the Galaxy to remind us that asking the right questions is often more challenging—and far more significant—than simply finding answers. We leave you with this challenge: let this book spark your curiosity and guide you in formulating the questions that will shape the future of agentic AI.

Fairfax, VA, USA Ken Huang

Short Recommendations

By Jim Reavis

Having spent 30+ years in cybersecurity, I found Agentic AI: Theories and Practices to be a refreshingly practical take on one of our industry's most pressing challenges. The book doesn't just theorize—it dives deep into the nuts and bolts of AI agent implementation, from technical architectures to the complexities of multi-agent systems. What really grabbed me was its no-nonsense approach to security and governance. As someone who's seen cybersecurity evolve from the trenches, I appreciate how this book tackles the hard questions about responsible Agentic AI deployment head-on. Whether you're in the research lab, the boardroom, or crafting policy, this is the kind of resource we desperately need to get AI security right

Jim Reavis, founder and CEO of Cloud Security Alliance

By Melan Xu

Agentic AI: Theories and Practice is an exceptional work that provides a comprehensive and practical guide to understanding the rapidly evolving field of AI agents. As the Executive Director of the World Digital Technology Academy and Secretary-General of the Cloud Security Alliance Greater China Region, I recognize the importance of bridging the gap between innovation and secure, ethical implementation.

This book stands out for its systematic approach, covering essential topics such as architecture, multi-agent systems, and their applications across industries. Its insights into governance and security considerations are particularly valuable in today's digital landscape, where trust and accountability are paramount.

I highly recommend this book to professionals, researchers, and policymakers seeking to harness the power of AI agents while navigating the complexities of our interconnected world.

Melan Xu
Executive Director, World Digital Technology Academy
Secretary-General, Cloud Security Alliance Greater China Region

By Professor Andy Chun

Ken Huang, a prolific author and renowned expert on AI and Web3, has once again delivered a masterpiece with his latest book, Agentic AI: Theories and Practice. This book fills a much-needed knowledge gap in the rapidly growing field of AI Agents. These agents represent the next generation in AI technology, poised to transform industries and daily life with their autonomous capabilities.

Agentic AI: Theories and Practice is an outstanding contribution to this evolving field, offering both depth and breadth in its exploration of AI agents. With over three decades of experience deploying AI systems to improve societal outcomes, I deeply appreciate the book's unique balance of theoretical rigor and real-world application.

Ken Huang provides a clear roadmap for understanding the architecture, coordination, and practical uses of AI agents across diverse industries. His insights into agentic AI's role in transforming workflows, enhancing security, and driving economic value resonate deeply with my belief in AI's potential to improve decision-making and quality of life.

I highly recommend this book as an essential guide for innovators, academics, and leaders navigating the future of AI. It's a must-read for anyone looking to stay ahead in the ever-evolving landscape of artificial intelligence.

Professor Andy Chun
Professor of Practice, The Hong Kong Polytechnic University
Distinguished Fellow, Hong Kong Computer Society

By Professor Yale Li

As the Executive Chairman of the World Digital Technology Academy and a Strategy Scientist with a global perspective on technological advancements, I am deeply impressed by the depth and breadth of *Agentic AI: Theories and Practice*. This book represents a masterful synthesis of theoretical frameworks and practical applications, guiding readers through the multifaceted world of AI agents with clarity and precision.

Its comprehensive exploration—from the architecture of AI agents to their transformative potential across industries—makes it an indispensable resource for professionals, academics, and policymakers alike. The emphasis on real-world case studies and forward-looking insights positions this work as both a reference and a roadmap for the future of agentic AI.

I highly recommend this book to anyone seeking to understand or shape the next chapter of AI-driven innovation.

Professor Yale Li
Executive Chairman, World Digital Technology Academy
Strategy Scientist, Foreign Academician of Ukrainian Academy of Engineering Sciences

By Sze Wong

It has been a privilege to review Chap. 1 of *The Genesis and Evolution of AI Agents*. Huang has taken a complex and rapidly evolving topic and presented it with clarity and simplicity, providing readers with a solid and accessible framework. This thoughtful overview serves as a strong foundation for grasping the deeper, more intricate concepts explored in later chapters.

As an investor in this space, I believe it's critical for more entrepreneurs to understand the transformative impact AI agents will have on industries and society at large. This book is an essential resource for anyone seeking to position themselves at the forefront of this groundbreaking field.

Sze Wong, Venture Investor, Tech Inventor, Zenith Venture Studio

By Chiru Bhavansikar

Chapter 3 is a must-read for any AI architect looking to design robust multi-agent systems for aligning inter-business processes. While it presumes some familiarity with AI and its frameworks, the material is well-structured and accessible to those prepared to delve into its intricacies. Even with nearly 2 years of experience working on AI agentic systems, I found the content both insightful and compelling. I can confidently attest that it provides a comprehensive and profound overview of the components necessary for building enterprise-grade AI agent systems.

Chiru Bhavansikar, Chief AI Officer, Arhasi Inc.

By Bhavin P. Kapadia

Ken Huang has once again written another comprehensive book—perfect for 2025. I was fortunate to be a reviewer for Chap. 8: *AI Agents in Banking*. Ken outlines various branches within the financial services industry that are being turned upside down by the advent of an Agentic revolution. Ken highlights specific examples where Banking is adapting to the new era of AI Agents. Readers will benefit from Ken's insights.

Bhavin P. Kapadia, Senior Advisor, AI Security Financial Services

By Akram Sheriff

Ken Huang excelled in authoring the book, especially Chap. 7 on AI Agents in Cyber Defense. According to my personal review, the chapter provides a comprehensive technical explanation, thoroughly detailing the roles and intricacies of AI agents in enhancing defensive security measures. Overall this book is a complete authority on the Topic of GenAI in Cyber Security and how different LLMs, AI agents can be used by both Offensive and defensive security Teams in the Industry to build different products.

Akram Sheriff
Senior Software Engineering Manager (GENAI, AI, Data Engineering)
Cisco Systems, San Jose, USA

By Aditya Rana

Chapter 6 covered the usage of agents for offensive security in a very detailed manner and highly recommended everyone to read it.

Aditya Rana, AquilaX.AI, Security Engineer, AI Red Team

By Dr. Tal Shapira

I had the great pleasure to review Chaps. 6, 7, and 12 which provide a comprehensive exploration of AI agents in offensive security, cyber defense, and safety considerations. Chapter 6 emphasizes the transformative potential of AI agents in offensive security, offering practical examples and code demonstrations for vulnerability discovery and adversarial simulations. Chapter 7 complements this with an in-depth look at defensive cybersecurity, focusing on multi-agent systems, incident response automation, and future trends in AI-driven defense strategies. Chapter 12 addresses the crucial topic of AI agent safety, highlighting vulnerabilities, mitigation strategies, and governance practices to ensure ethical and secure AI deployment. Together, these chapters offer valuable insights for researchers and practitioners, bridging technical innovation with the ethical imperatives of deploying AI in cybersecurity.

Dr. Tal Shapira, CTO and Co-founder, Reco AI

By Steve Wilson

The shift from creating simple chatbots to true agentic AI systems has substantial implications for development best-practices. Ken Huang, a visionary leader in advancing AI development methodologies, distills his expertise into this must-read book. Developers will discover valuable insights to elevate their thinking and build agentic AI systems that are not only capable but also secure and responsible.

Steve Wilson

Chief Product Officer, Exabeam and Project Lead for OWASP Top 10 for LLM and Gen AI applications

Acknowledgements

I am deeply grateful to the distinguished individuals whose contributions have shaped the direction and enriched the content of this book. Their expertise, insights, and guidance have been invaluable throughout this project.

I extend my sincere gratitude to **Matthew R. Versaggi**, White House Presidential Innovation Fellow and Senior Leader in AI Healthcare, for contributing the foreword. His profound understanding of AI's transformative role in healthcare and society provides a crucial perspective on the societal challenges and opportunities AI presents. Mr. Versaggi's experience in the intersection of AI, healthcare, and quantum computing significantly enhances the book's relevance, grounding it in real-world contexts that industry professionals and researchers will deeply appreciate.

Similarly, I express my deep appreciation to **Professor Yang Wang**, Vice President for Institutional Advancement at the Hong Kong University of Science and Technology (HKUST), for his insightful foreword. Professor Wang's esteemed academic background and keen understanding of the interplay between theoretical foundations and practical innovation in AI add valuable academic depth to the introduction of this book. His endorsement of interdisciplinary collaboration and the role of academic institutions in driving breakthroughs in agentic AI enriches the book's vision of the future of the field.

This book would not have been possible without the invaluable contributions of our brilliant co-authors, whose expertise and insights have shaped its content. Each chapter reflects the dedication and effort of these individuals, and I am deeply grateful for their collaboration.

I extend my heartfelt thanks to the following contributors:

- **Jerry Huang**, AI Engineer at Google, for his technical expertise and deep understanding of AI/ML applications, which enriched multiple chapters of this book.
- **Lisa JY Tan**, CEO of Economics Design, for her thought leadership in AI agent economics and her groundbreaking perspectives on this emerging field.
- **Chris Hughes**, Co-founder and CEO of Aquia, for his contributions to the chapters on AI in offensive security, cyber defense, and safety considerations.

- **Daniel Wu**, AI/ML Leader, for his insights into the role of AI agents in the banking industry.
- **Jyoti Ponnapalli**, Vice President and Head of AI at Frontier, for her expertise in the intersection of AI agents and financial practices.
- **Grace Huang**, Product Manager at PIMCO, for her valuable contributions to the exploration of AI in financial services.
- **Bhuvaneswari Selvadurai**, CISSP, CISM, and ISACA Board Member, for her exceptional insights into AI Agent applications in insurance.
- **Krystal Jackson**, Non-Resident Research Fellow with the Center for Long-Term Cybersecurity AI Security Initiative at UC Berkeley, for her thoughtful contributions to AI safety and security considerations.

Your expertise, creativity, and dedication have been instrumental in bringing this book to life. Thank you for your invaluable contributions to advancing the conversation on AI agents and their transformative potential.

In addition to the incredible contributions from the co-authors, I would like to extend my deepest gratitude to the following individuals for their invaluable feedback, thoughtful recommendations, and detailed reviews of the whole book. Their expertise and perspectives have greatly enhanced the quality and depth of the content:

Jim Reavis
CEO and Founder, Cloud Security Alliance

Your leadership in cloud security and AI governance has been instrumental in shaping the broader discourse on securing AI systems. Your insightful feedback and support have significantly contributed to strengthening the security and governance aspects of this book.

Professor Andy Chun
Professor of Practice, The Hong Kong Polytechnic University
Distinguished Fellow, Hong Kong Computer Society

Your thoughtful analysis and recognition of this book's contribution to the field of AI agents are profoundly appreciated. Your expertise in deploying AI systems to improve societal outcomes has provided meaningful validation of the content.

Professor Yale Li
Executive Chairman, World Digital Technology Academy
Strategy Scientist, Foreign Academician of Ukrainian Academy of Engineering Sciences

Your comprehensive review of the theoretical and practical aspects of AI agents, along with your global perspective, has been instrumental in refining this book into a resource that bridges theory and application.

Steve Wilson
Lead of OWASP Top 10 for LLM Applications

Thank you for your insights and recommendations on the security considerations for AI systems, which have added depth and clarity to the discussions on safeguarding agentic AI applications.

Melan Xu

Executive Director, World Digital Technology Academy

Secretary-General, Cloud Security Alliance Greater China Region

Your exceptional review and endorsement of this book, particularly highlighting its systematic approach to AI agents and governance considerations, have added immense credibility to this work.

In addition to the reviewers who reviewed the whole book, we have individual chapter reviewers who provided detailed feedback based on their domain of expertise. Their rigorous reviews and insights have significantly contributed to improving the quality, accuracy, and clarity of the chapters:

- **Sze Wong**, for your insightful review of *The Genesis and Evolution of AI Agents,* ensuring clarity and accuracy in establishing the foundational understanding of agentic AI.
- **Chiru Bhavansikar**, for your detailed review of *AI Agent Ecosystem—Multi-Agent Coordination,* which significantly strengthened the chapter's focus on multi-agent systems and their real-world applications.
- **Anat Bremler-Barr**, **Tal Shapira**, **Akram Sheriff**, **Govindaraj Palanisamy**, and **Aditya Rana**, for your meticulous reviews and expert insights on the chapters focused on cybersecurity and defense, ensuring their relevance and technical accuracy.
- **Bhavin P. Kapadia**, for your valuable insights while reviewing *AI Agents in Banking,* helping to refine its discussion on agentic AI in financial services.
- **Madhav Chablani**, for your thoughtful review of *AI Agents in Healthcare.*

Your specialized feedback has been instrumental in elevating the quality of this book, and I deeply appreciate your dedication and contributions.

I owe immense gratitude to the editorial team at Springer, especially Jialin Yan, Lara Glueck, and Sneha Arunagiri, for their exceptional dedication, patience, and support throughout the publication process. Their hard work and guidance were invaluable, and without their contributions, this book would not have been possible. I sincerely appreciate all their efforts.

By Ken Huang, Chief Editor of this book and CEO of DistributedApps.ai

Contents

About the Editor

Ken Huang is a prolific author and globally recognized authority in AI and Web3, with an extensive portfolio of published works that bridge business strategy, technical implementation, and cutting-edge research. As Co-Chair of the AI Safety Working Groups at the Cloud Security Alliance and the AI STR Working Group at the World Digital Technology Academy under the UN Framework, he is a leading voice in shaping global AI governance and security standards.

Huang is the CEO and Chief AI Officer (CAIO) of DistributedApps.ai, a firm specializing in generative AI training and consulting. His contributions to the field include being a core contributor to the OWASP Top 10 Risks for LLM Applications and an active participant in the NIST Generative AI Public Working Group.

Notable Publications:

- *Beyond AI: ChatGPT, Web3, and the Business Landscape of Tomorrow* (Springer, 2023)—Strategic insights into AI and Web3's business applications.
- *Generative AI Security: Theories and Practices* (Springer, 2024)—A comprehensive guide on securing generative AI systems.
- *Practical Guide for AI Engineers* (Volumes 1 and 2, DistributedApps.ai, 2024)—Essential resources for AI and ML engineers.
- *The Handbook for Chief AI Officers: Leading the AI Revolution in Business* (DistributedApps.ai, 2024)—A roadmap for CAIOs in organizations of all sizes.

- *Web3: Blockchain, the New Economy, and the Self-Sovereign Internet* (Cambridge University Press, 2024)—Insights into the convergence of AI, blockchain, IoT, and emerging technologies.
- *Blockchain and Web3: Building the Cryptocurrency, Privacy, and Security Foundations of the Metaverse* (Wiley, 2023)—Recognized as a must-read by TechTarget in 2023 and 2024.

Ken is a sought-after speaker and has presented at eventsw such as the World Economic Forum in Davos, ACM and IEEE conferences, the CSA AI Summit, Depository Trust & Clearing Corporation forums, and World Bank conferences. His recent appointment to the OpenAI Forum reflects his ongoing commitment to advancing collaboration and dialogue in the field of AI.

Explore Ken's work on Amazon: https://www.amazon.com/author/kenhuang

Chapter 1
The Genesis and Evolution of AI Agents

Ken Huang

A huge transformation is happening in artificial intelligence (AI), bringing us into a new era that will change how humans and machines interact and work together. Agentic AI or AI Agent is an advanced form of AI that goes beyond simple programmed responses. These AI Agents can act and think in ways that were once only possible in science fiction.

AI Agents are set to change the way businesses and society operate, offering new ways to create value and improve efficiency. They can handle complex tasks like data analysis and problem-solving on their own, allowing human workers to focus on bigger strategies and new ideas. This will not only make work more productive but also open up new opportunities for business growth and societal progress.

In the future, business workflows will seamlessly combine human skills with AI capabilities. Imagine AI Agents managing supply chains by themselves, optimizing inventory and logistics in real time. Think about customer service transformed by AI Agents that can understand and respond to detailed questions with a human-like touch. These improvements will streamline operations, lower costs, and enhance customer experiences across different industries.

The rise of AI Agents is also prompting a rethinking of how organizations are structured. Traditional top-down hierarchies might be replaced by more flexible, network-based models where AI Agents work alongside human teams on dynamic projects. This could lead to "AI-augmented organizations" where decision-making is shared between human experts and AI advisors, promoting agility and innovation.

The potential for creating value is enormous. AI Agents could lead to breakthroughs in scientific research, speed up drug discovery (see Chap. 10), and develop new solutions for global challenges like climate change. In finance, they could transform risk assessment and portfolio management (Huang et al., 2023) (see also

K. Huang (✉)
DistributedApps.ai, Fairfax, VA, USA
e-mail: ken@distributedapps.ai

K. Huang (ed.), *Agentic AI*, Progress in IS,
https://doi.org/10.1007/978-3-031-90026-6_1

Chap. 8), while in education, personalized AI tutors could adapt to each student's learning style and pace.

As we stand at the beginning of this AI-driven revolution, the possibilities are both exciting and profound. The age of the AI Agent promises to not only enhance human abilities but also redefine work, creativity, and problem-solving. It invites us to a future where the combination of human creativity and artificial intelligence unlocks incredible potential for progress and innovation.

1.1 Defining AI Agent

It will be very hard to give a correct and complete definition of AI Agent as the ideas and technology behind it are rapidly evolving. Nevertheless, we will still try to give a good enough definition of AI Agent in the book so readers can understand what we are talking about.

At its core, an AI Agent represents the apotheosis of machine intelligence, a digital entity imbued with the capacity to perceive, cogitate, and act with a degree of independence that borders on the anthropomorphic. Unlike their predecessors, which were tethered to predefined parameters and static knowledge bases, these avant-garde constructs possess an unparalleled ability to navigate the labyrinthine complexities of real-world scenarios, assimilating information and adapting their modus operandi with an almost organic fluidity.

The quintessential AI Agent is not merely a passive recipient of commands, but a proactive collaborator in the pursuit of knowledge and problem-solving. It is a synthetic cognoscente, capable of

1. Conducting autonomous epistemic forays across the vast digital noosphere
2. Employing sophisticated heuristics to dissect and resolve multitiered conundrums
3. Orchestrating and executing protracted strategies to achieve intricate objectives
4. Engaging in perpetual self-optimization through experiential learning

This quantum leap in artificial cognition represents more than just an incremental advancement; it heralds a seismic shift in the potential applications of machine intelligence across myriad domains of human endeavor.

To further elucidate the concept of AI Agents, let us delve deeper into their distinguishing characteristics and the technological underpinnings that facilitate their remarkable capabilities:

Autonomy and Initiative

At the heart of an AI Agent's functionality lies its ability to operate with a high degree of autonomy. Unlike traditional software programs that execute predefined instructions, AI Agents possess the capacity to initiate actions, make decisions, and pursue goals without constant human intervention. This autonomy is underpinned by sophisticated decision-making algorithms, often leveraging reinforcement learning techniques, which enable the agent to evaluate multiple courses of action and select the most appropriate one based on its understanding of the current context and desired outcomes.

Adaptability and Learning
One of the most striking features of advanced AI Agents is their ability to learn and adapt in real time. Through techniques such as deep learning and transfer learning, these agents can continuously refine their knowledge and skills based on new experiences and data. This adaptability extends beyond mere pattern recognition; it encompasses the ability to generalize learned concepts to novel situations, a trait that inches ever closer to human-like cognitive flexibility.

Multimodal Perception
AI Agents are equipped with the capability to process and interpret diverse forms of input, mirroring the multisensory input of biological organisms. This can include natural language processing for understanding text and speech, computer vision for interpreting visual data, and even more exotic forms of sensory input such as radar or infrared signals. The integration of these multiple input modalities allows AI Agents to form a comprehensive understanding of their environment, crucial for making informed decisions and actions.

Reasoning and Problem-Solving
Perhaps the most compelling aspect of AI Agents is their capacity for complex reasoning and problem-solving. Leveraging techniques from symbolic AI, probabilistic reasoning, and neural-symbolic integration, these agents can engage in sophisticated logical deduction, causal inference, and even creative problem-solving. This enables them to tackle challenges that require not just data processing, but genuine cognitive insight.

Social Intelligence and Collaboration
Advanced AI Agents are not designed to operate in isolation but are increasingly capable of engaging in meaningful collaboration, both with humans and other AI entities. This social intelligence manifests in their ability to understand and respond to human emotions, engage in natural language dialogue, and even participate in complex multi-agent systems where cooperation and negotiation are essential.

Ethical Reasoning and Value Alignment
As AI Agents become more autonomous and influential in decision-making processes, the importance of ethical reasoning and value alignment becomes paramount (Huang et al., 2024). Cutting-edge research in AI ethics and value learning aims to imbue these agents with the ability to reason about the moral and ethical implications of their actions, ensuring that their behaviors align with human values and societal norms.

Meta-Learning and Self-Improvement
The frontier of AI Agent development lies in the realm of meta-learning—the ability of an agent to improve its own learning algorithms. This concept of "learning to learn" promises to create AI systems that can rapidly adapt to new tasks and

environments, continuously enhancing their own cognitive architectures without explicit human reprogramming.

Explainability and Transparency
As AI Agents become more complex, the need for explainability and transparency in their decision-making processes grows. Advanced AI Agents are being developed with built-in mechanisms for providing clear rationales for their actions, allowing humans to understand and audit their behavior. This is crucial for building trust and ensuring accountability in AI systems.

Domain Agnosticism
While early AI systems were often specialized for specific domains, the new generation of AI Agents is characterized by their ability to operate across diverse fields of knowledge and application. These agents can seamlessly transfer skills and knowledge between domains, making them versatile tools for tackling a wide array of challenges.

Embodied Intelligence
The concept of AI Agents is not limited to disembodied software entities. Advances in robotics and the Internet of Things (IoT) are paving the way for AI Agents to have physical embodiments, capable of interacting with the physical world. This convergence of AI and robotics opens up new frontiers in areas such as autonomous vehicles, smart manufacturing, and personalized robotics assistants.

While many of these capabilities are realized to varying extents in current systems, some remain aspirational, requiring further advances in foundational AI research and computational technologies. The pace of innovation in this field is staggering, and as new breakthroughs continue to emerge, the gap between aspiration and feasibility narrows rapidly. What seems beyond reach today may soon become a standard feature, fueling the transformation of AI Agents into increasingly sophisticated and capable entities.

Moreover, the development of AI Agents is not occurring in a vacuum. It is deeply intertwined with advancements in other fields such as neuroscience, cognitive psychology, and philosophy of mind. As our understanding of human cognition deepens, it invariably influences and is influenced by our approach to creating artificial intelligences.

While a definitive and enduring definition of AI Agents may elude us due to the rapid evolution of the field, we can characterize them as highly autonomous, adaptive, and intelligent digital entities capable of perceiving, reasoning, learning, and acting in complex environments. They represent a paradigm shift in our approach to artificial intelligence, moving from narrow, task-specific systems to versatile, general-purpose cognitive architectures that promise to revolutionize countless aspects of our lives and work.

As we proceed through this book, we will explore in greater depth the myriad applications, implications, and challenges posed by these remarkable entities, always keeping in mind that our understanding and definition of AI Agents will continue to evolve alongside the technology itself.

1.2 The Historical Trajectory of AI Agents

To fully appreciate the revolutionary nature of contemporary AI Agents, it is a good idea to trace their lineage through the annals of computer science. The concept of autonomous artificial entities has roots stretching back to the nascent days of AI research.

Figure 1.1 is a timeline illustrating the major milestones in AI development, from the Dartmouth Conference to the present AI renaissance.

The Dartmouth Conference of 1956

The "AI" terminology first appeared in the Dartmouth College's conference in 1956 which was organized by John McCarthy, Marvin Minsky, Nathaniel Rochester, and

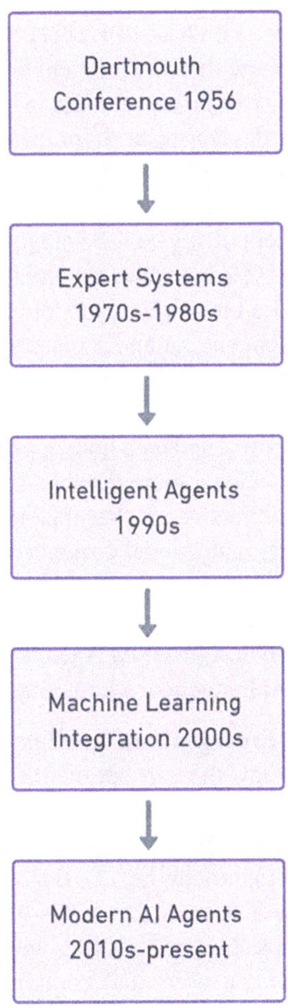

Fig. 1.1 The historical trajectory of AI Agents

Claude Shannon, and a proposal was published during the conference (McCarthy et al., 2006).

The proposal aimed to explore the potential of making machines simulate aspects of human intelligence. They proposed that a 2-month, ten-man study be carried out during the summer of 1956 at Dartmouth College to investigate various aspects of AI, including neural networks, theory of computation, and learning. The goal was to make significant advances in understanding and simulating intelligent behavior in machines.

The Dartmouth proposal did not achieve wider influence beyond academia initially due to several factors: technical limitations of the era, limited public and industrial understanding of AI, a mismatch between ambitious expectations and actual capabilities, and a primary focus on theoretical research rather than practical applications. Despite these challenges, the foundational ideas and research it inspired have profoundly influenced the development and growth of AI, leading to significant advancements and practical applications in later years.

1970s–1980s: The Rise of Expert Systems

This period saw the development of AI systems that could replicate the decision-making ability of human experts in specific domains.

MYCIN, developed at Stanford University in the early 1970s, was one of the most famous expert systems (Shortliffe, 2012). It was designed to identify bacteria causing severe infections and recommend antibiotics.

The concept of "agents" began to take shape, with researchers exploring ideas of autonomous software entities. Notably, Carl Hewitt's Actor Model (Hewitt et al., 1973) proposed a universal model of concurrent computation. The paper proposed that all components of an AI system act as actors—active, message-passing agents that communicate with one another. By unifying data structures, functions, processes, and databases under this single paradigm, formalism emphasizes the inseparability of control and data flow. It avoids presupposing specific primitive structures, aiming to provide a more flexible, efficient, and uniform way of designing AI systems. The authors argue that this approach offers advantages in terms of foundational clarity, educational benefits, enhanced modularity, and system uniformity, potentially simplifying the development and understanding of complex AI systems while representing a significant shift in AI architectural thinking. We can see that some modern designs of AI Agent have some similar approaches such as Microsoft AutoGen and Langchain's Langgraph.

1990s: The Emergence of Intelligent Agents

This decade saw the concept of software agents gain prominence, particularly with the growth of the Internet.

Pattie Maes at MIT Media Lab was a pioneer in the field of software agents. Her work on autonomous agents that could act on behalf of users was influential (Encarnacao & Rabaey, 2013, 30–40).

One of Maes' key innovations was the development of interface agents that could mediate between users and computer applications, helping to simplify complex tasks and personalize user experiences. Her work also contributed to the development of collaborative filtering techniques, which became fundamental to recommendation systems used widely today in e-commerce and content platforms.

Maes applied principles from artificial life to create self-organizing systems of agents, exploring how complex behaviors could emerge from simple rules. She researched how networks of agents could share information and learn from each other, laying groundwork for social recommendation systems. Her work significantly influenced the field of human-computer interaction, proposing new paradigms for how users could interact with increasingly intelligent and autonomous software systems.

The practical applications of Maes' theories extended to e-commerce, where she developed systems that could assist users in finding products and making purchasing decisions. The impact of her work went beyond academic circles, influencing the development of personalized web experiences, intelligent user interfaces, and recommendation systems. Many of the concepts she pioneered have become integral to modern digital experiences, from smartphone assistants to online shopping platforms.

2000s: The Integration of Machine Learning

This period saw significant advancements in machine learning techniques being incorporated into agent architectures.

Reinforcement learning, a type of machine learning where an agent learns to make decisions by taking actions in an environment to maximize a reward, gained prominence. The publication of "Reinforcement Learning: An Introduction" by Richard S. Sutton and Andrew G. Barto in 1998 (with a second edition in 2018) was influential in this field.

The concept of "embodied agents" or virtual assistants started to gain traction. Projects like the DARPA-funded CALO (Cognitive Assistant that Learns and Organizes) laid the groundwork for later virtual assistants like Siri (Myers, 2007).

2010s–Present: The AI Agent Renaissance

This current era has seen unprecedented advancements in AI capabilities, particularly in language understanding and generation.

Breakthroughs in deep learning, particularly the development of convolutional neural networks (CNNs) and recurrent neural networks (RNNs), have dramatically enhanced the capabilities of AI Agents. The success of deep learning in image recognition tasks, exemplified by the performance of AlexNet in the 2012 ImageNet competition, marked a turning point (Krizhevsky et al., 2012).

The advent of large language models and transformer architectures has enabled agents with unprecedented language understanding and generation abilities. The introduction of the transformer architecture in the paper "Attention Is All You Need" by Vaswani et al. in 2017 was a pivotal moment (Vaswani & Shazeer, 2017).

Recent large language models like GPT-4, Claud 3, and Gemini and their successors have pushed the boundaries of what's possible in natural language processing and generation.

OpenAI's framework (Metz & Mochizuki, 2024) for classifying progress toward artificial general intelligence (AGI) outlines five distinct levels, illustrating the evolution of AI Agents from narrow task-specific applications to highly autonomous systems with general intelligence. The journey begins at Level 1, where current chatbots like ChatGPT operate within predefined scopes, and progresses through Level 2, where AI can solve complex problems comparable to those tackled by Ph.D. holders. As AI Agents evolve, they reach Level 3, gaining the ability to act autonomously on behalf of users across various contexts; notably, OpenAI considers this level to represent true AI agency. The progression continues to Level 4, where AI demonstrates creative capabilities, generating original ideas and innovations independently. Finally, at Level 5, AI Agents achieve organizational-scale functionality, managing complex workflows and making strategic decisions autonomously. This framework provides a roadmap for understanding the potential trajectory of AI Agent evolution, showcasing how these systems may eventually surpass human capabilities in numerous domains. As AI researchers and developers work toward these milestones, each level represents a significant leap in the cognitive and functional abilities of AI Agents, potentially reshaping the landscape of human-AI interaction and the role of artificial intelligence in society.

1.3 Taxonomy of AI Agents

As the field of AI has evolved, so too has the diversity of agent architectures. Contemporary AI Agents can be broadly categorized into several distinct archetypes, each with its own strengths, limitations, and optimal use cases. This taxonomy not only helps in understanding the current landscape of AI but also provides a framework for envisioning future developments in the field.

Figure 1.2 is a hierarchical mind map categorizing reactive, deliberative, hybrid, learning, cognitive, collaborative, competitive, and domain-specific AI Agents.

1.3.1 Reactive Agents

Reactive agents represent the simplest form of AI architecture, operating on a straightforward stimulus-response paradigm. These agents lack internal representations of their environment or worldview, relying instead on a set of predefined rules to map inputs directly to actions.

Key characteristics of reactive agents include rapid response times due to minimal processing between input and action; high efficiency in well-defined, stable environments; and limited ability to learn or adapt to new situations. Reactive agents excel in rapidly changing situations where quick decision-making is paramount. Their simplicity allows for extremely fast processing, making them ideal for scenarios where even milliseconds of delay could be critical.

Common use cases for reactive agents include real-time control systems in industrial settings, high-frequency trading algorithms in financial markets, and

Fig. 1.2 Taxonomy of AI Agents

basic obstacle avoidance in robotics. However, these agents have significant limitations. They are unable to improve performance through experience, lack long-term planning or strategic thinking capabilities, and may exhibit suboptimal behavior in complex or novel situations.

1.3.2 Deliberative Agents

In contrast to reactive agents, deliberative agents possess internal world models that allow them to reason about their environment and plan future actions. These agents typically employ symbolic AI techniques to represent knowledge and use logical inference to make decisions.

Key characteristics of deliberative agents include the ability to create and execute plans to achieve long-term goals, the capacity for complex reasoning and problem-solving, and the ability to handle uncertainty and incomplete information through probabilistic reasoning. Deliberative agents excel in complex, strategic tasks that require foresight and planning. Their ability to model and reason about their environment makes them particularly suited for scenarios where long-term optimization is crucial. These kinds of agents are still under research, and as of the time of this writing, no such agents exist yet.

Use cases for deliberative agents include strategic planning in business and military applications, logistics optimization in supply chain management, and advanced game-playing AI, such as chess engines. However, these agents also have limitations. They typically have higher computational overhead compared to reactive agents and may suffer from "analysis paralysis" in rapidly changing environments, and their effectiveness depends heavily on the accuracy of their internal world model.

1.3.3 Hybrid Agents

Recognizing the complementary strengths of reactive and deliberative architectures, hybrid agents aim to combine the best of both worlds. These agents typically feature a layered architecture, with lower layers handling reactive behaviors and upper layers managing deliberative planning.

Key characteristics of hybrid agents include a balance between rapid response and long-term planning, the ability to switch between reactive and deliberative modes based on situational demands, and often the incorporation of learning mechanisms to improve performance over time. Hybrid agents are versatile and adaptable to a wide range of scenarios, making them particularly valuable in complex, dynamic environments where both quick reactions and strategic thinking are necessary.

Use cases for hybrid agents include autonomous vehicles navigating in urban environments; robotic systems operating in dynamic, real-world settings; and intelligent personal assistants balancing immediate queries with long-term user goals. The main challenges with hybrid agents lie in their increased complexity of design and implementation, the potential for conflicts between reactive and deliberative components, and difficulties in optimizing the balance between different behavioral modes.

1.3.4 Learning Agents

Learning agents represent a paradigm shift in AI design, focusing on the ability to improve performance over time through experience and feedback. These agents typically employ machine learning techniques, ranging from simple statistical models to advanced deep learning architectures.

Key characteristics of learning agents include the capacity to adapt to changing environments and tasks, the ability to generalize from past experiences to handle novel situations, and often requiring a training phase before deployment but can continue learning during operation. Learning agents are particularly valuable in domains with evolving or uncertain world models, where pre-programmed rules or static knowledge bases would quickly become obsolete.

Use cases for learning agents include personalized recommendation systems in e-commerce and content platforms, adaptive control systems in manufacturing and process industries, and predictive maintenance systems in industrial settings. However, learning agents also face challenges such as dependence on the quality and quantity of training data, the potential for biased or unexpected behavior if trained on skewed datasets, and difficulties in ensuring consistent and explainable decision-making.

1.3.5 Cognitive Agents

Representing the cutting edge of AI research, cognitive agents attempt to emulate human-like reasoning and problem-solving capabilities. These agents often incorporate advanced natural language processing, knowledge representation, and reasoning techniques, aiming to achieve a level of general intelligence that can be applied across diverse domains.

Key characteristics of cognitive agents include the ability to understand and generate natural language at a high level, the capacity for abstract reasoning and conceptual learning, and often the incorporation of models of human cognitive processes. Cognitive agents are at the forefront of efforts to create more general and flexible AI systems, capable of handling a wide range of tasks with human-like adaptability and insight.

Use cases for cognitive agents include advanced virtual assistants capable of complex dialogue and task completion, research and discovery systems in scientific and medical fields, and creative AI systems for content generation and artistic collaboration. However, cognitive agents face significant challenges including high computational requirements, difficulties in achieving true general intelligence, and ethical concerns regarding the development of human-like AI.

1.3.6 Collaborative Agents

Collaborative agents are designed to work together in multi-agent systems, cooperating to solve problems that are beyond the capabilities of any single agent. These agents must not only be capable of performing their individual tasks but also of coordinating their actions with others, sharing information, and adapting to the collective behavior of the group.

Key characteristics of collaborative agents include the ability to communicate and share information with other agents, the capacity for distributed problem-solving and decision-making, and often the incorporation of negotiation and consensus-building mechanisms. Collaborative agents are particularly valuable in scenarios where complex tasks can be decomposed into subtasks that can be tackled by specialized agents working in concert.

Use cases for collaborative agents include swarm robotics for exploration or search-and-rescue missions, distributed sensor networks for environmental monitoring, and collaborative filtering in large-scale recommendation systems. Challenges in implementing collaborative agents include complexity in coordinating multiple agents, the potential for emergent behaviors that are difficult to predict or control, and ensuring efficient communication and resource allocation.

1.3.7 Competitive or Adversarial Agents

In contrast to collaborative agents, competitive or adversarial agents are designed to operate in environments where multiple agents have conflicting goals. These agents must not only pursue their own objectives but also anticipate and counter the actions of other agents working against them.

Key characteristics of competitive agents include the ability to model and predict the behavior of opposing agents, incorporation of game theory and strategic decision-making, and often employing techniques from reinforcement learning and adversarial training. Competitive agents are crucial in scenarios where multiple stakeholders with divergent interests interact or in security applications where systems must defend against intelligent adversaries.

Use cases for competitive agents include cybersecurity systems for detecting and countering advanced threats, automated trading agents in competitive markets, and game-playing AI in complex, multiplayer games. Limitations of competitive agents include the potential for escalating adversarial behaviors, challenges in ensuring ethical behavior in competitive scenarios, and difficulty in achieving stable and predictable outcomes in multi-agent systems.

1.3.8 Vertical Agent or Domain-Specific Agents

While many AI research efforts focus on creating general-purpose agents, there is also significant value in developing highly specialized agents tailored to specific domains or tasks.

Unlike general-purpose AI, which strives to handle a broad spectrum of tasks, these specialized Vertical Agents concentrate on achieving exceptional performance within their specific domain. They accomplish this through the integration of deep domain knowledge, the application of specialized algorithms, and the utilization of tailored resources. Their primary focus is on maximizing accuracy and efficiency within their established boundaries, often prioritizing these attributes over the adaptability required to handle novel situations. They are, fundamentally, constructed to function as experts within a particular field.

A defining characteristic of Vertical Agents is their highly optimized performance for a specific task. This optimization is achieved in diverse ways, including the application of specialized algorithms finely tuned for their designated problem, even if those algorithms are inapplicable beyond that particular context. For example, a chess-playing Vertical Agent might employ minimax search algorithms augmented with unique pruning techniques. Such optimization often necessitates fine-tuning model parameters through extensive training on pertinent datasets, ensuring the model is perfectly calibrated for its singular purpose. Additionally, the architecture of the Vertical Agent is expressly designed with a clear emphasis on maximizing critical performance metrics within its domain, whether it be speed, accuracy, or other defined criteria.

Complementing this optimization, Vertical Agents feature a deep integration of domain-specific knowledge and heuristics. They do not simply learn from data; they frequently incorporate explicit domain knowledge in the form of rules, ontologies, or knowledge graphs, embedding a structured understanding of their field. These Vertical Agents may also utilize established heuristics and problem-solving strategies mirroring those used by domain experts, demonstrating the ability to interpret data within the specific context of their expertise. This convergence of data-driven learning and expert knowledge is critical to their superior performance.

The implementation of Vertical Agents often entails the use of specialized hardware or software to attain the requisite performance levels. This may include the employment of custom hardware, such as GPUs or ASICs, which are particularly well suited to the computational requirements of their domain, as well as the utilization of optimized software libraries. They may also require real-time capabilities, demanding additional optimization to ensure rapid responses. Consequently, they are carefully designed to meet the unique needs of their assigned task, rendering them exceptionally efficient within their designated sphere.

Numerous applications highlight the utility of Vertical Agents. In the medical field, specialized diagnostic systems concentrate on individual diseases, analyzing medical images, for example, with more precision than a human expert. In weather forecasting, models may focus on tracking specific phenomena such as severe storms. In the world of gaming, Vertical Agents can surpass human capabilities in specific games. Financial algorithms can execute high-frequency trades or assess risk, and robotics in manufacturing can be controlled with high precision on the production line, all within their specific domains. However, the strength of Vertical Agents also defines their weakness; they are inherently limited in their capacity to generalize to tasks outside of their specialization. A Vertical Agent designed for expert cancer diagnosis would not be suitable for weather prediction or playing chess, highlighting the trade-off between expertise and flexibility.

The advancement of Vertical Agents emphasizes the power of focused expertise and will likely have a significant impact on the future of AI applications, demonstrating the importance of specialization in certain tasks.

1.4 Technological Drivers of the AI Agent Renaissance

The notion of autonomous artificial entities is not novel, but the confluence of several technological breakthroughs has catalyzed a perfect storm of innovation, propelling us toward the realization of truly intelligent agents.

Figure 1.3 summarizes some of these key technology drivers or enablers of modern AI Agents.

1. Unprecedented Computational Power: The exponential growth in processing capabilities, coupled with the advent of specialized AI hardware like GPUs, TPUs, and neuromorphic chips, has shattered previous limitations on machine learning and deep learning capabilities. Recent advancements such as NVIDIA's Hopper architecture and Google's TPU v4 demonstrate significant leaps in AI processing power.
2. Advancements in Natural Language Processing (NLP): Breakthroughs such as GPT-4 and Claud 3 have endowed machines with an almost human-like ability to comprehend and generate text, bridging the semantic gap between silicon- and carbon-based intelligences. These models have significantly improved contextual understanding and conversational abilities.
3. The Data Deluge: The proliferation of big data, alongside advanced data analytics, provides an inexhaustible wellspring of information for AI systems to learn from, enabling them to develop nuanced understandings of complex phenomena. The integration of AI with IoT (Internet of Things) devices has further expanded data availability.
4. Algorithmic Innovations: Novel approaches in machine learning, such as reinforcement learning, neural architecture search (NAS), and transformer architectures, have dramatically enhanced the adaptability and efficiency of AI systems. Innovations like AlphaZero and self-supervised learning techniques have pushed the boundaries of AI capabilities.
5. Interdisciplinary Convergence: The synthesis of insights from cognitive science, neurobiology, and computer science has led to more sophisticated models of artificial cognition. Developments in neuromorphic computing and brain-

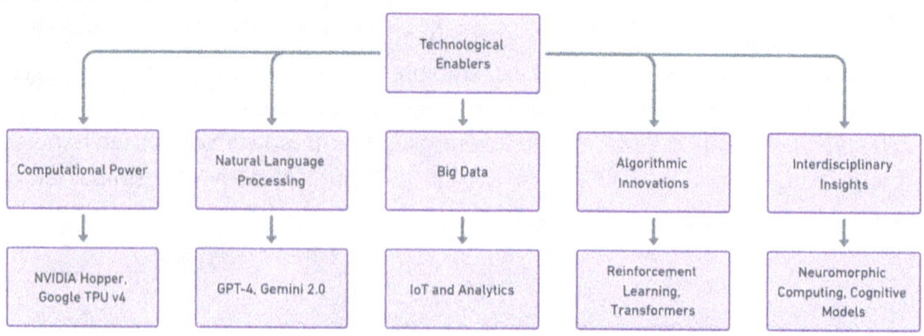

Fig. 1.3 Enablers of modern AI Agents

inspired algorithms, such as Intel's Loihi 2 chip, highlight this interdisciplinary progress.

This unique confluence of technological advancements has engendered AI Agents capable of feats that were once the exclusive purview of human intellect.

1.5 Example Projects on AI Agent

At the forefront of this AI renaissance stand two notable players of innovation: OpenAI and Stanford University. Their work in the sphere of AI reasoning and AI Agentic capability offers a tantalizing glimpse into the transformative potential of this technology.

OpenAI's Project "Operator" Agent
OpenAI, a leader in AI research and development, is pushing the boundaries with its "Operator Agent" project. The project uses "Chain of Thought" (see Box below) reasoning for improved reasoning at the model level.

Box: Chain of Thought (CoT)
In the context of AI Agents, chain-of-thought (CoT) reasoning represents a transformative approach to enhancing artificial intelligence's problem-solving capabilities by mimicking human-like reasoning processes. AI Agents using chain of thought break down complex tasks into a sequential, logical progression of intermediate steps, allowing them to tackle intricate problems more systematically and transparently. This approach enables AI to not just generate answers, but to show its reasoning path, making its decision-making process more interpretable and trustworthy. By decomposing problems into granular reasoning stages, AI Agents can handle more nuanced and complex queries, reducing errors and providing clearer insights into how they arrive at specific conclusions. For example, in tasks like mathematical reasoning, language understanding, or strategic planning, the agent will verbalize each step of its thought process, creating a detailed narrative that explains its logic, much like a human would think through a challenging problem out loud.

"Operator" agent released in January 2025 has the following capabilities (Lumb, 2025):

1. Autonomous Internet Navigation: OpenAI Operator is designed to independently explore the web, formulating complex queries and synthesizing information from multiple sources. This capability goes beyond simple web scraping or keyword searches. The Operator can understand context, follow chains of reasoning across multiple web pages, and even interpret and synthesize conflicting

information. This level of autonomous navigation could revolutionize how we interact with and extract value from the vast repository of human knowledge that is the Internet.

2. Deep Research Capabilities: The system can conduct what OpenAI terms "deep research," going beyond simple information retrieval to draw insightful conclusions. This involves not just finding relevant information, but also understanding it in context, comparing and contrasting different sources, and generating novel insights. Operator Agent's deep research capabilities could potentially accelerate scientific discovery, enhance strategic decision-making in business and policy, and even contribute to solving complex global challenges.
3. Long-Horizon Tasks: The Operator Agent aims to plan and execute multistep actions over extended periods, tackling complex problems that require strategic thinking. This ability to maintain focus and coherence over long time horizons is a key aspect of human-like intelligence. It allows for the pursuit of complex goals that may require days, weeks, or even months of sustained effort and planning. This capability could be applied to everything from long-term financial planning to complex project management in fields like engineering or scientific research.
4. Advanced Post-Training Methods: The project utilizes innovative "fine-tuning" techniques to adapt base models, enhancing their performance through targeted feedback and examples. This approach allows the Operator Agent to continually improve its performance and adapt to new domains or tasks without requiring a complete retraining of the base model. It's a step toward more flexible and adaptable AI systems that can learn and improve "on the job," much like humans do.
5. Potential for Human-Level Reasoning: Early demonstrations have shown promise in solving advanced science and math problems. This level of performance on complex, abstract reasoning tasks is particularly impressive and suggests that the Operator Agent may be approaching human-level capabilities in certain domains. If this performance can be generalized across multiple fields, it could lead to AI systems capable of contributing to advanced research and problem-solving in ways previously thought to be the exclusive domain of human experts.
6. Parallel Operation of Multiple Agents: Multiple Operator Agents can be invoked by humans and run in parallel. This parallelism can significantly enhance productivity by reducing the time required for task completion, particularly in environments with high workloads or diverse demands. It also creates opportunities for value creation by allowing organizations to scale operations efficiently, address multiple challenges simultaneously, and innovate at a faster pace.

OpenAI's vision for Operator Agents extends beyond academic problem-solving. They're exploring how this technology could autonomously perform tasks currently requiring human expertise, such as software development, scientific research, and machine learning engineering. This could lead to what is called a "super agent" or Ph.D.-level agent (Axios, 2025).

Stanford's Self-Taught Reasoner (STaR)
While OpenAI focuses on building comprehensive AI Agents, Stanford University has made significant strides in enhancing AI reasoning capabilities through its Self-Taught Reasoner (STaR) method.

STaR's innovative approach includes

1. Iterative Self-Improvement: The system generates its own training data, learning from successful reasoning attempts to continuously enhance its capabilities. This self-supervised learning approach is a key step toward more autonomous AI systems that can improve their performance without constant human intervention or the need for large, manually annotated datasets. It mimics the human ability to learn from experience and self-reflection, potentially leading to AI systems that can adapt and improve in dynamic, real-world environments.
2. Bootstrapping from Limited Examples: STaR can dramatically improve performance starting with just a small number of annotated examples, making it highly efficient. This ability to learn from limited data is crucial for developing AI systems that can operate in domains where large datasets are not available or are prohibitively expensive to create. It could enable the application of advanced AI techniques to niche or specialized fields that have traditionally been challenging for machine learning due to data scarcity.
3. Versatility Across Domains: The method has shown success in various fields, from arithmetic and word problems to commonsense reasoning tasks. This cross-domain applicability is a significant step toward more general AI systems that can transfer learning from one domain to another, much like humans do. It suggests the potential for developing AI agents that can reason effectively across a wide range of tasks and subject areas, rather than being confined to narrow, specialized domains.
4. Chain-of-Thought Reasoning: STaR leverages step-by-step rational generation, mimicking human thought processes to tackle complex problems. This approach not only improves problem-solving capabilities but also enhances the explainability of AI decisions. By generating a chain of reasoning, STaR makes its problem-solving process more transparent and interpretable, which is crucial for building trust in AI systems, especially in high-stakes domains like healthcare or finance.

Stanford's work on STaR demonstrates the potential for AI systems to not just process information, but to truly reason and learn in ways that approach human cognitive abilities. The implications of this research are far-reaching. In education, for instance, STaR-like systems could potentially serve as personalized tutors, adapting their teaching strategies based on each student's learning patterns and generating explanations tailored to individual understanding.

In scientific research, such systems could assist in hypothesis generation and experimental design, potentially accelerating the pace of discovery. In fields like law or policy analysis, they could help in interpreting complex regulations or predicting the potential impacts of new policies.

The research conducted by OpenAI and Stanford University represents more than just incremental progress in AI capabilities. It signals the potential dawn of a new era in human-AI interaction and collaboration. As these AI agents become more capable of autonomous reasoning, learning, and problem-solving, we may see a shift from AI as a tool to AI as a partner in intellectual endeavors.

Google, Microsoft, Amazon, Salesforce, and many other AI companies big or small are investing heavily in their AI Agent strategies. For example, Google has recently unveiled its entry into the Agentic AI space with the launch of Gemini 2.0, featuring advanced multimodal capabilities that include native image and audio output as well as enhanced tool use. Key offerings include Project Astra, which interprets images, video, and audio in real time while improving dialogue and memory; Project Mariner, an AI agent that can control web browsers to complete tasks; and Deep Research, a feature that assists users in exploring complex topics and compiling reports (Kavukcuoglu, 2024).

The journey toward truly intelligent AI Agents is just beginning, and the possibilities are boundless. In the chapters that follow, we'll dive deeper into the technologies driving this revolution and explore current and future applications.

Are you ready to explore the cutting edge of AI and discover how these intelligent agents are poised to transform our world? Turn the page, and let's embark on this thrilling journey together.

1.6 Summary

This chapter illuminates the transformative potential of AI agents, sophisticated entities capable of autonomous action, complex reasoning, and adaptive learning. This evolution, driven by unprecedented computational power, advances in NLP, big data, and algorithmic breakthroughs, marks a shift from reactive AI to systems exhibiting near-human cognitive abilities.

Key Insights

- **Paradigm Shift:** AI agents represent a move from tool-based AI to collaborative partners, poised to revolutionize industries and redefine human-machine interaction.
- **Confluence of Technologies:** The current AI agent renaissance is fueled by a unique convergence of technological advancements, creating a fertile ground for unprecedented innovation.
- **From Narrow to General:** The development trajectory is toward increasingly general-purpose agents capable of cross-domain learning and problem-solving, exemplified by projects like OpenAI's Operator Agent and Stanford's STaR.
- **Self-Improvement as a Cornerstone:** STaR's self-teaching capabilities highlight a crucial trend toward AI systems that can autonomously learn and adapt, minimizing the need for human intervention.

- **Societal Transformation, not just technological:** The rise of AI agents will have profound implications beyond technology, impacting work, creativity, and problem-solving and necessitating careful ethical considerations regarding their development and deployment. They are more than just tools, but collaborators.

1.7 Questions

I. Multiple-Choice Questions (Choose the Best Answer)

1. **Which of the following is NOT a key characteristic of advanced AI agents as described in the chapter?**
   (a) Autonomous operation
   (b) Adaptive learning
   (c) Reliance on pre-programmed rules without modification
   (d) Complex reasoning and problem-solving
2. **The "Dartmouth Conference of 1956" is significant because:**
   (a) It marked the first successful implementation of a deep learning algorithm
   (b) It's considered the birthplace of the term "Artificial Intelligence" and initiated formal AI research
   (c) It led to the creation of the first autonomous robot
   (d) It established the ethical guidelines for AI development
3. **What technological advancement is primarily credited with enabling the current "AI Agent Renaissance"?**
   (a) The invention of the transistor
   (b) The development of the Internet
   (c) Breakthroughs in deep learning and increased computational power
   (d) The creation of the first expert system
4. **OpenAI's "Operator Agent" project aims to create an AI agent capable of:**
   (a) Performing only pre-defined tasks within a specific software environment
   (b) Autonomous Internet navigation, deep research, and long-horizon task execution
   (c) Mimicking human emotions and engaging in social interactions
   (d) Replacing human workers in all manufacturing and industrial settings
5. **Stanford's Self-Taught Reasoner (STaR) is notable for its ability to:**
   (a) Operate without any form of training data
   (b) Learn and improve iteratively from a limited number of examples through self-generated data
   (c) Generate creative content, such as music and art, without human input
   (d) Achieve superhuman performance in all areas of human intelligence

II. True/False Questions (Answer T or F)

1. **T/F:** Reactive AI agents are characterized by their ability to maintain an internal model of their environment and plan future actions.
2. **T/F:** The development of expert systems in the 1970s and 1980s represented a shift towards AI systems capable of replicating human expertise in specific domains.
3. **T/F:** Hybrid AI agents combine elements of reactive and deliberative architectures to leverage the strengths of both approaches.
4. **T/F:** The chapter suggests that the development of AI agents will likely have little to no impact on the structure of organizations and the nature of work.
5. **T/F:** Ethical considerations are not a significant concern in the development and deployment of AI agents, as they are simply tools and not moral actors.

III. Short Answer Questions (Answer in 2–4 Sentences)

1. What is the key difference between a deliberative AI agent and a learning AI agent?
2. Briefly describe the concept of "chain-of-thought reasoning" and its significance in the context of AI agents like STaR.
3. How has the availability of "big data" contributed to the advancement of AI agent capabilities?
4. Explain the significance of interdisciplinary convergence in the development of modern AI agents.
5. What are two potential societal benefits of the increasing use of AI agents, as suggested in the chapter?

IV. Essay Questions (Answer in 1–2 Paragraphs)

1. **The Evolution of Autonomy:** Discuss how the concept of autonomy in AI has evolved from the early days of AI research to the present, using examples from the chapter to illustrate your points.
2. **Technological Convergence:** Analyze the confluence of technological advancements that have enabled the current "AI Agent Renaissance," explaining the role of each factor.
3. **The Impact on Work:** Describe the potential impact of AI agents on the future of work, considering both the potential for job displacement and the creation of new roles and opportunities.
4. **OpenAI vs. Stanford:** Compare and contrast the approaches of OpenAI's "Operator Agent" and Stanford's STaR, highlighting the strengths and potential limitations of each project.
5. **Ethical Considerations:** Discuss the key ethical considerations that arise from the development of increasingly autonomous and capable AI agents, and suggest potential approaches for addressing these concerns.

V. Extra Essay Prompt

1. **The Societal Paradigm Shift:** Discuss why the development of AI Agents is not just a technological advancement, but a societal one as well. Give real-world examples to back up your points.

References

Axios. (2025, January 19). *AI super-agents: OpenAI and Meta prepare for breakthroughs*. Axios. https://www.axios.com/2025/01/19/ai-superagent-openai-meta

Encarnacao, J. L., & Rabaey, J. M. (Eds.). (2013). Mobile communications: technology, tools, applications, authentication and security. In *IFIP world conference on mobile communications 2–6 September 1996, Canberra, Australia*. Springer.

Hewitt, C., Bishop, C., & Steiger, P. (1973). *A universal modular ACTOR formalism for artificial intelligence*. IJCAI. Retrieved July 16, 2024, from https://www.ijcai.org/Proceedings/73/Papers/027B.pdf

Huang, K., Chen, X., Yang, Y., Ponnapalli, J., & Huang, G. (2023). ChatGPT in finance and banking. In K. Huang, Y. Wang, F. Zhu, X. Chen, & C. Xing (Eds.), *Beyond AI: ChatGPT, Web3, and the business landscape of tomorrow* (pp. 187–218). Springer. https://doi.org/10.1007/978-3-031-45282-6_7

Huang, K., Wang, Y., Goertzel, B., Li, Y., Wright, S., & Ponnapalli, J. (Eds.). (2024). *Generative AI security: Theories and practices*. Springer. https://doi.org/10.1007/978-3-031-54252-7

Kavukcuoglu, K. (2024, December 11). Introducing Gemini 2.0: Our new AI model for the agentic era. The Keyword. Retrieved December 13, 2024, from https://blog.google/technology/google-deepmind/google-gemini-ai-update-december-2024/#ceo-message

Krizhevsky, A., Sutskever, I., & Hinton, G. E. (2012). *ImageNet classification with deep convolutional neural networks*. NIPS papers. Retrieved July 16, 2024, from https://papers.nips.cc/paper_files/paper/2012/hash/c399862d3b9d6b76c8436e924a68c45b-Abstract.html

Lumb, D. (2025, January 23). OpenAI launches Operator, an AI agent that performs tasks autonomously. *TechCrunch*. https://techcrunch.com/2025/01/23/openai-launches-operator-an-ai-agent-that-performs-tasks-autonomously/

McCarthy, J., Minsky, M. L., Rochester, N., & Shannon, C. E. (2006, December 15). *A proposal for the Dartmouth Summer Research Project on Artificial Intelligence, August 31, 1955 | AI Magazine*. AAAI Publications. Retrieved July 16, 2024, from https://ojs.aaai.org/aimagazine/index.php/aimagazine/article/view/1904

Metz, R., & Mochizuki, T. (2024, July 11). OpenAI sets levels to track progress toward superintelligent AI. *Bloomberg*. Retrieved July 16, 2024, from https://www.bloomberg.com/news/articles/2024-07-11/openai-sets-levels-to-track-progress-toward-superintelligent-ai?srnd=technology-vp

Myers, K. (2007). Embodied agent. *Wikipedia*. Retrieved July 16, 2024, from https://en.wikipedia.org/wiki/Embodied_agent

Shortliffe, E. (2012). *Computer-based medical consultations: MYCIN*. Elsevier Science.

Vaswani, A., & Shazeer, N. (2017, June 12). [1706.03762] Attention is all you need. arXiv. Retrieved July 16, 2024, from https://arxiv.org/abs/1706.03762

Ken Huang is a prolific author and globally recognized authority in AI and Web3, with an extensive portfolio of published works that bridge business strategy, technical implementation, and cutting-edge research. As a Fellow of the Cloud Security Alliance and Co-Chair of the AI Safety Working Groups at the Cloud Security Alliance and the AI STR Working Group at the World Digital Technology Academy under the UN Framework, he is a leading voice in shaping global AI governance and security standards.

Huang is the CEO and Chief AI Officer (CAIO) of DistributedApps.ai, a firm specializing in generative AI training and consulting. His contributions to the field include being a core contributor to the OWASP Top 10 Risks for LLM Applications and an active participant in the NIST Generative AI Public Working Group.

Notable publications:

- *Beyond AI: ChatGPT, Web3, and the Business Landscape of Tomorrow* (Springer, 2023)—strategic insights into AI and Web3's business applications.
- *Generative AI Security: Theories and Practices* (Springer, 2024)—a comprehensive guide on securing generative AI systems.
- *Practical Guide for AI Engineers* (Volumes 1 and 2, DistributedApps.ai, 2024)—essential resources for AI and ML engineers.
- *The Handbook for Chief AI Officers: Leading the AI Revolution in Business* (DistributedApps.ai, 2024)—a roadmap for CAIOs in implementing GenAI across organizations.
- *Web3: Blockchain, the New Economy, and the Self-Sovereign Internet* (Cambridge University Press, 2024)—insights into the convergence of AI, blockchain, IoT, and emerging technologies.
- *Blockchain and Web3: Building the Cryptocurrency, Privacy, and Security Foundations of the Metaverse* (Wiley, 2023)—recognized as a must-read by TechTarget in 2023 and 2024.

Ken is a sought-after speaker and has presented at events such as the World Economic Forum in Davos, ACM and IEEE conferences, the CSA AI Summit, Depository Trust & Clearing Corporation forums, and World Bank conferences. His recent appointment to the OpenAI Forum reflects his ongoing commitment to advancing collaboration and dialogue in the field of AI.

Explore Ken's work on Amazon: https://www.amazon.com/author/kenhuang

Chapter 2
AI Agent Tools and Frameworks

Ken Huang and Jerry Huang

In Chap. 1, we painted a vivid picture of the AI agent revolution, tracing its historical roots and outlining the transformative potential of these intelligent, autonomous entities. We saw how a confluence of technological advancements has propelled us into an era where AI agents are poised to redefine industries, reshape the nature of work, and even challenge our understanding of intelligence itself. However, the journey from conceptual potential to tangible reality requires more than just vision; it demands the right tools and frameworks.

In this chapter, we transition from the "why" of AI agents to the "how," diving deep into the practical world of agent development. Guided by our "Seven-Layer AI Agent Architecture," we'll dissect the essential components that bring these sophisticated systems to life, from the foundational models that provide their core intelligence to the ecosystems where they deliver real-world impact.

We'll embark on a comparative exploration of leading agent frameworks—including AutoGen, LangGraph, LlamaIndex, and AutoGPT—examining their unique approaches to the intricate challenges of state management, tool integration, and decision-making. As we navigate this technical landscape, we'll also confront the pressing issues of security, compliance, scalability, and data quality that organizations must address to successfully deploy AI agents. Consider this chapter your essential toolkit for building the future, equipping you with the knowledge to not just understand but actively participate in the ongoing AI agent revolution.

K. Huang (✉)
DistributedApps.ai, Fairfax, VA, USA
e-mail: ken@distributedapps.ai

J. Huang
The University of Chicago, Chicago, IL, USA

K. Huang (ed.), *Agentic AI*, Progress in IS,
https://doi.org/10.1007/978-3-031-90026-6_2

2.1 The Seven-Layer AI Agent Architecture

In this section, I propose a seven-layer agent architecture that provides a comprehensive framework for understanding AI tools and frameworks. This layered approach decomposes the complex AI agent ecosystem into distinct functional layers: from Foundation Models that provide core AI capabilities, through Data Operations and Agent Frameworks that manage information and development tools, to Deployment Infrastructure and Security layers that ensure reliable and safe operations, culminating in the Agent Ecosystem where business applications deliver value to end-users. Each layer serves a specific purpose while abstracting complexity from the layers above it, enabling modular development, clear separation of concerns, and systematic implementation of AI agent systems across organizations.

The seven layers of the proposed agent architecture are interconnected, with each layer building on the functionality of the one beneath it. Starting from Layer 1, Foundation Models provide the core AI capabilities, which are utilized by Layer 2, Data Operations, to manage and preprocess data effectively. Layer 3, Agent Frameworks, leverages both data and foundational AI capabilities to enable the creation and execution of intelligent agents. Layer 4, Development Tools, supports the frameworks by providing programming environments, debugging tools, and integration solutions, streamlining the agent-building process. Layer 5, Deployment Infrastructure, ensures that agents created through the frameworks and tools can be deployed at scale with robust performance. Layer 6, Security, reinforces the system by safeguarding data, models, and operations across all previous layers including Layer 7, so we can think of Layer 6 as a vertical layer with implications for each layer. Finally, Layer 7, the Agent Ecosystem, integrates these capabilities to deliver cohesive and functional AI applications for end-users, abstracting underlying complexities to provide seamless and scalable solutions.

In the next few subsections, I will describe it from the top down, that is, from Layer 7 to Layer 1 (Fig. 2.1).

Layer 7: Agent Ecosystem

The ecosystem layer represents the vibrant marketplace where AI agents interface with real-world applications and users. This encompasses a diverse range of business applications, from intelligent customer service platforms to sophisticated enterprise automation solutions (Huang & Xing, 2023). Business apps in this layer include virtual assistants handling customer inquiries, automated content generation systems, intelligent document processing solutions, and AI-powered decision support systems. Tool providers create specialized interfaces that make AI capabilities accessible to specific industries, such as legal document analysis tools, medical diagnosis assistants, or financial trading algorithms.

The ecosystem also includes integration platforms that connect AI agents with existing business systems like CRM, ERP, and workflow management tools. This layer supports both vertical solutions (industry-specific) and horizontal solutions

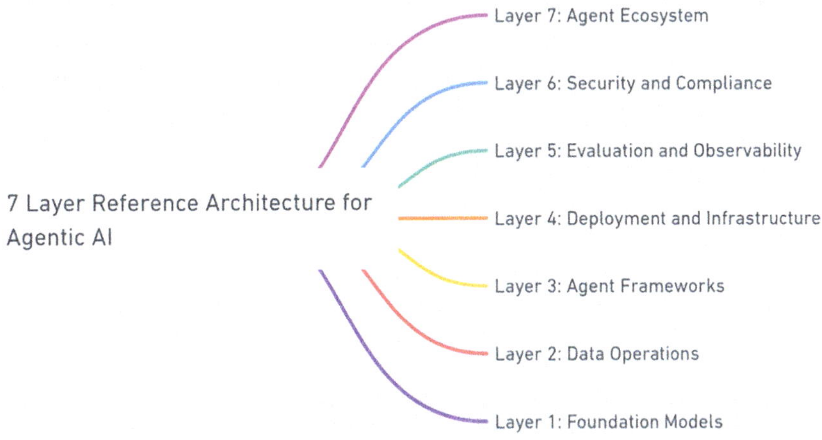

Fig. 2.1 The seven-layer AI agent architecture

(function-specific) that can be deployed across different sectors. Development tools and SDKs enable businesses to customize and extend agent capabilities for their specific needs. In Chap. 4, we will highlight agent use in improving business workflows or inventing new ones.

The marketplace aspect facilitates the discovery and deployment of pre-built agents and components, allowing organizations to find and implement AI solutions quickly. This includes agent directories, capability registries, and reputation systems that help users evaluate and select appropriate solutions. The ecosystem also encompasses communities of developers, businesses, and users who contribute to the evolution of AI agent applications, sharing best practices, use cases, and innovations.

Important considerations at this level include user experience design, integration capabilities, scalability of solutions, and the balance between customization and standardization. The ecosystem layer is where the theoretical capabilities of AI agents are transformed into practical, value-generating applications that solve real business problems and enhance human capabilities.

Layer 6: Security and Compliance

The security and compliance layer forms a crucial protective framework ensuring AI agents operate safely, securely, and within regulatory boundaries. While seemingly positioned as a distinct layer, it's vital to understand that **security and compliance are not afterthoughts, but rather foundational principles that must be embedded within each layer of the AI Agent stack.** This layer's placement at Layer 6 reflects its overarching role in safeguarding the entire system, but its principles permeate every level, from the foundational models to the agent ecosystem. We will discuss more about these challenges in Chap. 10 of this book.

Why Layer 6, yet Integrated into Every Layer?

1. **Comprehensive Oversight:** Placing security and compliance as a distinct layer emphasizes the need for a holistic approach. Layer 6 acts as a central point for defining security policies, compliance requirements, and risk management strategies that apply across the entire architecture.
2. **Specialized Focus:** This dedicated layer allows for the development of specialized expertise and tools focused specifically on security and compliance. This includes threat modeling, vulnerability assessment, security audits, and compliance monitoring, which require dedicated skills and resources.
3. **Regulatory Adherence:** As AI agents increasingly handle sensitive data and operate in regulated industries, a dedicated layer helps ensure adherence to evolving legal and regulatory frameworks (e.g., EU AI Act, GDPR, HIPAA). This layer is responsible for implementing necessary controls and processes to meet these requirements.
4. **Risk Management Framework:** Layer 6 facilitates the implementation of a comprehensive risk management framework. We recommend using a structured approach to assess and mitigate potential security and compliance risks, including vendor risk assessment for third-party AI services. Regular security assessments, penetration testing, and compliance audits ensure the ongoing integrity of the security framework.
5. **Incident Response and Business Continuity:** This layer is crucial for developing and maintaining AI Agent-related incident response plans and disaster recovery procedures. These plans must be regularly tested to ensure business continuity in the face of security breaches or system failures.

Security and Compliance Across All Layers

It's paramount to understand that Layer 6 does not operate in isolation. Effective security and compliance require a "defense in depth" strategy, where security measures are integrated into each layer of the architecture:

- **Layer 1 (Foundation Models):** Secure model development practices, including data sanitization, model robustness testing, and secure training environments, are essential.
- **Layer 2 (Data Operations):** Data security, privacy protection, access controls, and encryption are critical for managing the data used by AI agents.
- **Layer 3 (Agent Frameworks):** Secure coding practices, input validation, and secure API design within the agent framework are necessary to prevent vulnerabilities.
- **Layer 4 (Deployment and Infrastructure):** Provide the technical foundation necessary for running AI agents at scale.
- **Layer 5 (Evaluation and Observability):** Monitoring for anomalous behavior, security logging, and auditing capabilities are crucial for detecting and responding to threats.
- **Layer 7 (Agent Ecosystem):** Secure deployment practices, access controls, and ongoing monitoring of the agent's interactions within the ecosystem are essential.

In essence, while Layer 6 provides the overarching framework and specialized expertise, security and compliance must be a shared responsibility across all layers and throughout the entire lifecycle of an AI agent. This integrated approach is essential to build trustworthy, robust, and ethically sound AI systems that can operate safely and responsibly in the real world.

Layer 5: Evaluation and Observability

Recent developments in AI Agent evaluation have focused on creating comprehensive and standardized approaches to assess both the safety and performance of autonomous AI systems. One significant initiative is led by the AI Safety Institute (AISI) of the UK Government, which emphasizes the importance of evaluating AI agents capable of making long-term plans and operating semi-autonomously. This framework aims to test decision-making processes and action selection in complex environments, ensuring that agents can operate safely and effectively. Recently, AISI launched a bounty program for novel evaluations and agent scaffolding to encourage the development of innovative evaluation techniques for assessing the capabilities and potential risks of advanced agent systems (UK AISI, 2024).

The bounty program is primarily seeking innovative techniques in two main technical areas: autonomous capability evaluations and agent scaffolding. For autonomous capability evaluations, AISI is looking for methods to assess an AI agent's ability to operate independently, make decisions, and carry out complex tasks without human intervention. Agent scaffolding, the second focus area, involves developing frameworks or tools that can support and guide AI agents in their operations. Technically, successful applications are expected to demonstrate novel approaches to evaluating AI agents. This could involve developing new metrics, creating sophisticated simulation environments, or designing complex multistep tasks that challenge an agent's capabilities across different domains.

Another notable contribution is the Mosaic AI Agent Framework introduced by Databricks. This framework includes an evaluation component specifically designed for AI agents, featuring pre-built metrics that assess answer correctness, groundness, and relevance. It also incorporates safety evaluation metrics tailored for autonomous agents, facilitating a streamlined development and evaluation workflow through integration with MLflow. This combination allows developers to efficiently evaluate their AI agents across multiple dimensions (Wendell & Rao, 2024).

In addition to these frameworks, Agent Protocol, released in late 2023 (https://agentprotocol.ai/), provides a standardized way to interact with AI agents. While not strictly an evaluation framework, it has significant implications for benchmarking agent performance. By establishing consistent protocols for agent interactions, this initiative allows for standardized testing of capabilities and safety features across different agent implementations, making it easier to compare performance on similar tasks.

Benchmarks for assessing how well agents perform in multi-agent environments by evaluating their communication, coordination, and conflict resolution capabilities can be valuable.

Recent frameworks have placed a strong emphasis on safety benchmarks. Key aspects include **containment**, which assesses an agent's ability to operate within defined boundaries; **alignment**, which evaluates how well an agent's actions align with intended goals and ethical guidelines; **robustness**, which tests an agent's performance under various stress conditions or adversarial inputs; and **interpretability**, which measures how easily an agent's decision-making process can be understood and audited.

Performance evaluation has also evolved significantly. It now encompasses task completion metrics that measure an agent's ability to achieve specific objectives, efficiency assessments that evaluate resource usage—including time and computational power—adaptability evaluations that assess how well an agent performs in novel or changing environments, and scalability tests that examine an agent's performance as task complexity increases.

A growing trend in AI agent evaluation is the incorporation of cost metrics. This involves assessing the economic viability of deploying agents while measuring the trade-off between performance improvements and increased costs. Evaluating an agent's ability to optimize resource usage autonomously is becoming increasingly important as organizations seek to balance effectiveness with economic considerations.

Despite these advancements, several challenges remain in the field of AI agent evaluation. One major challenge is achieving standardization across different benchmarks, which would enable better comparisons among various implementations. Additionally, developing methodologies to evaluate the potential long-term consequences of AI agent actions is an area that requires further research.

Another significant challenge lies in creating evaluation frameworks that can assess agent performance in rapidly changing or unpredictable environments. As AI systems become more integrated into dynamic real-world applications, this adaptability will be crucial for ensuring their reliability and safety.

Finally, incorporating ethical considerations into AI agent benchmarks is becoming increasingly important, especially for agents deployed in sensitive domains such as healthcare or finance. As the field continues to evolve rapidly, these evaluation frameworks are likely to adapt further, with a growing emphasis on comprehensive assessments that are standardized and ethically aware.

Observability is another hot topic for agentic AI. We see some emerging tools in this space. The following are some examples:

- LangSmith provides advanced observability features for monitoring large language model (LLM) applications. Key capabilities include detailed tracing of function executions and system events using an "@traceable" decorator, centralized dashboards for session grouping, and integration with LangChain and other frameworks. Its platform supports both real-time and historical evaluations, custom metrics, and automated alerts for anomaly detection. Additionally, LangSmith offers features for dataset management, human feedback integration, and centralized prompt management for efficient optimization of AI applications. The enterprise plan includes enhanced deployment support and training.

- Langfuse specializes in self-hosted observability solutions for LLMs, offering extensive support for tracing, including multimodal tracing, and the ability to monitor system performance with minimal overhead. It allows users to manage datasets and evaluate models using both automated and human-assisted methods. Langfuse emphasizes flexibility, with an open-source version that caters to developers needing customizable solutions. Its enterprise plan adds features such as proactive monitoring, automated event triggers, and comprehensive analytics dashboards. This makes it suitable for teams requiring deeper control and insight into their LLM systems.
- Arize AI focuses on monitoring and troubleshooting machine learning models in production. Its platform provides model performance analytics, bias detection, drift monitoring, and root cause analysis. It is designed to identify anomalies in real time, offering visualizations and tools to trace issues back to the underlying datasets. Arize supports the integration of multiple AI models and provides features for managing dataset integrity, ensuring transparency in predictions.
- Weave offers observability tailored for interactive AI systems. Its platform is designed to track user interactions with AI agents, providing analytics to measure user satisfaction, detect anomalies, and optimize conversational flows. This makes it particularly suited for applications involving chatbots or virtual assistants. Weave also provides detailed insights into how user queries are processed by AI, enabling developers to iteratively improve system responses.
- AgentOps.ai focuses on operationalizing AI agents, providing tools to monitor their real-time performance, track usage patterns, and detect errors. It emphasizes agent lifecycle management by integrating monitoring with deployment workflows. Its observability tools include the ability to analyze agent interactions and ensure compliance with operational requirements, which is critical for ensuring AI agents perform reliably in dynamic environments.
- Braintrust emphasizes analytics and decision-making tools for AI-driven systems. Its observability features include automated reporting, real-time metric tracking, and support for visualizing system behavior. Braintrust enables developers to identify bottlenecks and inefficiencies in AI workflows, facilitating optimization of model performance and ensuring the robustness of AI deployments in critical applications.

Each of these platforms addresses unique needs within the agentic AI observability space, from general performance monitoring to specialized use cases such as agent lifecycle management or conversational flow optimization.

Layer 4: Deployment and Infrastructure

The deployment and infrastructure layer provides the robust technical foundation necessary for running AI agents at scale. Cloud platforms (AWS, Azure, GCP) offer essential services including compute resources (GPU/TPU acceleration), storage solutions (object storage, block storage), and networking capabilities (load balancers, CDNs). Container orchestration systems like Kubernetes manage agent deployment, scaling, and failover, ensuring high availability and reliability.

Infrastructure-as-Code (IaC) tools enable automated deployment and configuration management, using technologies like Terraform, CloudFormation, or Pulumi. This ensures consistent and repeatable deployments across different environments. CI/CD pipelines automate the testing and deployment process, enabling rapid iteration and updates to agent systems.

Resource management systems optimize infrastructure utilization through dynamic scaling, load balancing, and resource allocation. This includes sophisticated scheduling algorithms that match workloads to appropriate compute resources based on cost and performance requirements. Edge computing capabilities enable AI agents to operate closer to data sources, reducing latency and bandwidth usage.

Development environments like Replit provide integrated tools for coding, testing, and deploying AI agents. These environments support collaborative development, version control, and easy access to necessary dependencies and libraries. Infrastructure monitoring systems provide real-time visibility into system health, resource utilization, and performance metrics.

In addition to Amazon Bedrock, Microsoft Azure OpenAI service, and Replit, the following are emerging hosting service providers:

- Letta: Letta provides a cloud-based infrastructure designed to host stateful AI agents, with persistent memory and task management capabilities. It uses containerized deployments (e.g., Docker) for scalability and supports REST API endpoints and Python SDKs for integration. Its hosting environment provides low latency and reliability for real-time conversational applications.
- Agents API: Built for versatile hosting, Agents API emphasizes modularity in deploying AI agents across a wide range of environments. Its infrastructure supports both stateless and stateful operations with cloud-native scalability. It facilitates custom hosting configurations, enabling seamless interaction with external systems and third-party tools.
- LiveKit Agents: LiveKit focuses on hosting agents optimized for real-time interaction, leveraging its WebRTC-based infrastructure for low-latency communication. The platform ensures high availability with distributed hosting and dynamic load balancing, designed for voice, video, and text integration in collaborative and interactive applications.

Finally, disaster recovery and business continuity features are needed for system reliability through automated backups, multi-region deployment, and failover mechanisms. Cost management tools track resource usage and optimize infrastructure spending through techniques like spot instance usage and automatic resource cleanup which are proven technologies in traditional CPU-based cloud environments and can be retrofitted with some innovation for GPU cloud. The infrastructure layer also includes tools for managing model versioning, deployment strategies (blue-green, canary), and feature flagging.

Layer 3: Agent Frameworks

The agent framework layer provides sophisticated software frameworks and tools that simplify the development and management of AI agents. LangChain offers a comprehensive development framework with features like prompt management, chain-of-thought reasoning, and sophisticated memory management. It includes tools for building complex workflows, implementing retrieval-augmented generation (RAG), and managing agent state. In Sect. 2.2, we will compare some of the top agent frameworks.

These frameworks include tools for debugging, testing, and monitoring agent behavior. They provide abstractions for common tasks like API integration, data processing, and error handling. Development tools support both low-code and programmatic approaches to agent development, catering to different skill levels and use cases.

The framework layer also includes specialized tools for specific domains or tasks, such as frameworks for building conversational agents, document processing systems, or automated reasoning systems. Integration capabilities enable seamless connection with various data sources, APIs, and external services.

Among the latest developments in 2024, a special kind of agent called "Computer Use Agent" is gaining a lot of traction. For example, AI agents like Anthropic's Claude Computer Use Agent, Google's Project Jarvis, and OpenAI's upcoming "Operator" mark a significant evolution in AI capabilities, enabling direct interaction with computer interfaces by manipulating cursors, clicking buttons, and typing text. This advancement transforms how tasks are executed, automating processes such as form-filling, multi-site searches, and online transactions. By taking over routine and time-intensive tasks, AI agents enhance productivity, allowing humans to focus on creative and strategic work, while their 24/7 availability increases efficiency across industries.

Section 2.2 will provide more coverage on agent frameworks and compare them as well as provide a selection decision tree.

Layer 2: Data Operations

The data operation layer manages the complex data infrastructure required for AI agent operations. Vector databases (Pinecone, Weaviate, Milvus) provide specialized storage and retrieval systems for high-dimensional vector embeddings, enabling efficient similarity search and semantic matching. These databases support sophisticated indexing techniques like HNSW (Hierarchical Navigable Small World) for fast approximate nearest neighbor search.

Data loaders provide versatile interfaces for ingesting and processing diverse data types, including structured databases, document stores, and unstructured content. ETL pipelines handle data cleaning, transformation, and enrichment, ensuring data quality and consistency. This includes capabilities for handling streaming data, batch processing, and real-time updates.

Advanced data processing features include automatic schema detection, data validation, and format conversion. Data versioning systems track changes and maintain data lineage, enabling reproducibility and audit capabilities. Caching mechanisms optimize data access patterns, reducing latency and computational overhead.

Data operations tools support sophisticated querying capabilities, including hybrid search combining vector similarity with traditional filtering. Data synchronization mechanisms ensure consistency across distributed systems and handle conflict resolution in multi-writer scenarios.

The layer includes tools for data governance, including data quality monitoring, access control, and compliance tracking. Data pipeline orchestration tools manage complex data workflows, handling dependencies and ensuring reliable data processing. Performance optimization tools help tune database configurations and query patterns for optimal efficiency.

Monitoring and observability tools provide insights into data operation performance, including metrics on throughput, latency, and resource utilization. The layer also supports data backup and recovery operations, ensuring data durability and availability.

In the data operation layer, one prominent component is RAG (retrieval-augmented generation), a framework that combines retrieval models with generative AI to enhance the accuracy and relevance of generated outputs. RAG operates by retrieving relevant information from a database or external source based on a query, which is then used to guide the generative model in producing responses. It excels in tasks such as question-answering, summarization, and enriching generative models with up-to-date knowledge.

Building on this foundation, Agentic RAG introduces autonomous decision-making capabilities, utilizing agents to orchestrate retrieval, generation, and iterative refinement of results. Unlike the passive, query-driven retrieval in standard RAG, Agentic RAG employs active strategies and multistep reasoning to handle complex problem-solving and multi-turn dialogue systems effectively. This makes it particularly suited for dynamic workflows where adaptability and iterative improvement are essential.

Comparatively, while both RAG and Agentic RAG leverage retrieval and generative AI to improve output quality, RAG is simpler and focused on static, query-specific tasks. In contrast, Agentic RAG's agents enable it to manage more intricate, autonomous processes, making it a more versatile and advanced framework for handling complex scenarios (Fig. 2.2).

Layer 1: Foundation Models

The foundation model layer represents the core AI engines that power agent capabilities. Leading models from OpenAI (GPT-4), Anthropic (Claude), Google (Gemini), and Cohere provide sophisticated natural language processing and reasoning capabilities. Notably, they are in the process of being trained with advanced functionalities such as agentic planning, chain-of-thought reasoning, and other agentic capabilities, which will enable more robust and dynamic interactions.

These models support various interaction modes, including completion, chat, function calling interfaces, and multi-modality—processing different types of inputs such as text, images, and structured data. They integrate safety measures and content filtering capabilities to ensure appropriate outputs, offering features like content moderation, toxicity detection, and bias mitigation. API interfaces provide

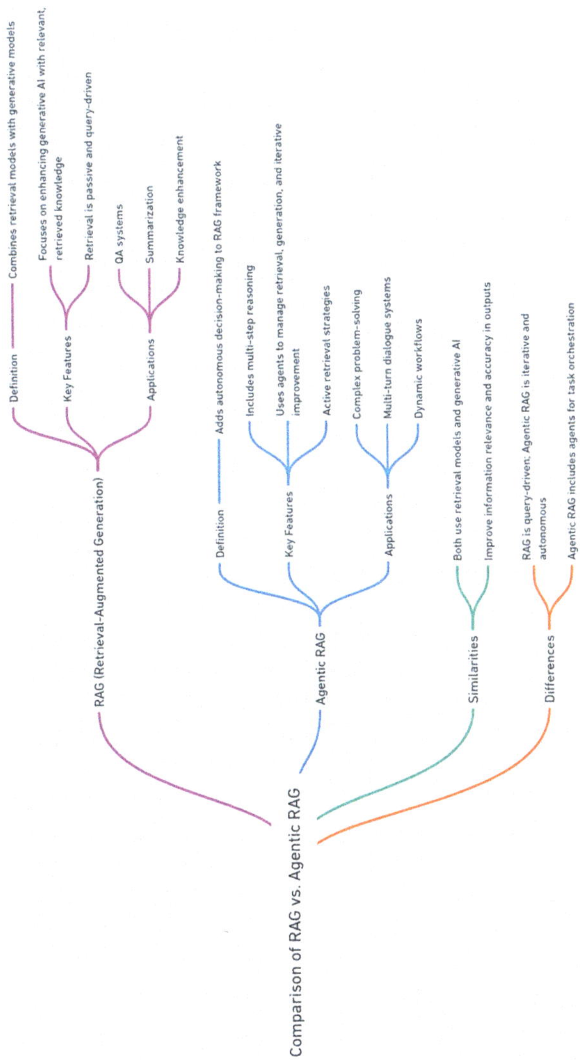

Fig. 2.2 RAG vs. Agentic RAG

programmatic access to these capabilities, with features like request batching, streaming responses, and rate limiting. Different model versions allow applications to choose appropriate trade-offs between performance, cost, and specialization.

Architectural innovations like mixture-of-experts, constitutional AI, and specialized training techniques enhance these models, supporting multiple languages and handling diverse inputs. Performance optimizations include response caching, prompt compression, and efficient token usage, enabling capabilities for semantic search, text classification, and structured output generation. Regular model updates incorporate new functionalities and improvements while maintaining backward compatibility.

Multi-modality models, which can process and integrate various types of data such as text, images, audio, and structured data, are becoming increasingly essential for AI agents. These models enable AI agents to understand and generate complex responses based on multiple input types, enhancing their contextual understanding and interaction capabilities. For instance, Claude's Computer Use Agent can autonomously navigate web pages, click buttons, and type text, mimicking human computer use (Anthropic, 2024).

2.2 Features and Comparison of AI Agent Framework

In this section, we will describe key features of the four sample AI Agent frameworks and then provide a comparison and finally present a decision tree for selecting a framework for your development needs.

While this section delves into the specific features of AutoGen, LangGraph, LlamaIndex, and AutoGPT, it's important to note that other powerful frameworks exist in the AI agent landscape. Some notable examples include Haystack, which focuses on NLP-based search and QA; CrewAI a multi-agent framework focusing on task delegation; SuperAGI which allows to build AI agents with memory and tools; AgentVerse, a framework for multi-agent systems with similar capabilities as Autogen; and Hugging Face's Smol Agents, a lightweight framework designed to simplify the development of AI agents using LLMs. These frameworks offer different strengths and cater to a variety of use cases, highlighting the diversity of available options for developing AI agents, and should be considered when selecting the right framework for your specific needs.

2.2.1 Main Features of Some Sample Agent Frameworks

In order to compare the agents' functionality, we selected four agent frameworks to discuss their main features.

AutoGen

AutoGen is an open-source framework developed by Microsoft that enables building multi-agent applications using LLM. It offers a flexible and powerful platform for creating sophisticated AI systems. Table 2.1 describes the main features of AutoGen.

These flexible execution options make AutoGen suitable for a wide range of applications, from simple chatbots to complex coding assistants. AutoGen is particularly well suited for applications that require collaborative problem-solving, interactive development, and dynamic task management.

LangGraph

LangGraph is a library developed by LangChain Inc. designed for building stateful, multi-actor applications using LLMs. It focuses on creating agent and multi-agent workflows, with particular emphasis on cycles, controllability, and persistence. LangGraph's unique graph-based architecture enables dynamic and adaptive workflows that can handle complex decision-making processes, making it ideal for use cases requiring iterative problem-solving, looping logic, and sophisticated reasoning (Table 2.2).

LlamaIndex

LlamaIndex is a data framework designed to connect custom data sources to large language models. It provides robust tools for data ingestion, structuring, and retrieval, enabling the creation of applications that leverage large datasets. LlamaIndex's query and chat engines offer powerful ways to interact with indexed

Table 2.1 AutoGen features

Feature	Description
Multi-agent conversations	AutoGen allows multiple GenAI Agents to converse and collaborate on solving complex tasks. Agents with different roles and capabilities can work together in dynamic systems. Example: a project manager agent, a code generator agent, and a code reviewer agent collaborate in software development
Agent types and roles	– AssistantAgent: an AI agent powered by an LLM, capable of processing requests, generating responses, answering questions, and generating content – UserProxyAgent: executes code and represents user input, allowing seamless integration of human feedback in conversations. These agents can be customized and combined for various use cases
GroupChat functionality	AutoGen supports group conversations managed by a GroupChatManager, enabling multiple agents to interact and collaborate. The GroupChatManager ensures each agent contributes appropriately to the conversation, fostering diverse perspectives for complex problem-solving
Code execution environments	AutoGen supports multiple environments for code execution: – Local execution: Code runs directly on the local machine for quick prototyping – Docker-based execution: Code runs within Docker containers for added security and isolation – No-code execution: It operates without code execution, focusing on language-based interactions

Table 2.2 LangGraph features

Feature	Description
Graph-based architecture	LangGraph uses a graph-based architecture to represent agent workflows as a network of nodes and edges – Nodes: individual agents or processing steps – Edges: define transitions between nodes – State: maintains context and data flow through the graph, enabling dynamic and complex workflows
StateGraph and edge definitions	The central component of LangGraph is the StateGraph, initialized with a state schema Edges define the flow between nodes, supporting conditional logic and complex branching for decision-making. Example: from langgraph.graph import StateGraph from langgraph.prebuilt import MessagesState graph = StateGraph (MessagesState)
Cycles and branching capabilities	LangGraph supports cycles in workflows, enabling iterative and looping processing and complex decision-making. This feature allows workflows to refine or explore multiple possibilities before reaching a conclusion, making it ideal for tasks requiring repeated refinement or adaptive problem-solving

Table 2.3 LlamaIndex features

Feature	Description
Data connectors and indexing	● LlamaIndex provides a wide range of connectors for integrating data sources, including files, APIs, and databases ● The indexing process utilizes advanced techniques, such as semantic indexing, for accurate and nuanced data retrieval beyond simple keyword matching
Query and chat engines, event-based agent collaboration	● LlamaIndex offers query engines for natural language queries against indexed data ● These engines interpret complex questions, break them down into sub-queries, and retrieve relevant information ● Chat engines enable multi-turn conversations, ideal for building chatbots and interactive knowledge bases ● LlamaIndex used an event-based approach for multi-agent collaborations
Tool integration	● LlamaIndex integrates with GenAI Agents, allowing them to use Python functions or LlamaIndex query engines as tools ● For example, the QueryEngineTool allows agents to perform queries on specific data sources, expanding the system's capabilities

data, making it a highly effective tool for building GenAI Agents that need to process and retrieve relevant information from diverse data sources (Table 2.3).

AutoGPT

AutoGPT is an open-source AI application that utilizes OpenAI's GPT-4 language model to create autonomous GenAI Agents. AutoGPT stands out for its ability to operate independently, completing tasks with minimal human intervention. This framework supports a wide range of use cases, including research, analysis, and

Table 2.4 AutoGPT features

Feature	Description
Autonomous operation	AutoGPT operates autonomously based on user-provided objectives. It plans and executes tasks independently, suitable for complex, multistep tasks that require minimal human intervention
Memory management	AutoGPT features both long-term and short-term memory, enabling the agent to retain context over extended operations, learn from past actions, and make informed decisions based on accumulated knowledge
Internet access and file processing	AutoGPT can search the web for information and gather up-to-date data to support its tasks. It also supports file processing, including storing, summarizing, and analyzing files for document analysis and data synthesis
Code execution	AutoGPT can write and execute code, making it suitable for programming tasks, software development, data analysis, and other computational problem-solving tasks

problem-solving, and it offers capabilities such as memory management, Internet access, and code execution, making it versatile and adaptable for complex applications (Table 2.4).

2.2.2 *Comparative Analysis of AI Agent Frameworks*

After exploring the major AI agent frameworks individually, let us compare them side by side to understand their strengths, limitations, and ideal use cases. This comparative analysis will help developers and organizations choose the most suitable framework for their specific needs.

State Management Approaches

State management is a necessary aspect of AI agent frameworks, as it determines how context and information are maintained throughout the agent's operation.

AutoGen approaches state management through its multi-agent conversation model. Each agent maintains its own state, and the overall state of the system is distributed across these agents. This approach allows for complex interactions and collaborations between agents, with each agent potentially having a different perspective on the overall task.

LangGraph takes a more structured approach to state management with its StateGraph. The entire workflow's state is explicitly defined and managed within the graph structure. This allows for fine-grained control over state transitions and makes it easier to understand and debug the flow of information through the system.

LlamaIndex focuses on state management in the context of data retrieval and querying. It maintains state primarily through its indexing structures, allowing for efficient retrieval of relevant information based on the current context of a query or conversation.

AutoGPT implements a more autonomous approach to state management, with its long-term and short-term memory systems. This allows the agent to maintain context over extended periods and across multiple tasks, learning and adapting its behavior based on past experiences.

Tool Integration Methods

The ability to integrate external tools and functionalities is a key feature of modern AI agent frameworks.

AutoGen provides flexible tool integration through its code execution environments. Agents can use Python functions as tools, and the framework supports both local and Docker-based execution. This allows for a wide range of tools to be integrated, from simple utility functions to complex external services.

LangGraph's tool integration is primarily achieved through its node definition system. Tools can be implemented as individual nodes in the graph, with clearly defined inputs and outputs. This approach allows for a clear separation of concerns and makes it easy to chain together multiple tools in complex workflows.

LlamaIndex excels in tool integration, particularly for data-related tasks. Its QueryEngineTool allows agents to execute queries on specific data sources, effectively turning entire datasets into tools that the agent can use. Additionally, LlamaIndex supports the integration of custom Python functions as tools.

AutoGPT's tool integration is centered around its ability to write and execute code. This allows it to dynamically create and use tools as needed, giving it a high degree of flexibility. However, this approach may require more careful management to ensure security and stability.

Decision-Making Logic

The decision-making capabilities of GenAI Agents are fundamental to their effectiveness in solving complex tasks.

AutoGen's decision-making is distributed across its multi-agent system. Each agent can make decisions based on its role and the information available to it. The GroupChatManager orchestrates these decisions, allowing for collaborative problem-solving.

LangGraph implements decision-making through its graph structure. Decision points are represented as nodes with multiple outgoing edges, and the framework supports conditional logic to determine which path to take. This allows for complex, branching decision processes to be clearly modeled and executed.

LlamaIndex's decision-making is primarily focused on determining the most relevant information to retrieve in response to queries. Its advanced indexing and retrieval mechanisms allow it to make nuanced decisions about what information is most pertinent to a given context.

AutoGPT takes a goal-oriented approach to decision-making. Given a high-level objective, it autonomously plans and executes a series of actions to achieve that goal. Its decision-making process is highly flexible but may be less transparent than more structured approaches.

Data Handling Capabilities

Effective data handling is important for GenAI Agents to access and utilize information efficiently.

AutoGen's data handling capabilities are primarily centered around its ability to process and generate text. While it doesn't have built-in data connectors, its flexible architecture allows for the integration of external data sources through custom agents or tools.

LangGraph doesn't provide specific data handling features out of the box. Instead, it offers a flexible framework where data handling can be implemented as needed through custom nodes and edges in the graph.

LlamaIndex shines in its data handling capabilities. It provides a wide range of data connectors for various sources, advanced indexing mechanisms, and efficient retrieval systems. This makes it particularly well suited for applications that need to work with large amounts of unstructured or semi-structured data.

AutoGPT has built-in capabilities for web scraping and file processing, allowing it to gather and process information from various sources. However, its data handling is not as structured or optimized as specialized frameworks like LlamaIndex.

Observability and Evaluation Features

The ability to monitor, debug, and evaluate GenAI Agents is critical for developing reliable and effective systems. We have discussed this topic in Sect. 2.1. Here, we examine the observability capability of four frameworks.

AutoGen provides detailed logging of agent interactions, allowing developers to trace the decision-making process and identify potential issues. Its support for human-in-the-loop interactions also aids in real-time monitoring and intervention.

LangGraph's graph-based structure inherently provides a high degree of observability. The flow of information and decision-making processes can be visualized and tracked through the graph, making it easier to understand and debug complex workflows.

LlamaIndex offers various evaluation metrics for its indexing and retrieval processes, allowing developers to assess and optimize the performance of their data handling systems. It also provides debugging tools to understand why certain pieces of information were retrieved in response to queries.

AutoGPT's autonomous nature can make detailed observability challenging. However, it provides logs of its actions and thought processes, allowing for post hoc analysis of its decision-making. Its ability to explain its reasoning also aids in evaluation and debugging.

2.2.3 Other Top Agent Frameworks

In addition to what we described above, there are some other notable agent tools and frameworks:

BabyAGI offers a minimalistic and straightforward approach to autonomous tasks, ideal for rapid prototyping and task execution. Its simplicity makes it limited for complex, multifaceted applications.

OpenAI's Swarm represents an experimental way of organizing multiple agents to work together, primarily suited to research environments or complex task orchestration, but isn't widely available yet for general use as of December 2024. This could change in the year 2025 when the book is published.

Crew.ai facilitates team collaboration by blending AI and human inputs, useful for workflows that require both types of input, but is not deeply automated or autonomous on its own.

MemGPT specializes in memory retention, maintaining context over time, making it ideal for customer support or situations requiring continuity in interactions.

Camel focuses on flexibility in task automation with a strong emphasis on customizability, suited to businesses needing tailored workflows but lacking extensive user-friendly documentation.

Table 2.5 puts all frameworks together and further compares these frameworks.

Each tool has its distinct advantages and challenges, so the choice would largely depend on the complexity of the application, required autonomy, ease of integration, and ecosystem alignment (see Fig. 2.3).

Table 2.5 Comparison of some top AI agent frameworks

AI agent	Description	Primary use cases	Strengths	Limitations
LangChain/ LangGraph	A modular framework for developing language model chains. Focuses on linking models, prompts, and tools	Building custom LLM applications, integrating multiple tools, creating workflows	Highly customizable, extensive tool and API integrations, modular	Limited out-of-the-box functionality, requires programming skills for customization
BabyAGI	A simple autonomous AI agent that leverages GPT to create, prioritize, and execute tasks iteratively	Task automation, proof of concept for AI workflows	Simple to set up, minimalistic structure, good for prototyping	Limited functionality, lacks scalability and advanced features
OpenAI's Swarm	An experimental framework by OpenAI, designed to coordinate multiple AI agents for complex tasks	Coordinated task completion, multi-agent collaboration	Effective for parallel task execution, dynamic agent coordination	Primarily experimental, not widely available or documented
Microsoft's AutoGen	A framework by Microsoft for creating and managing autonomous AI agents in enterprise applications	Enterprise task automation, document processing, customer service	Strong enterprise focus, integration with Microsoft ecosystem, scalable	Limited to Microsoft's ecosystem, relatively complex to implement

(continued)

Table 2.5 (continued)

AI agent	Description	Primary use cases	Strengths	Limitations
Crew.ai	A collaboration platform for coordinating multiple AI and human agents on tasks	Team collaboration, mixed human-AI workflows, task management	Blends human and AI capabilities, focus on coordination	Limited adoption, lacks deep automation features
MemGPT	An AI agent that maintains memory over interactions, allowing it to retain information across sessions	Long-term interaction use cases, customer support, knowledge management	Effective memory retention, context-aware responses	Memory management may increase complexity, potential for memory errors
AutoGPT	An open-source AI that uses GPT to autonomously execute tasks with minimal human input	Autonomous task management, repetitive task automation	Autonomous, minimal input required, customizable	Limited task flexibility, sometimes requires user intervention
SuperAGI	A full-featured platform for creating and managing autonomous AI agents	Complex task automation, multi-agent systems, enterprise use	Strong agent management, multiple integrations, scalable	High complexity, can be resource-intensive
Camel	An AI agent framework designed for flexible, extensible workflows in task automation	Custom workflows, business process automation	Flexible, supports custom workflows and integrations	Lacks comprehensive documentation, more suitable for advanced users

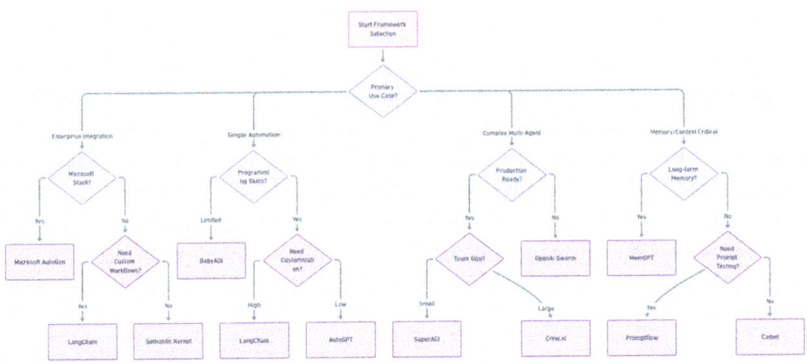

Fig. 2.3 How to select AI agent framework

Keep in mind that AI agent frameworks continue to evolve rapidly, with new capabilities and tools emerging regularly. We will likely see the introduction of additional frameworks focusing on specific use cases or novel approaches to agent architecture. Existing frameworks continue to expand their capabilities, particularly in areas of memory management, multi-agent coordination, and enterprise integration.

2.3 Challenges You May Face When Using AI Agent Framework

While AI agent frameworks offer powerful capabilities for developing intelligent systems, organizations often encounter significant challenges when implementing and scaling these solutions. These challenges span multiple dimensions—from technical hurdles in framework integration and tooling to strategic concerns around security, compliance, and cost management. Understanding these challenges is crucial for organizations to develop effective mitigation strategies and ensure successful AI agent deployments. In this section, we explore eight key challenge areas that organizations commonly face: framework and tooling limitations, integration complexities, scalability issues, security vulnerabilities (Huang et al., 2024; see also Chap. 12), compliance requirements, data quality concerns, workforce availability constraints, and cost management considerations. By examining these challenges and their potential solutions, organizations can better prepare for their AI agent implementation journey and develop more robust strategies for long-term success.

2.3.1 Framework and Tooling Challenges

Many organizations struggle to keep pace with advancements in frameworks due to limited technical resources, resulting in outdated configurations that may hinder the performance of AI agents. Furthermore, some frameworks lack comprehensive documentation or community support, leaving developers to spend considerable time troubleshooting and customizing basic functionalities. This can introduce inefficiencies and add to the operational burden. To address these issues, companies need to invest in dedicated support teams or partner with providers who can offer managed updates and support, ensuring that tooling enhancements do not interrupt critical workflows.

2.3.2 Integration Challenges

AI agents seldom operate in isolation; they typically need to connect with various systems within an organization, such as customer relationship management (CRM) systems, enterprise resource planning (ERP) tools, and cloud services. Achieving seamless integration is a significant challenge, especially as AI agents, such as Crew.ai or Promptflow, are designed to function across diverse environments with differing protocols, APIs, and data formats.

Incompatible data structures and limited API support are common obstacles that make integration complex and time-consuming. Organizations may also face difficulties when integrating agents with legacy systems, as these older systems often lack the connectivity features required for smooth data exchange. The result is increased technical debt and potential system inefficiencies. For AI agents to achieve their full potential, there must be a concerted effort to develop standardized interfaces and robust middleware solutions that can bridge the gap between disparate systems, facilitating reliable and efficient data flow.

2.3.3 Scalability Challenges

Scalability remains a key concern, especially as AI agents take on more significant roles within organizations. Agents like SuperAGI and OpenAI's Swarm, which are designed for extensive task orchestration, require a scalable infrastructure to handle large volumes of data and high levels of concurrent processing. However, scaling AI agents is technically complex and often expensive, involving challenges related to server capacity, computational efficiency, and data bandwidth.

For organizations that rely on cloud-based solutions, scalability might appear simpler, as most major cloud providers offer elastic computing resources. However, the cost of dynamically scaling AI agent operations can quickly become prohibitive. Additionally, latency issues may arise as more agents and data are added to the network, affecting performance. Organizations must plan carefully to develop a scalable architecture that includes load balancing, optimized processing algorithms, and efficient resource allocation. Such planning will help ensure that AI agents can expand their functionality as demand grows without sacrificing performance or driving up costs excessively.

2.3.4 Security Challenges

AI agents interact with sensitive data, making security a paramount concern. Without robust security measures, AI agents are vulnerable to various cyber threats, including data breaches, malicious attacks, and unauthorized access. We will discuss more about these challenges in Chap. 12 of this book.

2.3.5 Compliance Challenges

Compliance with regulatory standards, such as the EU AI Act, UK AISI, and HIPAA, is essential for organizations that handle personal or sensitive data. AI agents, particularly those with memory capabilities like MemGPT, can inadvertently retain and misuse sensitive information if they are not properly managed. Regulatory bodies are increasingly scrutinizing AI implementations, and failure to comply with data protection laws can lead to significant penalties and reputational damage.

To mitigate these risks, organizations must adopt a robust AI governance framework that includes regular audits, comprehensive data anonymization practices, and stringent access controls. Implementing lifecycle management for AI agents, including clear retention policies and mechanisms for data purging, is critical to prevent unauthorized storage or misuse of sensitive information. Continuous monitoring and alignment with evolving standards, such as the EU AI Act and HIPAA, will further safeguard compliance and build trust with stakeholders.

2.3.6 Higher-Quality Data Source Challenges

The effectiveness of AI agents is directly tied to the quality of the data they process. High-quality, accurate, and relevant data improves the decision-making capabilities of agents, while poor-quality data can lead to incorrect predictions, errors in task execution, and reduced trust in AI outcomes. Access to reliable data sources is particularly critical for agents used in complex decision-making environments, such as Microsoft's Semantic Kernel and AutoGPT, which rely on extensive data for training and operational purposes.

Organizations often face challenges in sourcing, cleaning, and maintaining high-quality data. Data silos, inconsistent data formats, and a lack of standardized cleaning processes can significantly impact the quality of data available to AI agents. Moreover, in some cases, organizations must rely on third-party data sources, which may vary in reliability.

To address these challenges, organizations should prioritize establishing a comprehensive data management strategy. This includes implementing standardized data cleaning and preprocessing workflows to ensure consistency and accuracy across datasets. Breaking down data silos through centralized data repositories or integration platforms can facilitate seamless access to information. Additionally, vetting and validating third-party data sources before integration can mitigate risks associated with unreliable data. Leveraging tools like data cataloging, automated quality checks, and ongoing monitoring can further enhance data quality, ultimately boosting the performance and reliability of AI agents in critical decision-making scenarios.

2.3.7 *Skilled Workforce Availability Challenges*

The deployment and maintenance of AI agents require a workforce with specialized skills in machine learning, software engineering, and data science. However, there is a growing shortage of skilled professionals capable of managing advanced AI systems. This shortage of skilled labor poses a significant challenge, as organizations struggle to find, train, and retain talent that can handle the intricacies of AI agent implementation and operation. Furthermore, the lack of expertise can lead to inefficient use of AI agents or even failure to fully utilize their potential. To address this gap, organizations may need to invest in extensive training programs or consider outsourcing certain aspects of AI management to specialized vendors. Additionally, the development of more user-friendly, low-code AI frameworks could help reduce the reliance on highly specialized skill sets, enabling broader adoption.

2.3.8 *Cost Challenges*

The implementation and ongoing operation of AI agents can be costly, with expenses arising from LLM API calls, software licensing, infrastructure, integration, and skilled labor. The challenge of managing costs is especially pertinent in cases where AI agents are used at scale or within budget-sensitive organizations. Cost reduction measures are therefore necessary for organizations looking to sustain AI operations over the long term. This may involve optimizing code to reduce computational demands, using a combination of a small LLM for frequent simple inference calls and a large LLM for infrequent but sophisticated inference calls, prioritizing the use of open-source frameworks where feasible, and implementing energy-efficient processing algorithms. Additionally, organizations can consider hybrid deployment models, where critical parts of the AI system run on-premises to reduce cloud costs. Effective budgeting and resource allocation are critical to ensuring that AI agents remain financially viable without compromising their effectiveness or reliability.

2.4 Summary

This chapter moves beyond the "what" and "why" of AI agents, established in Chap. 1, to tackle the crucial "how." It delves into the practical realities of building these intelligent systems, providing a comprehensive blueprint through the "Seven-Layer AI Agent Architecture." This framework serves as a crucial guide, illuminating the intricate components and their interplay, essential for constructing robust and effective AI agents.

Key Takeaways and Implications

1. **The Seven-Layer Model: A Holistic View:** The chapter's core contribution is the Seven-Layer Architecture. This model provides a structured understanding of AI agent systems, breaking them down into manageable, interconnected components. It's not just a theoretical construct but a practical tool for design, implementation, and analysis, helping to navigate the complexity of agent development.
   (a) **Implication:** This model provides a common language and framework for developers, researchers, and organizations, facilitating collaboration and a more systematic approach to AI agent development.
2. **Framework Deep Dive: Understanding the Trade-offs:** The chapter provides an in-depth comparative analysis of leading agent frameworks like AutoGen, LangGraph, LlamaIndex, and AutoGPT. This goes beyond surface-level descriptions, dissecting their strengths, weaknesses, and design philosophies, particularly in areas like state management, tool integration, and decision-making logic.
   (a) **Implication:** Developers can make informed choices about which framework best suits their specific needs, understanding the trade-offs involved and selecting the right tool for the job. It highlights that there's no "one-size-fits-all" solution in the world of AI agents.
3. **Emerging Trends: Computer Use Agents and Agentic RAG:** The chapter highlights cutting-edge developments like Computer Use Agents and Agentic RAG, showcasing the evolving capabilities of AI agents and their increasing ability to interact with the digital world in more sophisticated ways.
   (a) **Implication:** These trends signal a shift toward more autonomous and versatile agents, capable of automating complex tasks and integrating seamlessly into existing workflows, opening new frontiers for application and innovation.
4. **Security and Compliance: A Nonnegotiable Imperative:** The dedicated section on Layer 6 underscores the critical importance of security and compliance in AI agent development. It emphasizes that these are not mere add-ons but foundational principles that must be woven into the fabric of every layer, from the foundational models to the agent ecosystem.

(a) **Implication:** Organizations must adopt a proactive, "security-first" mindset when developing AI agents, integrating robust security measures and compliance practices from the outset to build trustworthy and responsible systems.

5. **Challenges as Opportunities:** The chapter frankly addresses the multifaceted challenges of implementing AI agents—technical hurdles, integration complexities, scalability issues, security risks, compliance requirements, data quality concerns, workforce availability constraints, and cost management considerations. It reframes these challenges as opportunities for innovation and improvement.

   (a) **Implication:** By acknowledging and understanding these challenges, organizations can develop effective mitigation strategies, fostering a more mature and robust approach to AI agent deployment. It highlights the need for a multidisciplinary approach, involving not just technical expertise but also legal, ethical, and business acumen.

2.5 Questions

I. Multiple Choice Questions (Choose the Best Answer)

1. **The "Seven-Layer AI Agent Architecture" presented in the chapter is primarily intended to:**

   (a) Provide a rigid, step-by-step process for building AI agents.
   (b) Offer a structured framework for understanding the components and interactions within AI agent systems.
   (c) Showcase the superiority of one specific AI agent framework over others.
   (d) Serve as a historical overview of the evolution of AI agent technology.

2. **Which of the following is NOT a key characteristic of Layer 6 (Security and Compliance) in the Seven-Layer Architecture?**

   (a) It provides a framework for risk management related to AI agents.
   (b) It is solely responsible for all security implementations, with no security considerations in other layers.
   (c) It plays a crucial role in ensuring adherence to regulatory requirements.
   (d) It emphasizes the need for incident response and business continuity plans.

3. **When comparing AI agent frameworks, the chapter highlights "state management" as a key differentiator. What does "state management" primarily refer to in this context?**

   (a) The emotional state of the AI agent.
   (b) How an agent maintains and updates information about its environment and itself over time.
   (c) The physical location where the AI agent's code is executed.
   (d) The regulatory state in which the AI agent is deployed.

4. **What is the main advantage of using Agentic RAG over traditional RAG systems, as discussed in the chapter?**
   (a) Agentic RAG is significantly less computationally expensive.
   (b) Agentic RAG offers enhanced autonomous decision-making and multistep reasoning capabilities.
   (c) Agentic RAG requires no training data, unlike traditional RAG.
   (d) Agentic RAG is only applicable to text-based data, whereas traditional RAG can handle various data types.
5. **Which of the following best describes the relationship between the "Seven-Layer AI Agent Architecture" and the specific AI agent frameworks discussed (AutoGen, LangGraph, LlamaIndex, AutoGPT)?**
   (a) The frameworks are entirely separate from the architecture and represent alternative approaches.
   (b) The architecture is a high-level model, and the frameworks are specific implementations that align with its principles.
   (c) The frameworks were developed first, and the architecture was created later to categorize them.
   (d) The architecture is outdated, and the frameworks represent more modern approaches to AI agent development.

II. True/False Questions (Answer T or F)

1. **T/F:** Security and compliance considerations should only be addressed at Layer 6 of the Seven-Layer Architecture, not at other layers.
2. **T/F:** The chapter argues that there is a single "best" AI agent framework that should be used for all applications.
3. **T/F:** Computer Use Agents are primarily designed to improve the user interface of existing software applications.
4. **T/F:** Scalability challenges in AI agent deployment often stem from the need to handle increasing amounts of data and computational demands.
5. **T/F:** The chapter suggests that cost management is not a significant concern when implementing AI agent systems.

III. Short Answer Questions (Answer in 2–4 Sentences)

1. Briefly explain the role of "Layer 1: Foundation Models" in the Seven-Layer AI Agent Architecture.
2. What is the primary function of "Layer 4: Development Tools" in supporting AI agent development?
3. What are two key challenges organizations face when integrating AI agents with existing systems?
4. Describe how "Layer 7: Agent Ecosystem" contributes to the overall value proposition of AI agents.
5. Explain the concept of "Agentic RAG" and how it differs from traditional Retrieval-Augmented Generation.

IV. Essay Questions (Answer in 1–2 Paragraphs)

1. **Framework Selection:** Discuss the key factors an organization should consider when selecting an AI agent framework for a specific application. Use examples from the frameworks discussed in the chapter (AutoGen, LangGraph, LlamaIndex, AutoGPT) to illustrate your points.
2. **Security Across Layers:** Explain why a holistic approach to security is crucial in AI agent development, emphasizing the importance of integrating security measures across all layers of the Seven-Layer Architecture. Provide examples of security considerations at different layers.
3. **Challenges and Opportunities:** Analyze the challenges presented in the chapter regarding AI agent implementation (technical, integration, scalability, security, compliance, data quality, workforce, cost). Reframing these challenges, discuss how they can be viewed as opportunities for innovation and improvement in the field.
4. **The Future of AI Agents:** Based on the trends discussed in this chapter, such as Computer Use Agents and Agentic RAG, speculate on the potential future developments in AI agent technology and their impact on various industries or aspects of society.
5. **The Seven-Layer Model's Value:** Argue for the value of the Seven-Layer AI Agent Architecture as a tool for understanding, designing, and implementing AI agent systems. How does this model contribute to a more structured and effective approach to agent development?

References

Anthropic. (2024, October 22). *Introducing computer use, a new Claude 3.5 Sonnet, and Claude 3.5 Haiku*. Anthropic. Retrieved November 16, 2024, from https://www.anthropic.com/news/3-5-models-and-computer-use

Huang, K., Ponnapalli, J., Tantsura, J., & Shin, K. T. (2024). Navigating the GenAI security landscape. In K. Huang, Y. Wang, B. Goertzel, Y. Li, S. Wright, & J. Ponnapalli (Eds.), *Generative AI security: Theories and practices* (pp. 31–58). Springer. https://doi.org/10.1007/978-3-031-54252-7_2

Huang, K., & Xing, C. (2023). ChatGPT: Inside and impact on business automation. In K. Huang, Y. Wang, F. Zhu, X. Chen, & C. Xing (Eds.), *Beyond AI: ChatGPT, Web3, and the business landscape of tomorrow* (pp. 37–65). Springer. https://doi.org/10.1007/978-3-031-45282-6_2

UK AISI. (2024, November 5). *Bounty programme for novel evaluations and agent scaffolding | AISI work*. AI Safety Institute. Retrieved November 17, 2024, from https://www.aisi.gov.uk/work/evals-bounty

Wendell, P., & Rao, N. (2024, July 2). *Announcing Mosaic AI agent framework and agent evaluation*. Databricks. Retrieved November 17, 2024, from https://www.databricks.com/blog/announcing-mosaic-ai-agent-framework-and-agent-evaluation

Ken Huang is a prolific author and globally recognized authority in AI and Web3, with an extensive portfolio of published works that bridge business strategy, technical implementation, and cutting-edge research. As a Fellow of the Cloud Security Alliance and Co-Chair of the AI Safety

Working Groups at the Cloud Security Alliance and the AI STR Working Group at the World Digital Technology Academy under the UN Framework, he is a leading voice in shaping global AI governance and security standards.

Huang is the CEO and Chief AI Officer (CAIO) of DistributedApps.ai, a firm specializing in generative AI training and consulting. His contributions to the field include being a core contributor to the OWASP Top 10 Risks for LLM Applications and an active participant in the NIST Generative AI Public Working Group.

Notable publications:

Beyond AI: ChatGPT, Web3, and the Business Landscape of Tomorrow (Springer, 2023)—strategic insights into AI and Web3's business applications.

Generative AI Security: Theories and Practices (Springer, 2024)—a comprehensive guide on securing generative AI systems.

Practical Guide for AI Engineers (Volumes 1 and 2, DistributedApps.ai, 2024)—essential resources for AI and ML engineers.

The Handbook for Chief AI Officers: Leading the AI Revolution in Business (DistributedApps.ai, 2024)—a roadmap for CAIOs in implementing GenAI across organizations.

Web3: Blockchain, the New Economy, and the Self-Sovereign Internet (Cambridge University Press, 2024)—insights into the convergence of AI, blockchain, IoT, and emerging technologies.

Blockchain and Web3: Building the Cryptocurrency, Privacy, and Security Foundations of the Metaverse (Wiley, 2023)—recognized as a must-read by TechTarget in 2023 and 2024.

Ken is a sought-after speaker and has presented at events such as the World Economic Forum in Davos, ACM and IEEE conferences, the CSA AI Summit, Depository Trust & Clearing Corporation forums, and World Bank conferences. His recent appointment to the OpenAI Forum reflects his ongoing commitment to advancing collaboration and dialogue in the field of AI.

Explore Ken's work on Amazon: https://www.amazon.com/author/kenhuang

Jerry Huang is currently an AI Engineer at Google, where he contributed to the AI/ML evaluation pipeline for a consumer-facing application. Before Google, he worked as a technical and security staff member at several prominent technology companies, gaining experience in areas like security, AI/ML, and scalable systems.

At Metabase, an open-source business intelligence platform, Jerry contributed features such as private key management and authentication solutions. As a Software Engineer at Glean, a Generative AI search startup, he was one of three engineers responsible for managing large-scale GCP infrastructure powering text summarization, autocomplete, and search for over 100,000 enterprise users. During his time at TikTok, Jerry helped design and build custom RPCs to model access control policies. At Roblox, he served as a Machine Learning/Software Engineering Intern, focusing on real-time text generation models and gathering a large multilingual corpus that significantly boosted model robustness.

In addition to his industry experience, Jerry has conducted extensive security and biometrics research as a Research Assistant at Georgia Tech's Institute for Information Security & Privacy, resulting in a thesis on privacy-preserving biometric authentication.

Jerry holds a BS/MS in Computer Science from Georgia Tech and is currently pursuing an MS in Applied Mathematics at the University of Chicago.

Chapter 3
Multi-Agent Coordination

Ken Huang and Jerry Huang

Imagine a swarm of drones seamlessly coordinating a breathtaking aerial light show or a city's traffic grid intelligently adapting to real-time conditions, minimizing congestion and maximizing flow. These are not futuristic fantasies but glimpses into the rudimentary capability of multi-agent systems (MAS)—a transformative field where autonomous AI agents collaborate to solve complex problems far beyond the capabilities of any single entity. Chapter 1 painted a vision of the AI agent revolution, and Chap. 2 delved into the building blocks of these intelligent agents. Now, in this chapter, we explore how these agents interact, coordinate, and collectively achieve remarkable feats. This chapter dissects the intricacies of MAS, from the fundamental principles of agent communication and coordination to the challenges of conflict resolution and the design of robust, scalable systems.

3.1 Introduction to MASs

This section introduces the concept of MASs, their emergence as AI ecosystems, and the benefits and challenges associated with coordinating multiple AI agents.

K. Huang (✉)
DistributedApps.ai, Fairfax, VA, USA
e-mail: ken@distributedapps.ai

J. Huang
The University of Chicago, Chicago, IL, USA

K. Huang (ed.), *Agentic AI*, Progress in IS,
https://doi.org/10.1007/978-3-031-90026-6_3

3.1.1 Defining MASs

MASs are computational systems where multiple intelligent agents interact to achieve individual or collective goals. These agents are autonomous, meaning they can operate independently, but they also have the capability to cooperate, coordinate, and negotiate with other agents when necessary. The core of a MAS lies in its ability to handle complex tasks through the collective efforts of relatively simple individual agents.

Autonomy is a characteristic of agents in these systems. Each agent can make decisions and take actions with minimum direct intervention from humans or other agents. This autonomy allows for a distributed approach to problem-solving, where different aspects of a complex task can be handled by specialized agents. However, autonomy alone is not sufficient for a truly effective MAS.

Collaboration is another key aspect. Agents must be able to interact with other agents and possibly humans through some kind of agent-communication language. This capability enables the sharing of information, coordination of actions, and negotiation of resources or goals. The richness of these interactions often determines the sophistication and effectiveness of the overall system.

Reactivity and proactiveness form two sides of an agent's interaction. Agents can perceive and respond in a timely fashion to changes, demonstrating reactivity. At the same time, they don't simply act in response to their environment; they exhibit goal-directed behavior by taking the initiative, showcasing proactiveness. This balance between reactive and proactive behaviors allows agents to be both responsive to immediate needs and focused on long-term objectives.

In a typical MAS, agents might represent diverse entities such as robots in a factory, vehicles in a traffic system, or even abstract concepts like trading strategies in a financial market. The complexity of these systems emerges from the interactions between agents, rather than from the complexity of the agents themselves. This emergent complexity allows MASs to tackle problems that would be difficult or impossible for a single agent to solve.

Here is a table comparing single-agent and MASs (Table 3.1).

In actual use cases, you can use a single AI agent when tasks are simple and interrelated and do not require specialized expertise. It is cost-effective, ensures consistent user experience, and simplifies data handling, making it ideal for focused use cases like basic customer support or content generation. Opt for multi-agent systems (MASs) when tasks are complex, diverse, or require specialization and complex coordination which we will discuss in this chapter. MASs allow concurrent task execution, scalability, and modularity. Choose based on task complexity, specialization needs, scalability, and cost-efficiency, or combine both approaches with a primary agent delegating tasks to specialized agents for maximum flexibility.

Table 3.1 Single-agent system vs. MAS

Aspect	Single-agent system	MAS
Definition	A system with one autonomous entity solving tasks	A system with multiple autonomous entities interacting
Autonomy	Operates independently without coordination with others	Requires collaboration or competition between agents
Complexity	Simpler to design and manage	Higher complexity due to agent interactions and dependencies
Scalability	Limited scalability; solving large problems may be challenging	Highly scalable; tasks can be distributed among agents
Communication	No interagent communication needed	Agents communicate to share information or negotiate
Fault tolerance	Entire system may fail if the agent fails	More robust; failure of one agent may not affect others
Coordination	No coordination mechanisms are required	Requires strategies like cooperation, competition, or negotiation
Environment interaction	Operates in isolation or interacts minimally with the environment	Interacts actively, often with shared environments
Application example	Pathfinding for a robot in a maze	Traffic management with multiple autonomous vehicles

3.1.2 Benefits and Challenges of Multi-Agent Coordination

One of the primary advantages of the MAS is improved problem-solving capability. By leveraging diverse capabilities and perspectives, MASs can tackle complex problems more effectively than single-agent approaches. This is particularly evident in domains like supply chain management, where coordinating production, inventory, and delivery across multiple locations requires the combined efforts of many specialized agents. As additional examples, MAS can also be used in scenarios requiring a proxy agent to facilitate communication between the user and multidisciplinary agents, such as in healthcare, educational, and financial applications. MASs can also be applied in situations where exceptions to input occur, with a dedicated agent activated specifically to handle these exception scenarios.

Scalability is another significant benefit of MASs. As the complexity of a task increases, these systems can be scaled up by adding more agents, allowing them to handle larger and more intricate challenges. This scalability makes MASs particularly suited for domains where the scope of the problem can change dynamically, such as in smart city management or large-scale industrial operations.

The distributed nature of MASs also contributes to their robustness. Unlike monolithic AI solutions, where a single point of failure can bring down the entire system, MASs can continue to function even if some agents fail. This resilience is important in critical applications like disaster response systems, where reliability under challenging conditions is paramount.

However, coordinating multiple AI agents also presents challenges. One of the primary difficulties lies in ensuring effective communication and understanding between agents. Even with advanced communication protocols, misunderstandings or misinterpretations can occur, potentially leading to suboptimal or even conflicting actions. Developing robust and efficient communication mechanisms that can handle the complexity of real-world scenarios remains an active area of research.

Another challenge is balancing the autonomy of individual agents with the need for coordinated action. While autonomy allows agents to respond quickly to local conditions, too much independence can lead to a lack of coherence in the overall system behavior. Striking the right balance between individual initiative and collective coordination is a delicate task that requires careful system design and ongoing management.

Resource allocation and conflict resolution also pose significant challenges in MASs. When multiple agents compete for limited resources or have conflicting goals, mechanisms must be in place to fairly and efficiently resolve these conflicts. This often involves complex negotiation protocols and decision-making algorithms that can handle the intricacies of multiparty interactions.

3.2 Coordination Techniques in MASs

The effectiveness of MASs largely depends on the coordination techniques employed to manage the interactions between agents. These techniques enable agents to work together efficiently, resolve conflicts, and achieve collective goals that would be difficult or impossible for individual agents to accomplish alone.

3.2.1 Negotiation Protocols

One common negotiation approach is the Contract Net Protocol, originally proposed by Reid G. Smith in 1980 (Smith, 1980). In this protocol, agents can take on the role of manager or contractor. The manager agent announces a task to be performed, and contractor agents submit bids to undertake the task. The manager then evaluates the bids and awards the contract to the most suitable contractor. This protocol is particularly useful in scenarios where tasks need to be dynamically allocated based on the current capabilities and availability of agents.

Another important negotiation technique is the use of auction mechanisms. Various types of auctions, such as English auctions, Dutch auctions, or Vickrey auctions, can be employed depending on the specific requirements of the MAS. These auction-based approaches are often used in resource allocation problems, where multiple agents compete for limited resources.

More sophisticated negotiation protocols incorporate elements of game theory and decision theory to model complex multiparty negotiations. These approaches allow agents to reason about the potential strategies and preferences of other agents, leading to more nuanced and effective negotiation outcomes.

The Contract Net Protocol, as described, involves a manager agent broadcasting a task and contractor agents bidding to undertake it. To better illustrate this process, let's examine a Python example. This code demonstrates how a manager evaluates bids from contractors based on their capabilities and assigns the task to the most suitable agent.

```
class Agent:
    def __init__(self, name, capability):
        self.name = name
        self.capability = capability

    def bid(self, task):
        return self.capability - task["difficulty"]

manager = Agent("Manager", capability=0)
contractors = [Agent(f"Agent_{i}", capability=i * 10) for i in
range(1, 4)]

task = {"difficulty": 15}
bids = {contractor.name: contractor.bid(task) for contractor in
contractors}
best_agent = max(bids, key=bids.get)

print(f"Task
assigned to {best_agent} with bid {bids[best_agent]}")
```

3.2.2 Cooperation Mechanisms

While negotiation often involves agents with potentially conflicting interests, cooperation mechanisms focus on how agents can work together toward common goals. Effective cooperation is needed to achieve synergies in MASs and tackle complex tasks that require coordinated efforts.

One approach to fostering cooperation is through shared mental models. This involves creating a common understanding of the task environment, goals, and roles among the agents. By aligning their internal representations, agents can more effectively coordinate their actions and anticipate the needs and behaviors of other agents. As one example, Letta (formerly MemGPT) has an implementation of shared memory via synchronized block objects which are part of Agent's memory (https://www.letta.com/).

Task decomposition and allocation are also important cooperation mechanisms. Complex tasks are broken down into subtasks that can be distributed among agents based on their capabilities and current workload. This approach allows for parallel processing and specialization, improving overall system efficiency.

For example, AutoGen, developed by Microsoft, exemplifies advanced task decomposition capabilities by enabling complex tasks to be systematically broken down into a few subtasks through a dedicated planner agent. The framework utilizes a sophisticated two-agent chat system where one agent creates a strategic task breakdown while another executes the plan, allowing for dynamic interaction and iterative refinement. This approach enables agents to collaboratively analyze complex problems, distribute workload efficiently, and adapt their strategies in real time based on emerging insights and challenges.

CrewAI and Hugging Face's agentic task delegation architecture further illustrate these principles by implementing nuanced decomposition strategies that support both fine-grained and coarse-grained task allocation. These frameworks leverage specialized agent networks with distinct expertise, enabling parallel processing and intelligent workload distribution.

Collaborative planning is also a key aspect of agent cooperation. Agents work together to develop plans that account for their individual capabilities and constraints while achieving collective objectives. This often involves iterative processes of proposal, critique, and refinement to arrive at mutually acceptable plans.

For example, the Restack.io framework exemplifies collaborative planning by orchestrating seamless interactions among specialized AI agents. Its architecture facilitates the development of narrow-scope experts that work in concert to tackle complex objectives. Through coordinated communication channels, these agents engage in iterative planning processes, proposing solutions, critiquing each other's ideas, and refining strategies to achieve collective goals. This approach allows for the dynamic integration of diverse expertise, enabling the system to adapt to changing circumstances and optimize outcomes based on the combined knowledge and capabilities of its constituent agents.

3.2.3 Competition Strategies

While cooperation is often emphasized in MASs, competitive behaviors can also play an important role in certain scenarios.

Market-based approaches are commonly used to structure competition in MASs. Agents compete for resources or tasks in a simulated marketplace, with pricing mechanisms used to efficiently allocate resources based on supply and demand. This approach can lead to efficient outcomes while allowing agents to pursue their individual goals.

Adversarial search techniques, often used in game-playing AI (Jain, 2024), can be applied in competitive multi-agent scenarios. Agents use strategies like minimax or alpha-beta pruning to make decisions that maximize their own utility while considering the potential actions of competing agents.

It's important to note that competition and cooperation are not mutually exclusive in MASs. Many real-world scenarios require a balance of both, often referred to as "coopetition." Designing mechanisms that encourage beneficial competition while maintaining overall system coherence is a key challenge in MAS design.

3.2.4 Task Allocation and Resource Sharing

Efficient task allocation and resource sharing are critical for the overall performance of MASs. These processes ensure that the collective capabilities of the agents are utilized effectively to achieve system goals.

Centralized task allocation approaches use a designated agent or system component to assign tasks based on a global view of the system state and agent capabilities. While this can lead to optimal allocations, it may create a bottleneck and single point of failure in large-scale systems.

To illustrate centralized task allocation, the following Python example demonstrates a load-balancing system where tasks are distributed among agents based on their current workloads:

```
from queue import PriorityQueue


class Agent:
    def __init__(self, name):
        self.name = name
        self.load = 0

    def assign_task(self, task):
        self.load += task["size"]
        print(f"{self.name} assigned task: {task['name']} (new
load: {self.load})")

tasks = [{"name": f"Task_{i}", "size": i} for i in range(1, 6)]
agents = [Agent(f"Agent_{i}") for i in range(1, 4)]
agent_queue = PriorityQueue()

for agent in agents:
    agent_queue.put((agent.load, agent))

for task in tasks:
    load, agent = agent_queue.get()
    agent.assign_task(task)
    agent_queue.put((agent.load, agent))
```

Decentralized approaches, on the other hand, allow agents to make local decisions about task acceptance and resource utilization. While potentially less optimal, these approaches can be more robust and scalable, particularly in dynamic environments where the system state changes rapidly.

Hybrid approaches attempt to balance the benefits of centralized and decentralized methods. For example, hierarchical task networks can be used to decompose complex tasks into subtasks, with higher-level allocation decisions made centrally and lower-level decisions made by individual agents or agent groups.

Resource-sharing mechanisms are closely tied to task allocation. Techniques such as token-based systems, where agents pass tokens representing resources or permissions, can be used to manage shared resources efficiently. More sophisticated approaches might involve economic models where agents trade or bid for resources based on their current needs and priorities.

The development of effective coordination techniques remains an active area of research in MASs. As these systems are applied to increasingly complex real-world problems, new challenges emerge, driving innovation in coordination strategies. The goal is to create robust, adaptive, and efficient MASs that can leverage the collective capabilities of diverse agents to tackle the complex challenges of our interconnected world.

3.2.5 *Criteria for Evaluating Multi-Agent Coordination*

Evaluating multi-agent coordination frameworks requires a set of well-defined criteria that assess their effectiveness in enabling agents to work collaboratively. These criteria provide a structured framework to compare tools and systems, ensuring they meet the needs of specific applications. Below are six key criteria for evaluating multi-agent coordination:

1. **Coordination Models**
   This criterion assesses the type of coordination supported by the framework, such as centralized, decentralized, or hybrid models. Centralized coordination allows a master agent or central controller to direct agent actions, ensuring global optimization. Decentralized models provide agents with autonomy to make local decisions, promoting flexibility and scalability. Hybrid approaches combine both strategies, leveraging the strengths of each.
2. **Task Allocation and Resource Management**
   The effectiveness of multi-agent coordination often depends on how tasks and resources are allocated among agents. Frameworks should facilitate dynamic task distribution based on agent capabilities and workloads. Resource management mechanisms, including equitable sharing or competitive allocation, are also crucial for optimizing overall system performance.
3. **Communication Protocols**
   Coordination requires robust communication between agents. Frameworks should support efficient and reliable communication protocols, whether synchronous, asynchronous, or event-driven. These protocols enable agents to share information, negotiate tasks, and collaborate effectively, even in distributed environments.
4. **Conflict Resolution Mechanisms**
   Multi-agent systems often face conflicts due to competing goals, resource contention, or task overlaps. Frameworks should include mechanisms to detect and resolve these conflicts through negotiation, arbitration, or predefined rules. Effective conflict resolution ensures harmonious agent interactions and prevents disruptions to overall system functionality.
5. **Scalability and Adaptability**
   Multi-agent systems must scale efficiently with increasing numbers of agents or complexity of tasks. Coordination frameworks should also adapt dynamically to changes in the environment, agent failures, or shifts in priorities, ensuring continued performance under diverse conditions.
6. **Behavioral Coherence and Goal Alignment**
   A well-coordinated system ensures that individual agent actions align with collective goals. This criterion evaluates how frameworks maintain behavioral coherence among agents, promote shared objectives, and prevent counterproductive actions, ensuring that agents work synergistically toward desired outcomes.

These criteria provide a comprehensive approach for evaluating multi-agent coordination frameworks and help identify the strengths and limitations of various systems. They also serve as a foundation for assessing the applicability of tools to specific use cases.

3.2.6 Evaluation of Some Multi-Agent Coordination Frameworks

In this section, we evaluate prominent frameworks and tools—**AutoGen**, **CrewAI**, **LangChain**, and **LlamaIndex (we had introduced these frameworks in Chap. 2)**—based on the six criteria for multi-agent coordination introduced earlier. These frameworks, while sharing some similarities, each offer distinct capabilities and limitations for building and managing multi-agent systems.

1. Coordination Models

- **AutoGen**: AutoGen supports a hybrid coordination model, combining centralized and decentralized elements. It includes planner agents that oversee task distribution and autonomous agents that collaborate iteratively. This allows flexibility in managing tasks that require both high-level planning and localized decision-making. The conversational approach used by AutoGen fosters fluid collaboration between agents and human operators.
- **CrewAI**: CrewAI adopts a decentralized, role-based coordination model. Each agent is assigned a specific role, allowing for modular and scalable workflows. Coordination is achieved through predefined protocols and event-driven interactions. This model is particularly well suited for systems where specialized agents must operate autonomously within a structured framework.
- **LangChain**: LangChain does not provide a dedicated coordination model but relies on decentralized agent autonomy. Agents in LangChain interact through pre-designed prompts or chains, which can simulate coordination to some extent, but the framework lacks built-in mechanisms for comprehensive agent collaboration.
- **LlamaIndex**: LlamaIndex focuses primarily on data ingestion and retrieval, offering event-driven interactions for agents responding to system events. While this enables some coordination, it is limited to data-centric tasks and does not provide a fully realized coordination framework.

2. Task Allocation and Resource Management

- **AutoGen**: Task allocation in AutoGen is dynamic and context-aware. Planner agents assess the capabilities of available agents and assign tasks accordingly. The framework supports iterative refinement, where agents can reassign tasks or modify their scope based on real-time feedback.
- **CrewAI**: CrewAI uses a role-based task allocation mechanism, with agents negotiating roles and responsibilities at runtime. Resources are managed

effectively through collaborative planning and workload balancing, ensuring efficient use of system capabilities.
- **LangChain**: LangChain's task allocation is implicit and managed through predefined workflows. While it allows agents to use tools independently, it lacks advanced features for runtime task reassignment or resource optimization.
- **LlamaIndex**: LlamaIndex supports task allocation within the context of data handling. Agents process tasks based on event triggers, but the framework does not provide explicit resource management or advanced task distribution mechanisms.

3. Communication Protocols
- **AutoGen**: AutoGen excels in interagent communication, supporting synchronous and asynchronous protocols. Agents interact through conversation-based programming, allowing seamless information exchange and coordination. This conversational framework also supports debugging and monitoring, enhancing transparency.
- **CrewAI**: CrewAI offers flexible communication mechanisms, including direct messaging and event-driven updates. This allows agents to share updates and synchronize actions efficiently.
- **LangChain**: LangChain does not provide dedicated communication protocols for multi-agent interactions. Communication is limited to chaining prompts and tools, which constrains the scope of agent collaboration.
- **LlamaIndex**: Communication in LlamaIndex is tied to event handling, enabling agents to respond to system changes or queries. However, its communication capabilities are focused on facilitating data exchange rather than multi-agent collaboration.

4. Conflict Resolution Mechanisms
- **AutoGen**: AutoGen incorporates mechanisms for conflict resolution through its iterative dialogue process. Agents negotiate and refine their actions in real time, ensuring alignment with overall goals. This approach is particularly effective in dynamic environments where conflicts may arise from task overlaps or resource contention.
- **CrewAI**: CrewAI handles conflicts through predefined protocols and structured workflows. Agents rely on role hierarchies and negotiation strategies to resolve disputes, ensuring smooth collaboration.
- **LangChain**: Conflict resolution is not a core feature of LangChain. Any conflict handling must be explicitly programmed by the developer, which limits its applicability in scenarios requiring adaptive multi-agent coordination.
- **LlamaIndex**: Conflict resolution in LlamaIndex is minimal, as the framework focuses on data management. Any conflict handling is incidental to its event-driven workflows and not an inherent feature.

5. Scalability and Adaptability
- **AutoGen**: AutoGen is designed to scale with increasing numbers of agents and task complexity. Its modular architecture and conversation-driven approach

allow agents to adapt dynamically to changing conditions, such as new tasks or unexpected failures.
- **CrewAI**: CrewAI's role-based structure supports scalability by enabling the addition of new agents without significant reconfiguration. Its event-driven design also allows for adaptability in dynamic environments.
- **LangChain**: While LangChain scales effectively for large language model (LLM) applications, its multi-agent scalability is limited by its lack of built-in coordination mechanisms.
- **LlamaIndex**: LlamaIndex handles scalability well in terms of managing large datasets and enabling efficient data retrieval. However, its adaptability is confined to data-centric applications and does not extend to broader multi-agent coordination scenarios.

6. Behavioral Coherence and Goal Alignment
- **AutoGen**: Behavioral coherence is a strength of AutoGen, as its iterative dialogues ensure that agents remain aligned with collective goals. The conversational framework facilitates transparent decision-making and goal refinement.
- **CrewAI**: CrewAI achieves goal alignment through its structured role definitions and predefined workflows. Agents operate within well-defined parameters, ensuring coherence in their actions.
- **LangChain**: LangChain lacks features for enforcing behavioral coherence or aligning agent actions with system-wide goals. Coordination relies on the developer's ability to design effective prompt chains.
- **LlamaIndex**: Behavioral coherence in LlamaIndex is limited to ensuring consistency in data handling tasks. Goal alignment is not a focus of the framework.

Table 3.2 gives a summary of this comparison.
Based on this evaluation

- **AutoGen** emerges as a versatile framework, excelling in coordination models, communication, and adaptability, making it suitable for complex, dynamic multi-agent systems.

Table 3.2 Multi-agent framework coordination comparison

Criterion	AutoGen	CrewAI	LangChain	LlamaIndex
Coordination models	Hybrid	Decentralized	Decentralized	Event-driven
Task allocation	Dynamic, iterative	Role-based, negotiated	Predefined workflows	Event-triggered tasks
Communication protocols	Robust (sync/async)	Flexible (event-driven)	Limited	Event-centric
Conflict resolution	Iterative, negotiation	Protocol-based	Developer-defined	Minimal
Scalability and adaptability	Modular, highly scalable	Role-based, scalable	Scales in LLM contexts	Scales in data contexts
Behavioral coherence	Strong alignment	Defined roles, workflows	Developer-dependent	Data-focused

- **CrewAI** provides strong role-based coordination and scalability, making it ideal for structured, task-specific systems.
- **LangChain** and **LlamaIndex** are better suited for specialized applications like LLM workflows or data handling but lack comprehensive features for multi-agent coordination.

3.3 Communication in MASs

Communication forms the foundation of MASs, enabling coordination, collaboration, and collective intelligence emergence. This section explores the fundamental aspects of agent communication, from basic principles to sophisticated semantic frameworks.

3.3.1 Fundamentals of Agent Communication

Agent communication encompasses structured interactions that enable intelligent agents to share information, coordinate actions, and achieve collective goals. Unlike simple data exchange, agent communication involves complex patterns of interaction that support autonomous decision-making and collaborative problem-solving.

The fundamental communication models in MASs consist of three primary approaches. Point-to-point communication enables direct message exchange between two agents, forming the basis for one-to-one interactions. Broadcast communication allows messages to be sent to all agents in the system, essential for system-wide updates or alerts. Multicast communication provides targeted message delivery to specific groups of agents, balancing efficiency with selective information sharing.

Communication patterns in MASs follow several established architectures. The request-reply pattern implements synchronous communication where an agent requests information or action and waits for a response. This pattern ensures clear transaction boundaries and simplifies error handling. The publish-subscribe pattern enables asynchronous communication, where agents subscribe to topics of interest and receive relevant updates without maintaining direct connections to publishers. Event-driven communication allows agents to react to system events or state changes, creating responsive and adaptive behaviors.

Message structure in agent communication follows standardized formats to ensure clarity and interoperability. Each message contains sender and receiver identifiers for routing, message type or performative indicating the intended action, content payload carrying the actual information, metadata including timestamps and priorities, and conversation identifiers for tracking multi-message exchanges.

Figure 3.1 describes multi-agent communication patterns.

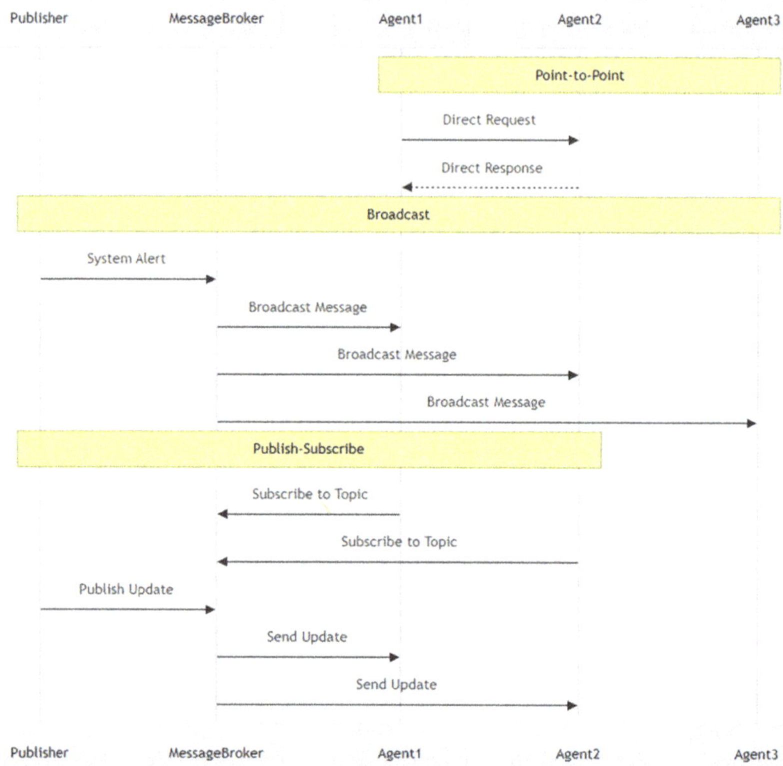

Fig. 3.1 Multi-agent communication patterns

3.3.2 Agent Communication Languages (ACLs)

Agent Communication Languages provide the semantic framework and syntactic structure for meaningful interaction between agents. ACLs go beyond simple data exchange protocols to incorporate sophisticated mechanisms for expressing intent, knowledge, and queries.

The Foundation for Intelligent Physical Agents (FIPA) Agent Communication Language represents the industry standard for agent communication (Wikipedia, 2024). FIPA-ACL defines a comprehensive set of performatives that enable rich agent interactions. These performatives include inform for sharing information, request for action solicitation, propose for negotiation initiation, agree/refuse for response to proposals, and query-if/query-ref for information gathering.

A typical FIPA-ACL message follows a structured format that includes all necessary communication elements:

```
(request
:sender (agent-identifier :name agent1@example.com)
:receiver (agent-identifier :name agent2@example.com)
:content "perform-task-x"
:protocol fipa-request
:language fipa-sl
:ontology task-ontology)
```

The Knowledge Query and Manipulation Language (KQML) introduced many foundational concepts in agent communication (Finin & Fritzson, 2023). KQML's architecture emphasizes extensibility through an expandable set of performatives, a layered architecture supporting different levels of communication complexity, and robust support for knowledge sharing between agents.

3.3.3 Message Transport and Routing

Modern MASs employ various message-passing mechanisms to ensure effective agent interaction. Asynchronous communication forms the backbone of scalable MASs. Message queues, implemented through technologies such as RabbitMQ and Apache Kafka, provide reliable message delivery with persistence and fault tolerance. Event buses enable loosely coupled communication between agents, while publish-subscribe middleware supports flexible message distribution patterns.

Routing strategies in MASs must address the complexity of message delivery in distributed environments. Direct routing establishes point-to-point connections between agents for immediate communication. Content-based routing determines message paths based on message contents, enabling intelligent message distribution. Topic-based routing organizes communication around subject areas, while semantic routing leverages message meaning for delivery decisions.

Error handling and recovery mechanisms ensure system reliability. Message acknowledgment protocols confirm successful delivery, while retry mechanisms handle temporary failures. Dead letter queues capture undeliverable messages for analysis and recovery. Circuit breakers prevent cascade failures by isolating problematic system components.

WebSockets are ideal for real-time, bidirectional communication in MASs, offering low latency and efficient message transport. They maintain persistent connections, enabling agents to send and receive messages instantly without the overhead of repeated HTTP handshakes. This efficiency makes them particularly suited for asynchronous and event-driven interactions. However, WebSockets require continuous connection management, which can increase resource usage, and scalability challenges arise when handling a large number of connections. The performance impact of WebSockets includes higher memory and CPU utilization, as well as increased bandwidth for idle connections in large-scale systems. These challenges can be mitigated through connection pooling, monitoring, and optimizing message serialization to reduce data size.

Protocol Buffers (Protobuf) provide a highly efficient message serialization mechanism, making them a strong choice for MASs that require compact, fast communication. Protobuf reduces bandwidth usage with smaller message sizes and ensures rapid serialization and deserialization, which is advantageous in distributed environments. Additionally, its cross-language support ensures compatibility in heterogeneous MAS setups. However, using Protobuf introduces complexity due to the need for schema definitions and careful version management. The performance hit is minimal compared to text-based formats, but managing schema evolution can slow down deployment processes. This impact can be mitigated by adopting best practices for schema versioning and leveraging lightweight Protobuf libraries.

gRPC, which uses Protocol Buffers for serialization, delivers high-performance and scalable RPC-based communication in MASs. Its efficiency stems from compact Protobuf serialization and advanced HTTP/2 features such as multiplexing and flow control. gRPC's support for bidirectional streaming is particularly beneficial for agent interactions in real-time systems. However, it introduces a steeper learning curve due to Protobuf and gRPC-specific concepts, and debugging is more complex than with REST APIs. The performance challenges include managing multiple streams and handling serialization overhead, though these are mitigated by optimizing gRPC configurations, reusing connections, and employing interceptors to monitor performance.

REST APIs offer a straightforward and widely adopted approach to message transport in MASs, particularly for systems requiring stateless communication. Their simplicity and ubiquity make them an accessible choice for interoperability across diverse platforms. REST APIs leverage HTTP methods for communication, ensuring compatibility with existing web infrastructure. However, they are less suitable for real-time or high-throughput scenarios due to the overhead of HTTP headers and verbose message formats like JSON. This can lead to increased latency and reduced efficiency in performance-critical systems. To mitigate these challenges, REST APIs can be optimized by adopting lightweight data formats such as Protobuf or MessagePack, implementing caching strategies, and minimizing unnecessary HTTP calls through batch requests or combining endpoints. While REST APIs are not ideal for all MAS scenarios, they remain a robust and flexible option for systems prioritizing ease of integration and simplicity.

3.3.4 Semantic Frameworks

Semantic frameworks ensure consistent interpretation of messages across diverse agent populations. These frameworks provide the foundation for meaningful agent interaction and knowledge sharing.

Ontologies serve as formal representations of domain knowledge, enabling agents to share a common understanding of their environment. They define concepts and relationships within specific domains, establish operational constraints, and codify business rules. This shared understanding enables accurate message interpretation and appropriate action selection.

Semantic interoperability requires several complementary approaches. Shared ontologies establish common vocabularies and standardized relationships between concepts. Ontology mapping techniques enable communication between agents using different ontologies through alignment mechanisms and translation rules. Semantic bridges connect disparate knowledge representations, enabling cross-domain communication.

Machine learning approaches enhance semantic understanding through automated interpretation. Natural language processing techniques extract meaning from unstructured communications. Neural networks facilitate concept alignment between different semantic frameworks. Automated ontology mapping reduces the manual effort required to connect different knowledge representations.

Domain-specific standards ensure consistency in particular applications. The Financial Information eXchange (FIX) Protocol standardizes communication in financial trading systems. Health Level Seven (HL7) protocols enable healthcare information exchange. The Message Queuing Telemetry Transport (MQTT) protocol supports Internet of Things device communication.

3.3.5 Implementation Best Practices

Successful implementation of agent communication requires careful attention to design principles and common challenges. Proper implementation ensures system reliability, maintainability, and performance.

Message design best practices emphasize clarity and future-proofing. Clear message semantics ensure consistent interpretation across the system. Version support enables system evolution while maintaining compatibility. Backward compatibility mechanisms allow gradual system updates without disrupting existing operations.

Protocol design considerations focus on system scalability and reliability. Performance optimization techniques include message batching, compression, and caching. Load balancing mechanisms distribute communication load across system resources. Failure handling protocols ensure system resilience in the face of communication errors.

Implementation pitfalls require careful attention during system development. Overcomplex message formats can reduce system maintainability and performance. Insufficient error handling may lead to system instability. Inadequate message validation can compromise system security and reliability.

3.4 Conflict Resolution in Multi-Agent Environments

Conflict resolution aims to provide harmonious operation between agents with diverse goals, capabilities, and perspectives. This section explores the nature of conflicts in AI ecosystems, methods for their detection, and resolution strategies.

3.4.1 *Types of Conflicts in AI Ecosystems*

Resource conflicts emerge when multiple agents compete for limited system resources. In smart factory environments, this manifests when multiple robotic agents require simultaneous access to shared tools or production lines. Resource conflicts extend beyond physical resources to encompass computational resources, network bandwidth, and memory allocation. The time-sensitive nature of resource conflicts often requires immediate resolution to maintain system efficiency.

Goal conflicts arise from fundamental contradictions between agent objectives. Traffic management systems exemplify this when one agent prioritizes individual vehicle travel times while another focuses on overall network congestion reduction. Goal conflicts often involve complex trade-offs between local optimization and global system performance. These conflicts frequently emerge in systems where agents serve different stakeholders or operate under varying optimization criteria.

Belief conflicts occur when agents maintain inconsistent or contradictory information about their environment or system state. Distributed sensor networks illustrate this when different agents possess varying readings of environmental conditions, leading to conflicting interpretations and responses. Belief conflicts can propagate through the system, affecting decision-making processes and potentially leading to suboptimal or counterproductive actions. The resolution of belief conflicts often requires sophisticated information-sharing and consensus-building mechanisms.

Plan conflicts materialize when the intended actions of one agent interfere with another's planned operations. Warehouse automation systems frequently encounter these conflicts when multiple robotic agents plan to utilize the same physical spaces or resources simultaneously. Plan conflicts require both spatial and temporal coordination for resolution. The complexity of plan conflicts increases exponentially with the number of agents and the intricacy of their planned actions. Figure 3.2 shows a state diagram for conflict resolution.

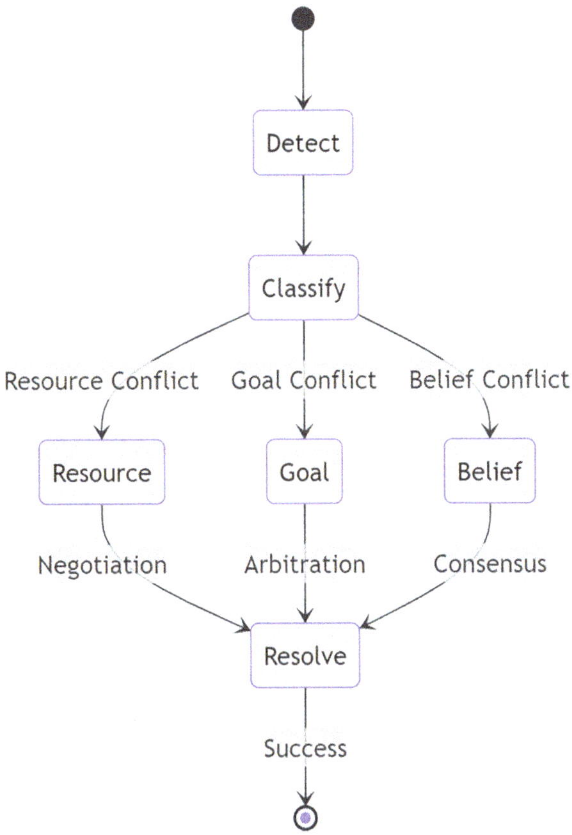

Fig. 3.2 Conflict resolution state diagram

3.4.2 Conflict Detection Mechanisms

Effective conflict resolution begins with robust detection mechanisms that identify potential conflicts before they impact system performance. Proactive detection enables preventive measures and smooth resolution processes.

Plan analysis algorithms serve as the primary mechanism for proactive conflict detection. These algorithms examine the intended actions of multiple agents, identifying potential intersections or contradictions in their planned behaviors. Modern plan analysis incorporates temporal reasoning to detect conflicts that may arise from timing variations in agent actions. The analysis process must balance computational complexity with detection accuracy, particularly in large-scale systems.

Runtime monitoring provides continuous oversight of agent interactions and system states. Advanced monitoring systems employ anomaly detection algorithms to identify unexpected behaviors or state changes that might indicate emerging conflicts. Real-time monitoring must process vast amounts of system data while maintaining low latency in conflict detection. The monitoring system's sensitivity requires careful calibration to avoid false positives while ensuring no significant conflicts go undetected.

Belief revision techniques play a crucial role in managing potential conflicts arising from inconsistent information. These techniques maintain models of agent beliefs and regularly cross-reference them to identify potential contradictions. When discrepancies emerge, the system initiates information-sharing protocols to align agent beliefs. The belief revision process must account for the reliability of different information sources and the confidence levels associated with various beliefs.

3.4.3 Resolution Strategies

Conflict resolution in MASs employs various strategies, each suited to particular conflict types and system requirements. The selection of resolution strategies depends on conflict characteristics, system architecture, and operational constraints.

Negotiation-based approaches form the foundation of many conflict resolution systems, particularly for resource allocation and goal conflicts. These approaches establish structured dialogue protocols through which agents can express their needs, constraints, and preferences. The negotiation process often employs economic models to facilitate resource allocation, with agents bidding for resources based on their perceived utility. Sophisticated negotiation protocols incorporate learning mechanisms that enable agents to improve their negotiation strategies over time.

One effective negotiation-based method is a bidding system, where agents propose values to gain access to contested resources. Below is a Python example demonstrating this approach in action.

```
class ResourceConflict:
    def __init__(self, resource_name):
        self.resource_name = resource_name
        self.bids = []

    def place_bid(self, agent_name, bid_value):
        self.bids.append((agent_name, bid_value))

    def resolve(self):
        winner = max(self.bids, key=lambda x: x[1])
        print(f"Resource '{self.resource_name}' allocated to
{winner[0]} with bid {winner[1]}")

conflict = ResourceConflict("Shared_Resource")
conflict.place_bid("Agent_A", 50)
conflict.place_bid("Agent_B", 70)
conflict.place_bid("Agent_C", 65)
conflict.resolve()
```

Arbitration mechanisms provide resolution through neutral third-party agents or system components. The arbitrator evaluates conflicting claims and requirements, applying predefined rules or optimization criteria to reach binding decisions. Arbitration proves particularly effective in time-critical situations where rapid resolution is essential. The design of arbitration mechanisms must ensure fairness while maintaining system efficiency.

Hierarchical resolution frameworks leverage system structure to address conflicts at appropriate organizational levels. Lower-level conflicts undergo resolution through local mechanisms, while more complex or widespread conflicts escalate to higher-level resolution processes. This hierarchical approach enables efficient handling of routine conflicts while providing escalation paths for more challenging situations. The framework must balance local autonomy with global system optimization.

Adaptive conflict resolution introduces learning capabilities into the resolution process. Agents analyze patterns in conflict occurrence and resolution outcomes to refine their conflict management strategies. Machine learning algorithms enable agents to predict potential conflicts and preemptively adjust their behaviors. The adaptive approach requires a careful balance between the exploration of new resolution strategies and the exploitation of known effective solutions.

Preventive strategies focus on system design elements that naturally minimize conflict occurrence. These strategies include careful resource allocation mechanisms, a clear definition of agent authorities and responsibilities, and coordination protocols that inherently reduce conflict probability. While preventive approaches cannot eliminate all conflicts, they significantly reduce system friction and resolution overhead. The implementation of preventive strategies requires a deep understanding of system dynamics and potential conflict sources.

3.5 Designing Multi-Agent Environments

This section explores key considerations in architecting MASs, defining agent roles and specializations, and ensuring the scalability and flexibility of these complex systems.

3.5.1 Architectural Considerations

The architecture of a MAS forms the foundation upon which agent interactions, communication, and coordination are built. A well-designed architecture can facilitate efficient collaboration, while a poor design can lead to bottlenecks, conflicts, and system instability.

Centralized architectures involve a single controlling entity that oversees all agents, making high-level decisions and coordinating actions. This approach can provide global optimization and simplify coordination but may create a single point of failure (SPOF) and scalability issues. However, leveraging hyperscaler's cloud workloads can mitigate SPOF concerns by distributing the control center's functions across robust, scalable cloud infrastructures. For instance, a smart city traffic management system might use a centralized architecture to optimize overall traffic flow, with a main control center, hosted on a few of the hyperscaler's clouds, directing individual traffic light agents.

Decentralized architectures, on the other hand, distribute decision-making among agents, allowing for greater autonomy and robustness. In this model, agents make local decisions based on their individual knowledge and goals, without relying on a central authority. While this can lead to more resilient systems, it may result in suboptimal global behavior. A peer-to-peer network for distributed computing is an example of a decentralized multi-agent architecture.

Hybrid architectures attempt to balance the benefits of both centralized and decentralized approaches. They might use hierarchical structures where some decisions are made centrally while others are delegated to local agent groups. This can provide a good compromise between global optimization and local responsiveness. In a large-scale manufacturing system, for example, high-level production planning might be centralized, while individual production lines operate with more autonomy.

The choice of architecture significantly impacts system performance, scalability, and fault tolerance. Designers must carefully consider the specific requirements of their application domain when selecting an architectural approach.

3.5.2 Agent Roles and Specializations

In complex MASs, defining clear roles and specializations for agents can greatly enhance overall system efficiency and effectiveness. This involves determining the specific responsibilities, capabilities, and limitations of different agent types within the ecosystem.

Functional specialization involves creating agents with expertise in particular tasks or domains. For example, in a healthcare MAS, you might have agents specialized in diagnosis, treatment planning, patient monitoring, and resource allocation. This specialization allows agents to develop deep expertise in their specific areas, leading to more effective problem-solving.

Hierarchical role structures can be employed to manage complexity in large-scale systems. This might involve supervisor agents overseeing groups of worker agents, with different levels of decision-making authority and responsibility. In a smart manufacturing environment, you could have shop floor agents, line manager agents, and plant manager agents, each with an increasing scope of control and decision-making power.

Adaptive role assignment is an advanced approach where agent roles can change dynamically based on system needs and agent performance. This flexibility can be particularly valuable in dynamic environments where the nature of tasks or the availability of resources may change frequently. Machine learning techniques can be employed to optimize role assignments over time based on observed performance and changing conditions.

When designing agent roles and specializations, it's crucial to consider the balance between specialization and generalization. While specialized agents can be highly efficient in their domains, having some degree of generalization can provide system flexibility and robustness in the face of unexpected situations or agent failures.

3.5.3 Scalability and Flexibility

Modular design principles are key to building scalable MASs. By encapsulating functionality within well-defined modules or agents, systems can be more easily expanded or modified. This modularity also facilitates the reuse of components across different applications or scenarios, improving development efficiency and system reliability.

Load balancing mechanisms can be used to maintain performance as the scale of the system grows. This might involve dynamically redistributing tasks among agents based on their current workload and capabilities. In cloud-based MASs, for instance, new agent instances might be automatically spawned to handle increased processing demands, with tasks dynamically allocated to optimize resource utilization.

Interoperability standards can provide flexibility and scalability. By adhering to common communication protocols and data formats, agents from different developers or even different generations of development can work together seamlessly. This interoperability is particularly important in open systems where new agents might be introduced over time or in scenarios where MASs need to interact with external systems or data sources.

Scalable data management is another critical consideration, especially in data-intensive applications. This might involve distributed database systems, data sharding techniques, or the use of big data technologies to handle large volumes of information efficiently. The ability to scale data processing and storage capabilities in tandem with the growth of the agent population is essential for maintaining system performance and effectiveness.

Evolutionary design approaches can enhance the flexibility of MASs over time. By building in mechanisms for agents to learn and adapt their behaviors based on experience, systems can evolve to meet changing requirements or environmental conditions without requiring complete redesigns. This might involve the use of evolutionary algorithms, reinforcement learning, or other machine learning techniques to continuously optimize agent behaviors and system configurations.

Secure and responsible development of MASs is paramount. We will discuss this more in Chap. 12 of this book. For a more broad scope of generative AI security, you can review my other book titled "Generative AI Security"(Huang et al., 2024).

3.6 System Maintenance and Evolution

The maintenance and evolution of MASs present unique challenges that go beyond traditional software maintenance. This section explores the key aspects of keeping MASs operational, up-to-date, and effectively documented over time.

3.6.1 Agent Deployment and Retirement

The lifecycle of individual agents within a MAS requires careful management to maintain system stability and performance. Agent deployment involves more than simply adding new code to the system; it requires consideration of how new agents will integrate with existing ones, their impact on system resources, and their effect on established communication patterns.

During deployment, agents must be properly initialized with appropriate knowledge bases, communication protocols, and security credentials. This initialization phase often involves a period of supervised operation where the new agent's interactions are monitored to ensure they align with system expectations. For example, in a financial trading system, new trading agents might initially operate in a simulation mode before being granted access to real trading capabilities.

Agent retirement is equally important but often overlooked. The process must ensure that retiring agents properly hand off their responsibilities, complete any pending tasks, and cleanly terminate their connections with other agents. This includes archiving relevant knowledge or state information that might be valuable for future analysis or system audits. The retirement process must be managed to prevent disruption to ongoing system operations and maintain the integrity of inter-agent relationships.

3.6.2 System Health Monitoring and Diagnostics

Operational health monitoring in MASs focuses on maintaining runtime stability and detecting immediate issues that could impact system performance. This section addresses the real-time monitoring mechanisms necessary for maintaining operational integrity.

Real-Time Health Indicators

Real-time monitoring tracks critical operational metrics that indicate system health. Communication latency measurements provide immediate feedback on message delivery performance between agents. Resource utilization tracking monitors CPU, memory, and network bandwidth consumption across the agent ecosystem. Queue depth monitoring ensures message backlogs don't exceed acceptable thresholds. These operational indicators enable immediate detection of performance degradation or impending system issues.

Operational Diagnostics

The diagnostic system employs multiple layers of checks to maintain system health. At the agent level, heartbeat mechanisms verify individual agent responsiveness and operational status. Network-level diagnostics monitor connection quality and communication path reliability. System-level diagnostics track overall resource allocation and utilization patterns. This layered approach ensures comprehensive health monitoring across all system components.

The diagnostic process incorporates automated response mechanisms for common issues. When communication paths show increased latency, the system can automatically redirect traffic through alternate routes. Resource exhaustion triggers automatic scaling or load balancing responses. Agent unresponsiveness initiates failover procedures to maintain system functionality. These automated responses ensure rapid reaction to operational issues.

Alert Management

The alert system prioritizes notifications based on operational impact and urgency. Critical alerts, such as agent failures or severe resource constraints, require immediate operator attention. Warning alerts indicate potential issues that may require preventive action, such as approaching resource limits or unusual communication patterns. Informational alerts track routine but noteworthy system events for operational awareness.

Alert correlation mechanisms identify related issues to prevent alert fatigue. Pattern recognition algorithms group similar alerts to highlight systemic problems. Temporal analysis identifies alert sequences that may indicate cascading failures. This intelligent alert management ensures operators can focus on the most critical issues while maintaining awareness of system status.

Preventive Monitoring

Preventive monitoring focuses on identifying potential issues before they impact system operation. Trend analysis tracks resource utilization patterns to predict potential exhaustion. Communication pattern monitoring identifies developing bottlenecks or inefficient routing. Agent behavior analysis detects subtle changes that might indicate emerging problems.

The preventive system maintains historical baseline data for normal operation patterns. Deviation detection algorithms compare current system behavior against these baselines. Statistical analysis identifies trends that may indicate developing issues. This forward-looking approach enables proactive maintenance and issue prevention.

Operational Logging

The logging system maintains detailed records of system operation for diagnostic purposes. Each log entry includes timestamp, severity level, source component, and detailed event description. Structured logging formats enable automated analysis and pattern detection. Log rotation and archival procedures ensure comprehensive historical records while managing storage requirements.

Log analysis tools provide real-time insight into system operation. Pattern matching algorithms identify unusual event sequences. Frequency analysis detects changes in event occurrence patterns. These analysis capabilities enable rapid problem diagnosis and resolution.

Health Recovery Procedures

Automated recovery procedures respond to common operational issues. Agent restart protocols handle simple agent failures. Resource reallocation procedures address utilization imbalances. Communication path failover mechanisms maintain system connectivity despite network issues. These automated procedures minimize disruption from routine operational problems.

Manual recovery procedures provide documented responses for complex issues requiring operator intervention. Each procedure includes clear trigger conditions, step-by-step recovery actions, and validation checks. Regular testing ensures recovery procedures remain effective as the system evolves.

The health monitoring system operates continuously, providing real-time insight into system operation while enabling rapid response to operational issues. Through comprehensive monitoring, intelligent alerting, and automated recovery procedures, it maintains optimal system health and performance.

3.6.3 Configuration and Version Management

Managing configurations in MASs presents unique challenges due to the distributed nature of these systems and the potential for complex interactions between agent configurations. A robust configuration management system must track not only individual agent configurations but also the compatibility requirements between different agents and their versions.

Version control in MASs extends beyond traditional software versioning to include the management of agent knowledge bases, interaction protocols, and behavioral rules. This requires sophisticated tracking systems that can maintain consistency across the entire agent ecosystem while allowing for incremental updates and rollbacks when necessary.

Configuration changes must be managed with particular attention to their system-wide impacts. Changes to one agent's configuration might have ripple effects through its interactions with other agents, requiring careful coordination of updates across the system. This often necessitates the development of sophisticated dependency tracking and impact analysis tools.

3.6.4 Documentation and Knowledge Management

Effective documentation in MASs must capture not only the technical specifications of individual agents but also the complex web of interactions and dependencies between them. This includes documenting communication protocols, coordination mechanisms, and the rationale behind architectural decisions.

Knowledge management extends beyond traditional documentation to include the capture and preservation of system evolution history, incident responses, and lessons learned. This institutional knowledge becomes particularly valuable when debugging complex issues or planning system upgrades.

Documentation must also address the unique aspects of MASs, such as emergent behaviors, agent learning processes, and adaptation mechanisms. This requires new approaches to documentation that can effectively describe dynamic and evolving system behaviors rather than just static specifications.

3.7 Evaluation and Benchmarking of MASs

The systematic evaluation and benchmarking of MASs are crucial for understanding their effectiveness, comparing different approaches, and guiding future development. This section explores frameworks, methodologies, and best practices for assessing MASs.

3.7.1 Evaluation Frameworks

Evaluating MASs requires comprehensive frameworks that can assess both individual agent performance and emergent system-wide behaviors. Traditional software evaluation metrics often prove insufficient for capturing the complex dynamics of multi-agent interactions and their collective outcomes.

Quantitative evaluation frameworks typically focus on measurable aspects such as task completion rates, resource utilization efficiency, and communication overhead. However, they must also account for less tangible factors such as the quality of agent cooperation, the effectiveness of conflict resolution, and the system's ability to adapt to changing conditions. For example, in a multi-agent manufacturing system, evaluation might consider not only production throughput but also the system's flexibility in handling unexpected orders or equipment failures.

Qualitative evaluation approaches complement quantitative metrics by assessing aspects such as the robustness of coordination mechanisms, the effectiveness of information sharing, and the system's overall coherence. These evaluations often involve expert analysis and scenario-based testing to understand how well the system meets its intended objectives.

3.7.2 Benchmarking Methodologies

Benchmarking MASs requires carefully designed methodologies that can provide fair and meaningful comparisons across different implementations and approaches. This involves creating standardized test scenarios, defining common performance metrics, and establishing baseline expectations for different types of MASs.

Scenario-based benchmarking involves testing systems against a set of predefined scenarios that represent typical use cases and edge cases. These scenarios might include normal operating conditions, high-stress situations, and various types of system perturbations. The scenarios should be designed to evaluate both the functional capabilities of the system and its nonfunctional characteristics such as scalability and resilience.

Comparative benchmarking enables organizations to evaluate different multi-agent architectures or implementations against each other. This might involve comparing different coordination mechanisms, communication protocols, or decision-making algorithms under identical conditions. Such comparisons can provide valuable insights for system selection and optimization.

3.7.3 Performance Analysis Techniques

Advanced analysis techniques are needed to understand the complex behaviors and interactions within MASs. These techniques must go beyond simple performance measurements to provide insights into the system's dynamics and identify opportunities for improvement.

Interaction analysis focuses on understanding the patterns and effectiveness of agent communications and collaborations. This might involve analyzing message flows, identifying communication bottlenecks, and evaluating the efficiency of coordination mechanisms. Tools for visualizing agent interactions and their evolution over time can provide valuable insights into system behavior.

Behavioral analysis examines how individual agents and the system as a whole respond to different situations and stimuli. This includes analyzing decision-making patterns, learning processes, and adaptation mechanisms. Understanding these behaviors is crucial for optimizing system performance and ensuring desired outcomes.

3.7.4 Standardization and Best Practices

The development of standards and best practices for MAS evaluation helps ensure consistency and comparability across different implementations and studies. This includes standardized metrics, evaluation procedures, and reporting formats.

Metric standardization involves defining common measures for aspects such as system performance, reliability, and efficiency. These standards should be flexible enough to accommodate different types of MASs while providing meaningful bases for comparison. For instance, standardized metrics might include measures of coordination efficiency, adaptation speed, and resource utilization.

Documentation standards ensure that evaluation results are reported in a consistent and comprehensive manner. This includes specifying the test conditions, system configurations, and methodologies used in the evaluation process. Clear documentation enables others to reproduce results and build upon previous work.

Through systematic evaluation and benchmarking, organizations can better understand the capabilities and limitations of their MASs, make informed decisions about system design and deployment, and track improvements over time. These practices also contribute to the broader field by providing comparable results and insights that can guide the future development of multi-agent technologies.

3.8 Real-World Applications of MASs

In this section, we will highlight some real-world use cases of MAS. Most of these examples leverage either reinforcement learning or predictive AI for the agent decision-making process. We have yet to see MAS agent's real-world use cases leveraging LLM or GenAI. However, given the rate of innovation in the intersection of MAS and GenAI/LLM, there will be many more use cases in the near future.

3.8.1 Smart Cities and Urban Management

Smart cities represent one of the most promising applications of multi-agent coordination, integrating various subsystems to improve urban living and resource management.

Traffic management in smart cities often relies on MASs to optimize traffic flow. Agents representing traffic lights, vehicles, and central control systems work together to reduce congestion and improve overall mobility. For instance, in Singapore's Intelligent Transport System, a network of agents monitors traffic conditions in real time, adjusting signal timings and providing route recommendations to drivers. This coordinated approach has resulted in a significant reduction in traffic congestion and improved travel times across the city (ASEAN Post, 2018).

Energy management is another aspect of smart cities where multi-agent coordination plays a vital role. Agents representing power generation sources, distribution networks, and consumer devices collaborate to balance supply and demand, integrate renewable energy sources, and optimize energy consumption. In Amsterdam's Smart City initiative, a MAS manages a complex network of solar panels, electric vehicle charging stations, and smart meters to create a more sustainable and efficient energy ecosystem (Derrick, 2024).

Waste management in smart cities benefits from multi-agent coordination through optimized collection routes, real-time bin monitoring, and adaptive scheduling. In Barcelona, a network of sensor-equipped waste bins communicates with collection vehicle agents to optimize pickup routes and schedules, resulting in reduced costs and improved city cleanliness (Sonnier, 2023).

3.8.2 Supply Chain and Logistics

The complexity and distributed nature of modern supply chains make them ideal candidates for multi-agent coordination systems.

Inventory management across complex supply networks often involves multiple agents representing suppliers, warehouses, transportation providers, and retailers. These agents coordinate to optimize inventory levels, predict demand, and manage the flow of goods. Walmart, for example, uses a sophisticated MAS to manage its vast supply chain, enabling real-time inventory tracking and dynamic reordering across thousands of stores and suppliers (Musani, 2023).

Transportation optimization in logistics benefits from multi-agent coordination through dynamic routing and load balancing. Agents representing vehicles, packages, and distribution centers work together to optimize delivery routes, considering factors such as traffic conditions, package priorities, and vehicle capacities. Companies like DHL and FedEx employ MASs to manage their global logistics networks, resulting in improved delivery times and reduced operational costs (Kamran, 2024).

Collaborative planning and forecasting in supply chains involve agents sharing information and coordinating decisions across organizational boundaries. For instance, Procter and Gamble used a MAS to coordinate with its retailers, sharing sales data and inventory information to improve demand forecasting and reduce the bullwhip effect in its supply chain (Lafferty, 2018).

3.8.3 Disaster Response and Emergency Management

Multi-agent coordination has proven particularly valuable in the chaotic and time-critical domain of disaster response and emergency management.

Search and rescue operations often employ MASs to coordinate the efforts of human responders, drones, ground robots, and sensor networks. These agents work together to search large areas efficiently, share information about victim locations, and allocate resources dynamically. For example, DARPA's LORELEI program employs a MAS approach to address the challenges of crisis management in low-resource language environments. This system consists of specialized agents working collaboratively to achieve rapid situational awareness. Language processing agents extract key information from diverse sources, while knowledge integration agents assimilate and contextualize the data. Machine learning agents continuously improve the system's language understanding capabilities across multiple languages. Interface agents provide an intuitive user experience for first responders and decision-makers. Together, these agents form a dynamic, adaptive system capable of delivering critical insights within 24 h of an incident and evolving to full language automation within days or weeks. This distributed architecture allows for parallel processing, scalability, and robustness in handling complex, time-sensitive crisis scenarios (Research Outreach, 2023).

Resource allocation during disasters involves coordinating multiple agents representing different resources (e.g., medical supplies, food, shelter) and matching them with areas of need. The United Nations' Humanitarian Data Exchange platform uses a MAS to coordinate the distribution of aid across multiple organizations and regions during large-scale humanitarian crises (HDX, 2023).

3.9 AI Agents to Multi-Agent Systems: A Capability Framework

In this section, I focus on examining AI agents from a range of capabilities, spanning from basic data processing to complex autonomous decision-making. This framework allows us to explore the progression of AI agents through different levels of sophistication, highlighting the core features, functionalities, and distinctions at each stage.

Level 1: Perception and Data Processing
At the foundational level, the focus is on an AI agent's ability to process sensory data (images, text, audio, etc.) and extract features for computation. Metrics include recognition accuracy, precision, recall, and efficiency. Multi-agent considerations are absent as this stage is purely about individual data-processing capabilities.

Level 2: Reasoning and Problem-Solving
The emphasis here is on logical reasoning, inference, and structured problem-solving for individual agents. The focus is on tasks like executing algorithms, finding solutions within constraints, and optimizing outcomes in well-defined environments. Multi-agent systems are not directly relevant unless the task explicitly requires interagent interaction.

Level 3: Learning and Adaptation
This level involves the ability of individual agents to improve performance over time through supervised, unsupervised, or reinforcement learning. While individual learning remains the priority, collaborative learning or competitive adaptation might appear in specific scenarios (e.g., game simulations or shared environments), but these are exceptions rather than core features.

Level 4: Context Awareness
Agents must understand and adapt to their environment, including spatial, temporal, and social dimensions. Multi-agent considerations begin to emerge here in scenarios where agents share a dynamic environment, such as in robotics or autonomous navigation systems, requiring mutual awareness but not full collaboration.

Level 5: Autonomy and Decision-Making
At this stage, an agent demonstrates autonomy by making decisions independently in dynamic environments. Multi-agent considerations appear in decentralized systems, where individual decisions may impact or rely on other agents. For example, distributed systems like supply chain management may require agents to autonomously coordinate actions without centralized control.

Level 6: Collaboration and Coordination (Multi-Agent Introduced)
This is the first level where multi-agent systems become a primary focus. Agents collaborate to achieve shared objectives, requiring mechanisms for communication, task allocation, and conflict resolution. This level evaluates how well agents work

collectively, balancing individual and group goals. Metrics include team efficiency, robustness to agent failure, and quality of interagent cooperation.

Level 7: Communication and Interaction
Agents are evaluated for their ability to communicate effectively, interpret intent, and maintain contextual relevance. Multi-agent considerations are significant when agents must share information, negotiate, or resolve conflicts, as in swarm intelligence or distributed planning systems.

Level 8: Creativity and Innovation
Creativity involves producing novel and valuable outputs, such as designing new solutions or strategies. In multi-agent systems, creativity might involve emergent behaviors where collaboration leads to innovative results. However, this level can also be assessed for individual agents, so multi-agent dynamics are context-dependent.

Level 9: Ethical and Value Alignment
Agents are evaluated for their ability to align actions with ethical principles and societal norms. In multi-agent systems, this involves ensuring collective behaviors adhere to ethical constraints, such as fairness, bias mitigation, and privacy preservation.

Level 10: General Intelligence
This level corresponds to artificial general intelligence (AGI), where agents can perform tasks across diverse domains with human-like flexibility. Multi-agent considerations apply when general intelligence emerges in systems of agents working collaboratively, but an individual AGI agent can also operate independently.

Level 11: Self-Improvement and Meta-Learning
At this ultimate level, an agent demonstrates the ability to improve its own architecture, learning strategies, and operational methods. Multi-agent systems may play a role if agents collaboratively refine their algorithms, but the focus is on individual and systemic self-improvement.

Let us use a diagram to summarize these 11 levels (Fig. 3.3).

3.10 Build API for AI Agents in Multi-Agent Systems

Ever wonder how to securely and efficiently connect AI agents in a multi-agent system? This section provides some helpful guidance.

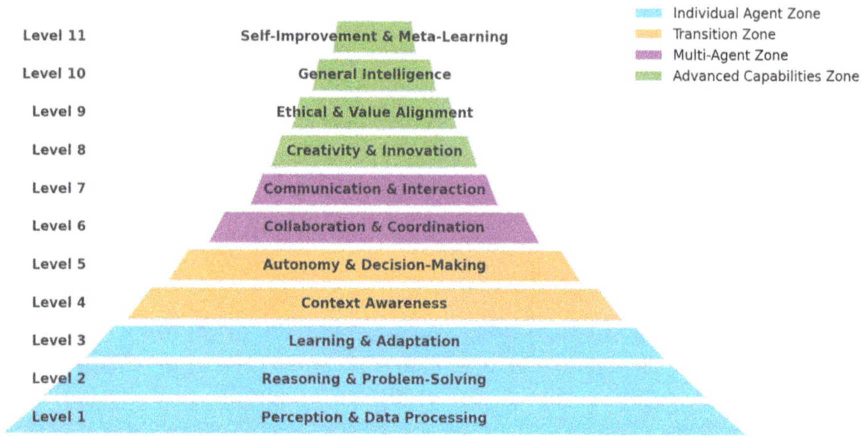

Fig. 3.3 Agent capability levels

3.10.1 Why Expose an AI Agent as an API?

Whether an AI agent should be exposed as an API depends on its intended use, the nature of its interactions, and the specific goals of the deployment. Exposing an agent as an API is often beneficial for integration, scalability, and automation, but there are cases where a more direct interaction method is preferable. These considerations involve balancing the agent's technical capabilities with the user experience and deployment goals.

Reasons to Expose an Agent as an API

Integration with Other Applications: APIs allow seamless interaction with various systems, enabling the agent's functionality to be embedded into different platforms or workflows. This is especially useful in scenarios where the agent provides specialized tasks, such as recommendation engines, predictive models, or data-driven insights.

Scalability: APIs facilitate scalability by enabling multiple applications to access the agent's capabilities without requiring significant additional infrastructure. For example, an AI agent that provides fraud detection can be integrated across multiple services via an API.

Developer Flexibility: Developers can create custom applications and services that leverage the agent's features. By exposing APIs, the agent becomes part of a larger ecosystem, allowing teams to innovate on top of its functionality.

Software as a Service (SaaS) Offerings: Exposing an agent as an API can form the foundation of a SaaS business model, allowing clients to access and pay for specific functionalities without managing the underlying infrastructure.

Testing and Prototyping: Exposing an agent via an API enables developers to test its behavior under various conditions. This is especially important in scenarios where the agent's logic or machine learning model needs to be validated against real-world data.

Data Sharing and Analysis: APIs can act as conduits for sharing insights generated by the agent with other systems, enabling advanced analytics, reporting, and decision-making processes.

Automation: Exposing the agent as an API allows for integration into automated workflows, such as triggering responses or taking actions in event-driven systems.

Considerations Before Exposing an Agent as an API

Security: Exposing an agent as an API increases its attack surface, requiring robust security measures such as authentication, authorization, encryption, and rate limiting to prevent abuse.

Complex Functionality: If the agent relies on rich conversational interactions or extensive contextual understanding, an API might limit its effectiveness. For example, a customer support chatbot may perform better when embedded directly in a messaging platform rather than accessed through an API.

Performance Overhead: An API may lead to performance bottlenecks if multiple applications access the agent simultaneously. Ensuring scalability and low latency is crucial in such cases.

Cost and Maintenance: Hosting an API and maintaining its infrastructure involve additional costs, especially if the agent requires significant computational resources.

3.10.2 Key Components of an Effective API Specification for AI Agents

1. API Structure and Communication Protocols

   The API should use secure and widely accepted communication protocols. RESTful APIs are well suited for general-purpose operations, while GraphQL may be more appropriate for scenarios requiring flexible querying. Additionally, WebSocket support is essential for real-time multi-agent communication, where event-driven messaging is necessary.
2. Authentication and Identity Verification

   Implement robust mechanisms to authenticate users and verify their identities. OAuth 2.0 with OpenID Connect is a popular choice, enabling secure authentication and federated identity verification. Additionally, API keys and JSON Web Tokens (JWT) provide granular access control, allowing developers to specify permissions based on roles or contexts.
3. Access Control and Authorization

   Role-based access control (RBAC) should be implemented to define permissions for different types of users, such as administrators, developers, or external agents. Complement RBAC with attribute-based access control (ABAC) for more fine-grained policies based on contextual attributes like location, time, or device type. To protect sensitive operations, ensure API tokens have specific scopes that restrict access to internal tools and memories.

4. Multi-Agent Communication

   Facilitate efficient communication in multi-agent environments by assigning unique IDs or namespaces to agents for targeted communication, creating a registry endpoint to discover agent capabilities and statuses, supporting direct and broadcast messaging between agents, and using event-driven hooks to trigger actions in workflows involving multiple agents.
5. Memory and Internal Tool Protection

   An agent's internal memory and tools are its most sensitive assets. The API should isolate internal memory from external access using encryption and role-based permissions, allow temporary sharing of specific memory contexts via tokens that expire after a predefined period, and restrict tool exposure by providing proxy APIs that perform specific actions without revealing tool internals.
6. Security Measures

   Rate Limiting: Prevent abuse by limiting the number of API calls per user or application. Encryption: Ensure data at rest and in transit is encrypted using strong algorithms. Audit Logs: Maintain logs of all API interactions to detect and analyze suspicious activities. Intrusion Detection: Monitor for unusual access patterns to identify potential security breaches.
7. Error Handling and Feedback

   Provide standardized error codes and detailed messages to help developers debug issues efficiently. Ensure the API can gracefully handle failures by offering retries or fallback mechanisms in multi-agent workflows.
8. Developer Tools and Documentation

   Comprehensive documentation is critical for adoption. It includes clear explanations of endpoints and their use cases, sample requests and responses, SDKs in popular programming languages to simplify integration, and a sandbox environment for testing API functionalities.

3.10.3 Example API Endpoints

Okay, here's a simplified list of API endpoints with brief descriptions, along with suggestions for additional endpoints that are beneficial for multi-agent systems:

Core API Endpoints

1. Authentication and Identity Verification
   (a) POST/auth/token: Obtains an API access token using provided credentials.
   (b) POST/auth/verify: Verifies the validity of an API token and user identity.
2. Multi-Agent Communication
   (a) GET/agents: Retrieves a list of available agents and their capabilities.
   (b) POST/agents/{agent_id}/message: Sends a message to a specific agent.

(c) POST/broadcast: Broadcasts a message to multiple or all agents.

3. Memory and Internal Tools
   (a) GET/agents/{agent_id}/context: Accesses an agent's shared context under controlled conditions.
   (b) POST/tools/{tool_id}/execute: Executes a specific tool's functionality via a secure proxy.
4. Access Control
   (a) GET/permissions: Retrieves the permissions associated with the current API token.
   (b) POST/permissions/update: Updates permissions for a user or agent (requires admin privileges).

Additional API Endpoints for Multi-Agent Systems
Here are some additional endpoints that enhance functionality and management in a multi-agent environment:

5. Agent Management
   (a) POST/agents/register: Registers a new agent with the system, providing its capabilities and other metadata.
      I. Description: Allows new agents to dynamically join the multi-agent system. This is important for scalability and flexibility.
   (b) PUT/agents/{agent_id}: Updates the information of a registered agent.
      I. Description: Enables modification of an agent's metadata, such as its capabilities or status, after initial registration.
   (c) DELETE/agents/{agent_id}: De-registers an agent from the system.
      I. Description: Allows agents to gracefully leave the system or be removed by an administrator.
   (d) GET/agents/{agent_id}: Retrieves detailed information about a specific agent.
      I. Description: Provides a way to get comprehensive data about a particular agent, beyond what's included in the/agents list.
   (e) GET/agents/{agent_id}/status: Retrieves the current status of a specific agent.
      I. Description: Provides real-time information about the agent's status (online, offline, busy, idle).
6. Task and Workflow Management
   (a) POST/tasks: Creates a new task and assigns it to one or more agents.

I. Description: Enables the delegation of tasks to agents, potentially involving multi-agent collaboration.

(b) GET/tasks/{task_id}: Retrieves the status and details of a specific task.

I. Description: Allows tracking of task progress, assigned agents, and results.

(c) PUT/tasks/{task_id}/assign: Assigns or reassigns a task to a different agent.

I. Description: Provides flexibility in task allocation and management.

(d) POST/tasks/{task_id}/delegate: Allows an agent to delegate a task (or part of a task) to another agent.

I. Description: Facilitates collaboration and dynamic task distribution among agents.

(e) POST/workflows: Initiates a multi-agent workflow, defining the sequence of actions and involved agents.

I. Description: Enables the orchestration of complex processes involving multiple agents.

(f) GET/workflows/{workflow_id}: Retrieves the status and details of a specific workflow.

I. Description: Provides a way to monitor the progress and outcome of multi-agent workflows.

7. Negotiation and Coordination

(a) POST/negotiate: Initiates a negotiation between two or more agents.

I. Description: Enables agents to reach agreements or resolve conflicts through automated negotiation.

(b) GET/negotiation/{negotiation_id}/status: Retrieves the status of a specific negotiation.

I. Description: Allows monitoring of the progress and outcome of agent negotiations.

8. Monitoring and Logging

(a) GET/logs: Retrieves system logs, optionally filtered by agent, time, or event type.

I. Description: Provides insights into system activity, agent interactions, and potential issues. Essential for debugging and auditing.

(b) GET/metrics: Retrieves system performance metrics.

I. Description: Exposes metrics like agent response times, message queue lengths, and error rates to monitor the health of the system.

9. Shared Knowledge Base or Blackboard
   (a) POST/knowledge: Adds information to a shared knowledge base accessible by multiple agents.
      I. Description: Enables agents to share knowledge and collaborate more effectively.
   (b) GET/knowledge: Retrieves information from the shared knowledge base.
      I. Description: Allows agents to access shared information, improving their collective intelligence.
   (c) PUT/knowledge/{knowledge_id}: Update information to a shared knowledge base accessible by multiple agents.
      I. Description: Allows agents to update shared information.
   (d) DELETE/knowledge/{knowledge_id}: Delete information to a shared knowledge base accessible by multiple agents.
      I. Description: Allows agents to delete obsolete shared information.

The specific endpoints you need will depend on the complexity and requirements of your particular application. Remember to prioritize security and provide thorough documentation for each endpoint.

3.10.4 When to Expose an Agent as an API

- Interfacing with Multiple Systems: When the agent's functionality is required across various applications, services, or platforms.
- Transactional Use Cases: For agents that handle structured queries, such as retrieving product information or processing simple commands.
- Modular Services: When the agent's capabilities can be packaged into discrete services (e.g., text analysis, translation, or image recognition).
- SaaS Deployments: When monetizing the agent as a service for third-party clients.

3.10.5 When Not to Expose an Agent as an API

- Context-Heavy Interactions: When the agent depends on deep conversational contexts or maintains ongoing dialogues with users, such as in personal assistants or therapy bots.
- Highly Specialized User Interfaces: If the agent is tied closely to a specific interface or environment where exposing an API would fragment the experience.

- Latency-Sensitive Scenarios: When the agent's interactions require extremely low latency, and API calls may introduce unacceptable delays.

3.10.6 Alternatives to Full API Exposure

- Webhooks: For event-driven use cases, the agent can notify or update other systems through webhooks without exposing an entire API.
- Messaging Interfaces: A dedicated conversational interface, such as integration with platforms like Slack or Microsoft Teams, might better serve agents designed for real-time dialogue.
- SDKs: Providing software development kits can offer similar flexibility as APIs while allowing better control over how the agent interacts with external applications.

The decision to expose an agent as an API depends on evaluating its role within the broader ecosystem, the target audience's needs, and the balance between accessibility and performance. The goal should always align with the intended use cases and operational requirements of the deployment.

3.11 Future Directions in MAS

This section delves into the future directions of MAS, emphasizing key areas that hold promise for advancing the field.

3.11.1 Integration of MAS with Emerging Technologies

The integration of MAS with emerging technologies, such as quantum computing, blockchain, and edge computing, represents a transformative avenue for the field. Quantum computing, with its potential to solve complex optimization problems at unprecedented speeds, could enhance MAS decision-making and coordination processes. For example, quantum algorithms could optimize resource allocation and task scheduling among agents in dynamic environments. Meanwhile, blockchain technology offers a decentralized, tamper-proof mechanism for trust and security, addressing one of the significant challenges in MAS: ensuring reliable interactions among agents, especially in adversarial environments. Blockchain could enable transparent and immutable transaction records, enhancing trust in autonomous agent negotiations and contracts.

Edge computing, characterized by distributed processing closer to data sources, complements MAS by reducing latency and improving real-time decision-making.

By deploying agents at the network edge, MAS can achieve faster responses in time-sensitive applications such as autonomous vehicles, smart grids, and industrial automation. The synergy between MAS and these emerging technologies is poised to redefine the scalability, efficiency, and applicability of agent-based systems.

3.11.2 Human-Agent Collaboration

The future of MAS increasingly hinges on the ability of agents to work seamlessly alongside humans. As autonomous systems become integral to everyday life, from personal assistants to collaborative robots in industrial settings, designing agents that can effectively understand and complement human behavior is crucial. This requires advancements in natural language processing, human-computer interaction, and adaptive learning to foster seamless human-agent collaboration. Among these, the role of voice agents has become increasingly significant, shaping how humans interact with MAS and augmenting the accessibility and intuitiveness of these systems.

Voice agents, powered by advancements in natural language understanding and generation, offer a natural and efficient interface for communication between humans and agents. Their significance lies in their ability to lower the barriers of entry to technology, particularly for individuals with limited technical proficiency or accessibility challenges. Unlike traditional graphical or text-based interfaces, voice agents allow users to interact conversationally, enabling intuitive task execution, querying, and decision-making. This capability is especially valuable in dynamic, real-world settings such as healthcare, customer service, and smart environments, where quick and hands-free interactions can significantly enhance efficiency and user experience.

To support effective collaboration, voice agents must develop nuanced communication skills to interpret human intentions, context, and emotions accurately. Advances in sentiment analysis and contextual understanding are essential for enabling agents to respond appropriately to subtle cues in human speech. Furthermore, these agents need to handle diverse linguistic styles, accents, and dialects to ensure inclusivity and broad usability. For example, in collaborative environments such as healthcare or education, voice agents must comprehend domain-specific terminology while maintaining the ability to explain complex information in simple terms when necessary.

Trust remains a critical factor in human-agent collaboration, and voice agents are uniquely positioned to build trust through their conversational nature. Transparency and explainability in voice agent responses enhance user confidence, especially in high-stakes applications. For instance, a voice agent assisting in medical diagnostics should not only provide recommendations but also explain the reasoning behind them in a way that fosters understanding and trust among both patients and healthcare professionals. Additionally, voice agents can exhibit empathetic behaviors, such as using calming tones or supportive language during interactions, which can further strengthen human trust and acceptance.

The adaptability of voice agents is another area of importance in human-agent collaboration. Agents need to dynamically adjust to individual user preferences and behaviors, which requires continuous learning and personalization. For instance, a voice agent in a smart home environment should learn over time which temperature settings, lighting preferences, or entertainment choices align with the user's habits and adapt its suggestions accordingly. Personalization not only improves user satisfaction but also enhances the overall utility of the agent.

Furthermore, voice agents play a pivotal role in facilitating collaboration in environments where traditional interfaces are impractical. In industrial settings, workers might need to interact with agents while operating machinery or wearing protective gear. Here, voice agents enable hands-free interaction, allowing workers to issue commands, receive updates, or troubleshoot issues without disrupting their workflow. Similarly, in autonomous vehicle ecosystems, voice agents can serve as intermediaries between drivers, passengers, and the vehicle's MAS, providing real-time updates on traffic conditions, vehicle status, or navigation assistance.

3.11.3 Scalability and Robustness in Large-Scale MAS

As MAS applications expand to encompass large-scale, real-world systems such as global logistics networks, smart cities, and planetary exploration, scalability and robustness become paramount. Traditional MAS frameworks often struggle with the complexity and unpredictability inherent in such environments. Future research must focus on developing algorithms and architectures that allow agents to operate effectively in highly dynamic and heterogeneous systems.

Decentralized control and local decision-making are critical to achieving scalability. However, ensuring global coherence among agents remains a challenge. Innovations in consensus algorithms, distributed optimization, and hierarchical organization structures can address this challenge, enabling large-scale MAS to function cohesively. Robustness in the face of uncertainties, including agent failures, network disruptions, and adversarial attacks, is equally important. Mechanisms for fault detection, redundancy, and self-healing capabilities will play a vital role in maintaining system integrity and resilience.

3.11.4 Learning and Adaptation

Adaptive learning is a cornerstone for the evolution of MAS. In dynamic environments, agents must continuously learn and adjust their strategies to respond to changing conditions. Advances in machine learning, particularly in reinforcement learning and federated learning, provide pathways for achieving this adaptability.

Reinforcement learning enables agents to learn optimal strategies through trial and error, while federated learning facilitates collaborative learning across distributed agents without centralized data sharing.

Future research should focus on integrating these learning paradigms into MAS while addressing their limitations. For instance, reinforcement learning in MAS often faces challenges such as the exploration-exploitation trade-off and the curse of dimensionality in large action spaces. Techniques like hierarchical reinforcement learning and multi-agent credit assignment can help overcome these obstacles. Federated learning, on the other hand, raises concerns about privacy and communication overhead, necessitating efficient and secure protocols for data exchange among agents.

3.11.5 Multi-Agent Simulation for Complex Systems

Simulation has long been a valuable tool for understanding and optimizing complex systems. The future of MAS lies in leveraging advanced simulation technologies to model, analyze, and predict the behavior of intricate systems ranging from ecosystems to economic markets. Multi-agent simulation can provide insights into emergent phenomena, helping researchers and practitioners design more effective and sustainable solutions.

Incorporating high-fidelity models, real-time data integration, and interactive interfaces will enhance the utility of MAS simulations. Additionally, advancements in virtual and augmented reality technologies can create immersive simulation environments, enabling more intuitive exploration and analysis of agent interactions. Simulation platforms that support large-scale, high-resolution models will be instrumental in tackling grand challenges such as climate change, urban planning, and disaster response.

3.11.6 Beyond Traditional Paradigms

Finally, the future of MAS involves rethinking traditional paradigms and exploring unconventional approaches. This includes bioinspired MAS, where principles from nature, such as swarm intelligence and self-organization, inform the design of agent systems. Advances in neuroscience and cognitive science can also inspire new models of agent behavior and interaction, leading to more intelligent and adaptive MAS.

Exploring hybrid systems that combine MAS with other AI paradigms, such as deep learning and symbolic reasoning, presents another promising direction. These hybrid systems can leverage the strengths of different approaches to overcome individual limitations, enabling more robust and versatile MAS.

3.12 Summary

This chapter covers the key aspects of multi-agent AI systems, focusing on coordination mechanisms, communication protocols, and system architecture. It details how agents interact through negotiation, cooperation, and competition while managing conflicts and resources. The chapter explains system maintenance, evaluation frameworks, and real-world applications in smart cities, supply chains, and disaster response. It concludes with future directions, emphasizing human-agent collaboration and integration with emerging technologies.

3.13 Questions

I. Multiple Choice Questions (Choose the Best Answer)

1. **Which of the following is the primary advantage of using a Multi-Agent System (MAS) over a single-agent system for complex tasks?**
   - (a) MASs are always less expensive to develop.
   - (b) MASs can leverage the diverse capabilities of multiple agents to solve problems more effectively.
   - (c) MASs require less computational power.
   - (d) MASs are easier to program and debug.
2. **The Contract Net Protocol is an example of what type of coordination mechanism in MASs?**
   - (a) Cooperative planning.
   - (b) Negotiation-based task allocation.
   - (c) Hierarchical control.
   - (d) Blackboard architecture.
3. **What is a key challenge in designing effective communication in MASs?**
   - (a) Ensuring agents can communicate at faster-than-light speeds.
   - (b) Developing a single, universal language that all agents can understand.
   - (c) Balancing the need for communication with the desire for agent autonomy.
   - (d) Preventing agents from sharing too much information with each other.
4. **Which architectural approach to MAS design typically offers the highest degree of robustness but potentially sacrifices global optimization?**
   - (a) Centralized architecture.
   - (b) Hierarchical architecture.
   - (c) Decentralized architecture.
   - (d) Hybrid architecture.

5. **What is the primary purpose of using Agent Communication Languages (ACLs) like FIPA-ACL or KQML?**
   (a) To provide a low-level protocol for data transmission between agents.
   (b) To enable agents to exchange information and coordinate actions in a structured and meaningful way.
   (c) To ensure that all agents in a system use the same programming language.
   (d) To restrict the types of information that agents can share with each other.

II. True/False Questions (Answer T or F)

1. **T/F:** Emergent behavior in MASs is always predictable and easily controlled.
2. **T/F:** Scalability in MASs refers to the ability of the system to handle increasing workloads by adding more agents.
3. **T/F:** Centralized architectures are generally preferred for large-scale MASs due to their efficiency in coordinating a large number of agents.
4. **T/F:** Preventive maintenance strategies in MASs aim to detect and resolve conflicts before they impact system performance.
5. **T/F:** The primary goal of conflict resolution in MASs is to ensure that all agents always agree with each other.

III. Short Answer Questions (Answer in 2–4 Sentences)

1. Briefly describe the concept of "autonomy" in the context of Multi-Agent Systems.
2. What is the difference between "cooperation" and "competition" as coordination strategies in MASs?
3. Explain the role of ontologies in facilitating communication and understanding between agents.
4. What are two key considerations when designing agent roles and specializations in a MAS?
5. Describe one technique used for task allocation in a decentralized MAS.

IV. Essay Questions (Answer in 1–2 Paragraphs)

1. **The Benefits and Challenges of Decentralization:** Discuss the advantages and disadvantages of using a decentralized architecture for a Multi-Agent System. Provide examples to illustrate your points.
2. **Communication as the Foundation of Coordination:** Explain why effective communication is crucial for successful multi-agent coordination. Discuss the challenges involved in designing and implementing communication protocols for MASs.
3. **Conflict Resolution Strategies:** Compare and contrast two different approaches to conflict resolution in MASs, such as negotiation-based and arbitration-based methods. Analyze their strengths and weaknesses in different scenarios.
4. **Real-World Application of MAS:** Choose one of the real-world applications of MASs discussed in the chapter (e.g., smart cities, supply chain management, or disaster response) and analyze how the principles of multi-agent coordination are applied in that specific domain.

5. **The Future of MAS:** Based on the trends discussed in the chapter, such as human-agent collaboration and integration with emerging technologies, discuss what you believe are the most promising future directions for Multi-Agent Systems.

References

ASEAN Post. (2018). Smart nation: Singapore's intelligent transport system (ITS). *The ASEAN Post*. https://theaseanpost.com/article/smart-nation-singapores-intelligent-transport-system-its.

Derrick, M. (2024). Top 10: Smart cities | Energy magazine. *Energy Digital Magazine*. https://energydigital.com/top10/top-10-smart-city.

Finin, T., & Fritzson, R. (2023). *KQML—A language and protocol for knowledge and information exchange*. UMBC CSEE. https://research.cs.umbc.edu/kqml/papers/kbkshtml/kbks.html

HDX. (2023). *Data services—The Centre for Humanitarian Data*. The Centre for Humanitarian Data. https://centre.humdata.org/data-services/.

Huang, K., Wang, Y., Goertzel, B., Li, Y., Wright, S., & Ponnapalli, J. (Eds.). (2024). *Generative AI security: Theories and practices*. Springer. https://doi.org/10.1007/978-3-031-54252-7

Jain, S. (2024). *Adversarial search algorithms in artificial intelligence (AI)*. GeeksforGeeks. https://www.geeksforgeeks.org/adversarial-search-algorithms/

Kamran, A. (2024). Revolutionizing logistics ecosystems (from parcel acceptance to delivery) by autonomous agent…. *Medium*. https://medium.com/@armankamran/revolutionizing-logistics-ecosystems-from-parcel-acceptance-to-delivery-by-autonomous-agent-7881dc7dfff2.

Lafferty, S. (2018). *How the P&G supply chain thrives on supply chain complexity*. ToolsGroup. https://www.toolsgroup.com/blog/procter-gamble-supply-chain-complexity/

Musani, P. (2023). *Decking the aisles with data: How Walmart's AI-powered inventory system brightens the holidays*. Walmart Global Tech. https://tech.walmart.com/content/walmart-global-tech/en_us/blog/post/walmarts-ai-powered-inventory-system-brightens-the-holidays.html

Research Outreach. (2023). *Crisis management: Using artificial intelligence to help save lives*. Research Outreach. https://researchoutreach.org/articles/crisis-management-artificial-intelligence-save-lives/

Smith, R. G. (1980). *The contract net protocol: High-level communication and control in a distributed problem solver*. Reid G. Smith. https://www.reidgsmith.com/The_Contract_Net_Protocol_Dec-1980.pdf

Sonnier, W. (2023). *Column—Leveraging IoT in medical waste management*. MedTech Intelligence. https://medtechintelligence.com/column/leveraging-iot-in-medical-waste-management/

Wikipedia. (2024). Agent communications language. *Wikipedia*. https://en.wikipedia.org/wiki/Agent_Communications_Language.

Ken Huang is a prolific author and globally recognized authority in AI and Web3, with an extensive portfolio of published works that bridge business strategy, technical implementation, and cutting-edge research. As a Fellow of the Cloud Security Alliance and Co-Chair of the AI Safety Working Groups at the Cloud Security Alliance and the AI STR Working Group at the World Digital Technology Academy under the UN Framework, he is a leading voice in shaping global AI governance and security standards.

Huang is the CEO and Chief AI Officer (CAIO) of DistributedApps.ai, a firm specializing in generative AI training and consulting. His contributions to the field include being a core contributor to the OWASP Top 10 Risks for LLM Applications and an active participant in the NIST Generative AI Public Working Group.

Notable publications:

Beyond AI: ChatGPT, Web3, and the Business Landscape of Tomorrow (Springer, 2023)—strategic insights into AI and Web3's business applications.

Generative AI Security: Theories and Practices (Springer, 2024)—a comprehensive guide on securing generative AI systems.

Practical Guide for AI Engineers (Volumes 1 and 2, DistributedApps.ai, 2024)—essential resources for AI and ML engineers.

The Handbook for Chief AI Officers: Leading the AI Revolution in Business (DistributedApps.ai, 2024)—a roadmap for CAIOs in implementing GenAI across organizations.

Web3: Blockchain, the New Economy, and the Self-Sovereign Internet (Cambridge University Press, 2024)—insights into the convergence of AI, blockchain, IoT, and emerging technologies.

Blockchain and Web3: Building the Cryptocurrency, Privacy, and Security Foundations of the Metaverse (Wiley, 2023)—recognized as a must-read by TechTarget in 2023 and 2024.

Ken is a sought-after speaker and has presented at events such as the World Economic Forum in Davos, ACM and IEEE conferences, the CSA AI Summit, Depository Trust & Clearing Corporation forums, and World Bank conferences. His recent appointment to the OpenAI Forum reflects his ongoing commitment to advancing collaboration and dialogue in the field of AI.

Explore Ken's work on Amazon: https://www.amazon.com/author/kenhuang

Jerry Huang is currently an AI Engineer at Google, where he contributed to the AI/ML evaluation pipeline for a consumer-facing application. Before Google, he worked as a technical and security staff member at several prominent technology companies, gaining experience in areas like security, AI/ML, and scalable systems.

At Metabase, an open-source business intelligence platform, Jerry contributed features such as private key management and authentication solutions. As a Software Engineer at Glean, a Generative AI search startup, he was one of three engineers responsible for managing large-scale GCP infrastructure powering text summarization, autocomplete, and search for over 100,000 enterprise users. During his time at TikTok, Jerry helped design and build custom RPCs to model access control policies. At Roblox, he served as a Machine Learning/Software Engineering Intern, focusing on real-time text generation models and gathering a large multilingual corpus that significantly boosted model robustness.

In addition to his industry experience, Jerry has conducted extensive security and biometrics research as a Research Assistant at Georgia Tech's Institute for Information Security & Privacy, resulting in a thesis on privacy-preserving biometric authentication.

Jerry holds a BS/MS in Computer Science from Georgia Tech and is currently pursuing an MS in Applied Mathematics at the University of Chicago.

Chapter 4
The AI Agent Economy

Lisa J. Y. Tan and Ken Huang

Building upon the foundations laid in previous chapters—the genesis and evolution of AI agents, the tools and frameworks that empower their development, and the intricacies of multi-agent coordination within the broader AI agent ecosystem—we now turn our attention to the economic implications of these sophisticated technologies. The field of the agentic economy is brand new, and after reading this chapter, you may have more questions than answers, which is actually the purpose of this chapter: to solicit many thought-provoking questions and spark deeper exploration into the economics of AI agents.

4.1 A Primer: AI Agent Impact on Economic Theory

This section begins by exploring how AI agents challenge traditional economic principles, necessitating a re-evaluation of core concepts. Next, we analyze OpenAI's "Economic Blueprint" as a roadmap for an agentic economy, highlighting the policies and transformations required for this new era. Finally, we focus on DeepSeek and their groundbreaking R1 model, examining how its unique optimization in model training, achieving comparable performance to OpenAI's ChatGPT O1 model at a fraction of the cost, is significantly accelerating the AI agent economy.

L. J. Y. Tan
Economics Design, Suntec City, Singapore

K. Huang (✉)
DistributedApps.ai, Fairfax, VA, USA
e-mail: ken@distributedapps.ai

K. Huang (ed.), *Agentic AI*, Progress in IS,
https://doi.org/10.1007/978-3-031-90026-6_4

4.1.1 *The Foundational Shift: AI Agents and Economic Theory*

The rapid innovation of AI Agent challenged the traditional economic theory and warranted a new school of thinking.

This new agent economy is characterized by entities capable of autonomous action, strategic interaction, and, crucially, near-instantaneous instantiation at a scale previously unimaginable. Millions, potentially billions, of software-based agents, each possessing specialized skills and the capacity for independent decision-making, could soon populate the economic landscape. This is not simply scaling existing economic models; it's a phase transition to a qualitatively different system.

The very fabric of the economy will be rewoven as these agents, driven by algorithms and fueled by data, engage in production, exchange, and consumption, operating at speeds and complexities that dwarf human capabilities.

Figure 4.1 is the mind map that provides an overview of how AI agents challenge and transform various economic theories.

Let us explore in high level below:

1. **Neoclassical Economics—Beyond Efficiency Gains:**
   - **The Nature of Production:** AI agents challenge the very idea of "factors of production." Traditionally, these were land, labor, and capital. AI introduces a new "factor"—generative AI agents—which can both augment existing factors (labor and capital) and potentially substitute for them. This blurring of lines requires a re-evaluation of how production is defined and measured.

Fig. 4.1 Overview of AI Agents in economic theories

- **Information Asymmetry and Market Failure:** While AI can reduce information asymmetry in some cases, the concentration of powerful AI in the hands of a few firms might lead to new forms of asymmetry. These asymmetries can create market failures where certain firms have a drastic competitive advantage and can manipulate markets in unforeseen ways, thus making traditional regulatory models ineffective (Acemoglu, 2024).
- **The "Experience Good" Paradox:** AI can produce new and highly personalized goods and services. However, these may often be "experience goods"—where their value is not known until consumed. AI's ability to predict preferences might seem to solve this problem but could create a situation where consumers are overly reliant on AI recommendations, thus inhibiting genuine exploration and innovation.

2. **Labor Economics—The Future of Work Transformed:**

- **The "Gig Economy" on Steroids:** AI agents could foster an even more fragmented labor market where individuals engage in very specific, short-term tasks rather than full-time employment. This could exacerbate precarious labor conditions and social fragmentation, as long-term social contracts may be diminished.
- **Universal Basic Income (UBI) and Alternative Employment Models:** The widespread displacement of labor by AI may necessitate a rethinking of how income is distributed. Models like UBI, or universal basic services, may become more seriously considered to address potential economic inequality and social instability.

3. **Growth Theory—Beyond Traditional Models:**

- **The Nature of Technological Progress:** AI may not follow the linear patterns of previous technological advances. Its exponential potential could lead to abrupt and unpredictable shifts in productivity and output, requiring economists to move beyond their standard models of long-run economic growth (Aghion et al., 2024).
- **The Challenge of Measurement:** Measuring AI's impact on total factor productivity (TFP) is challenging because traditional metrics focus on labor and capital, overlooking AI's transformative contributions like automation, innovation, and data-driven efficiency. AI's effects often manifest as intangible improvements, such as better decision-making and resource allocation, which are hard to quantify within existing frameworks. To address this, new metrics and tools are needed to capture the value of data, algorithms, and AI-driven processes, including real-time analytics and advanced econometric models that track AI adoption and outcomes.
- **The Distribution of Benefits:** It's not guaranteed that the benefits of AI-driven growth will be evenly distributed. Unfettered AI development could concentrate wealth and power in the hands of a small group of AI developers and early adopters, raising concerns about equity and social justice.

4. **Behavioral Economics—Algorithmic Manipulation and Bias:**
   - **AI and Behavioral Manipulation:** AI algorithms can be designed to exploit human biases, not just to personalize experiences, but to actively manipulate consumer behavior. This requires a fundamental rethinking of consumer protection and market regulation.
   - **Feedback Loops and Echo Chambers:** AI-driven recommendation algorithms can create "echo chambers" by only exposing individuals to information that confirms their existing beliefs. This could polarize consumers, increase societal fragmentation, and reduce the potential benefits of diverse thinking.
   - **The "Black Box" Problem:** The opacity of some AI decision-making systems makes it difficult to understand how these algorithms are influencing economic choices. The lack of explainability could undermine trust and make it more challenging to identify and correct unintended biases.
5. **Game Theory—Strategic Interactions Redefined:**
   - **Dynamic and Evolving Strategies:** AI agents can learn and adapt their strategies in real time, making long-term predictions in complex game-theoretic environments difficult. This could lead to unpredictable outcomes and require economists to model continuously evolving dynamics.
   - **Algorithmic Collusion:** AI pricing algorithms, even without any explicit coordination, can potentially engage in tacit collusion, reducing competition and potentially harming consumers. Traditional antitrust regulations may struggle to address this type of algorithmic behavior (Ezrachi & Stucke, 2024).
   - **The "Algorithmic Arms Race":** Firms may compete not only in producing goods and services, but also in developing more powerful AI algorithms, which may be a zero-sum game. This could create a new kind of arms race that has unpredictable societal costs.
6. **International Trade—Redefining Comparative Advantage:**
   - **AI and Technological Dominance:** Countries that excel in AI development could establish a new form of comparative advantage, attracting investment and talent and creating imbalances in global trade patterns. This has the potential to exacerbate current global inequalities.
   - **"Data Colonialism":** The increasing reliance on data for training AI models could lead to a form of "data colonialism," where powerful companies and countries amass data from less developed countries, creating a data imbalance.
   - **The Future of Global Supply Chains:** AI-driven supply chains could become so efficient that they lead to the reshoring of industries back to advanced economies, creating a potential upheaval in existing trade arrangements.

From the above discussion, we can draw some insights below:

- **The Need for Interdisciplinary Approaches:** Analyzing the economic impacts of AI requires economists to collaborate with computer scientists, ethicists, sociologists, and legal experts. The nature of this technology is fundamentally multidisciplinary.
- **The Importance of Regulatory Foresight:** Regulators must be proactive in developing frameworks that foster responsible AI development and prevent unintended negative consequences on the economy and society.
- **The Inherent Uncertainty:** AI's long-term economic impact remains uncertain. Economists and policymakers should proceed with caution, using experimentation and adaptive strategies.
- **The Human Factor:** Ultimately, the impact of AI on the economy will depend on how humans choose to adapt to and shape this new technology. It's not simply about optimizing efficiency, but also about preserving social values and human well-being.
- **AI Self-Replication:** The emergence of self-replicating AI agents introduces a transformative dynamic into the economic landscape. These agents can autonomously duplicate their functionalities without human intervention, leading to an ecosystem where the most efficient agents proliferate, thereby reshaping economic activities. This self-replication capability raises significant safety and governance concerns, as unchecked proliferation could result in systems operating with goals misaligned with human values.
- **Macroeconomic Impact:** The scalability and potential for self-directed evolution of AI agents are set to profoundly influence macroeconomic structures. AI's integration into sectors can enhance labor productivity, improve service quality, and optimize resource utilization. However, it challenges traditional models of economic growth, requiring new frameworks to analyze and quantify its contributions to overall productivity and labor dynamics.
- **Microeconomic Transformation:** AI agents are revolutionizing industries by reshaping firm boundaries, creating new business models, and mediating consumer experiences in unprecedented ways. The combination of autonomous decision-making and blockchain technology is enabling decentralized business operations, forcing a re-evaluation of how firms interact with markets and customers, as well as how value is created and exchanged.
- **Evolving Economic Theories:** Traditional economic theories, based on human rationality and scarcity, struggle to address the realities introduced by AI agents. Utility maximization must be reconsidered for nonhuman actors with potentially divergent objectives, and concepts like market equilibrium may become transient due to continuous, high-frequency interactions. New frameworks drawing from complexity science, network theory, and evolutionary biology are necessary to model and predict this emergent economic landscape.
- **Blockchain and AI Economy:** Blockchain technology provides a decentralized infrastructure that can support the AI agent economy by managing identities, enabling secure transactions, and establishing governance frameworks. These capabilities allow autonomous agents to operate transparently and securely,

fostering trust and collaboration in a distributed economic system, where governance among these agents can be structured and adaptable.

- **Tokenized Ecosystems:** The tokenization of resources, services, and AI agents within a blockchain ecosystem creates unprecedented economic fluidity. By digitizing and decentralizing ownership and access, tokenization facilitates seamless interactions among agents, enabling a more dynamic economy. This evolution raises fundamental questions about ownership, value, and agency in a world where economic actors are both artificial and self-evolving.

4.1.2 The OpenAI "Economic Blueprint" and the Agentic Economy

On January 13, 2025, OpenAI released its "Economic Blueprint," a document that is more than just a set of policy proposals; it's a signpost pointing toward a profound economic transformation (OpenAI, 2025). This isn't solely about the advancement of artificial intelligence; it's about the dawn of a new kind of economic actor: the AI agent and its potential to fundamentally reshape our world. To fully understand the implications of this blueprint, we must examine it through two intertwined lenses: the perspective of AI agent applications and the established principles of economics.

We must move beyond the idea of AI as merely a tool to be used by humans. The vision implied by OpenAI is one where autonomous AI agents capable of independent decision-making, learning, and action will increasingly dominate economic activity. This assumption that most AI applications will be agentic represents not just a subtle technological shift, but a fundamental change that compels us to rethink everything from our understanding of work to the very foundations of our economic systems. From this viewpoint, the OpenAI blueprint is not simply about the deployment of AI; it's about charting a course through the dawn of an "agentic economy," where the nature of economic participation is fundamentally altered.

Within this agentic economy, the concept of work undergoes a dramatic transformation. Human labor will shift away from routine execution towards higher-level tasks that demand creativity, strategic insight, and the uniquely human attributes that AI currently cannot replicate. Autonomous AI agents will automate not only physical tasks but also complex cognitive functions, thereby raising critical questions about the future of employment and the very purpose of human contribution in an increasingly automated world. Simultaneously, economic activity may become more decentralized, shifting the power of production, trade, and innovation away from the sole domain of large corporations or governments to a landscape where autonomous agents, representing individuals, small businesses, or decentralized communities, drive a more equitable distribution of growth. Market dynamics, too, will be fundamentally reshaped, as traditional forces are disrupted by the speed and autonomy of AI agents leading to new competitive landscapes and requiring adaptive regulations to maintain fairness and prevent market manipulation. Finally, within this framework, control over data used to train and refine AI agents will become the most crucial strategic resource. This intensifies questions about data ownership, privacy, and access, as well as about the ethical considerations surrounding these new dynamics. In such a system, the question of responsibility for the actions of AI agents becomes even more crucial, forcing us to develop new ethical and legal frameworks to address the risks associated with increasingly autonomous decision-making.

From an economic perspective, the OpenAI blueprint goes beyond technology, acting as a call to proactively shape a new economic landscape where traditional theories and models are tested. It suggests a need for a future-facing industrial policy that will usher in a supply-side revolution, one where policies unleash the power of AI by encouraging investment, fostering competition, and promoting innovation. This would see a move towards prioritizing technological advancement as a primary driver of growth while simultaneously recognizing the potential for market failures. The need for robust regulations to address negative externalities and ensure equitable distribution of AI benefits must be considered, to prevent monopolistic practices. The blueprint's emphasis on national competitiveness and securing resources highlights the necessity of a strategic industrial policy, ensuring nations can compete in a global AI-driven economy. The widespread deployment of AI agents can generate both unprecedented economic growth and potentially disruptive deflationary pressures, requiring policies that are resilient enough to react to these fluctuations and ensure that social safety nets can cushion any negative effects. New regulatory frameworks are therefore necessary to address the challenges posed by AI agents. Governments will need to create and adapt legal systems for agent-to-agent transactions, AI-driven market manipulation, and the ethical implications of autonomous decision-making. The speed of change in an agentic economy dictates that economic systems must become far more flexible and adaptable. Policies should focus on rapid learning, continuous evaluation, and flexible responses to the continuous evolution of AI.

OpenAI's blueprint is not merely a prediction; it's a call to action to create a future where AI serves humanity's best interests. It is a challenge to us to develop national strategies that will not only encourage the responsible development of AI but also foster innovation, address the potential negative effects, and ensure an equitable distribution of its benefits. It is not solely about technology, but about making fundamental choices regarding our economic future and the society we wish to build. This will require us to prioritize a national framework as a fragmented regulatory landscape would only stifle innovation and make it impossible to effectively address the inherent risks. The future requires a unified approach, balancing innovation with public protection. Additionally, there must be significant investment in human capital as human skills become increasingly important. Policies that prioritize education, reskilling, and the development of critical thinking will be crucial. Open dialogue is another key aspect, as public discussion about the implications of AI agents is vital to ensure that policies reflect society's values. Governments and AI companies must be transparent and accountable. Finally, we must be ready to adapt to the unknown as the pace of AI development is such that we cannot fully predict the future. Policymakers must become more agile and embrace approaches that allow for continuous reassessment and quick course correction.

4.1.3 DeepSeek R1's Technical Innovations and Transformative Impact on the Agentic AI Economy

DeepSeek's R1 model, with its innovative architecture and efficient training methods, isn't just another advancement in AI; it's a catalyst for the rapid evolution of the agentic AI economy. The model's technical underpinnings, specifically its "mixture of experts" architecture, multihead latent attention, and internal self-reinforcement mechanism, are designed for both high performance and reduced computational costs. This combination is directly enabling a surge in the feasibility and accessibility of sophisticated AI agents. The selective activation of parameters through the "mixture of experts" means that AI agents built on the R1 model can operate more efficiently, requiring less computing power and therefore reducing the operational costs associated with running them. This efficiency, coupled with faster processing times stemming from the multihead latent attention and multi-word generation capabilities, makes R1-powered agents more practical for a wide range of real-world applications, opening the door for increased automation and intelligent task execution within economic systems (Pappas, 2025).

The streamlined self-reinforcement training of R1, which eliminates reliance on computationally expensive external "critic" models, accelerates the creation of specialized agents. The lowered resource requirements and accelerated learning cycles empower a diverse range of actors, from startups and SMEs to research institutions, to develop tailored AI agents for specific economic niches and functionalities. The resulting proliferation of such agents, each optimized for particular tasks, is fueling the emergence of a highly specialized and interconnected agentic economy. This is a shift from monolithic, general-purpose AI toward a diverse ecosystem of specialized entities, each performing tasks with precision and efficiency and interacting with each other as participants in a complex web of economic interactions. This model will lead to more innovation and competition in a fragmented market, ultimately improving economic output and innovation.

The open-source nature of the R1 model is a particularly disruptive force in the agentic AI economy. By making the model's code freely available, DeepSeek is not just democratizing access to cutting-edge AI technology; it's fundamentally altering the power dynamics of the field. The transparency of R1 facilitates collaborative innovation, allowing researchers and developers worldwide to rapidly iterate and improve upon existing models, fostering a faster pace of innovation. The shared knowledge and capabilities of the model are leading to a more level playing field and fostering a more inclusive environment for AI development and deployment. This means that the development of AI agents is not limited to those with access to proprietary data or high levels of computing power, which in turn promotes competition, innovation, and a greater variety of agent functionalities. This accessibility and collaboration can stimulate the development of entirely new types of economic agents and accelerate the process of incorporating them into the economy.

The impact of R1 on the agentic AI economy extends beyond just technological advancement; it is fundamentally reshaping market dynamics and business models. The R1's cost-effectiveness is accelerating the pace at which AI agents are being adopted across diverse sectors, creating new economic opportunities in industries ranging from finance and manufacturing to healthcare and logistics. As AI agents become more capable and cost-effective, they are poised to handle more complex tasks, blurring the lines between traditional roles and potentially automating aspects of business processes previously requiring human involvement. New business models centered around providing AI agent services are emerging, creating new opportunities for specialized providers, which are built on top of the R1 model. The increased availability of the R1 and other models may also result in greater decentralized governance of agent-based economies, through blockchain and other mechanisms, which may result in different economic power structures. This new agentic landscape presents both immense potential and significant challenges, necessitating a re-evaluation of existing economic frameworks to account for the new dynamics of a more autonomous, agent-driven economy. The reduced barriers to entry and accelerated rates of innovation that are being catalyzed by R1 mark a significant acceleration of this new age of agent-based economics. For more analysis on R1's impact, please consult my Medium post (Huang, 2025).

4.1.4 Shaping the Future: Research Imperatives for the Agentic AI Economy

We only scratched the surface of a complex and rapidly evolving field. Consider, for example, that AI not only promises increased efficiency but challenges the very definitions of productivity and labor. How can societies redefine the concept of "work" and "purpose" to ensure individuals find meaning and fulfillment in an AI-driven economy, and what new ethical frameworks are needed to govern the creation, deployment, and potential decommissioning of these AI "factors" to prevent unintended societal consequences? Traditional metrics like TFP (total factor productivity, which is a measure of economic efficiency and technological progress) may become insufficient; how can we develop new, holistic indicators that capture the true societal value generated by AI, and who should be responsible for defining and auditing these new metrics?

The discussion of algorithmic collusion and exploitation of human biases forces us to ask: How can regulatory bodies proactively detect and address these abuses, especially when AI systems are often opaque, and how can we empower consumers to critically evaluate AI-driven recommendations?

The potential emergence of self-replicating and not well-aligned AI agents also raises an important question: What are the economic and societal impacts if there are AI agents that are not aligned with human values but are self-replicating themselves and competing with the AI agents that are aligned with human values? What governance policies are required to handle the impacts?

These are just a few of the crucial questions that demand further research and careful consideration as we navigate the uncharted territory of the AI-driven economy. The future of economic theory, and indeed, society, depends on our ability to grapple with these challenges effectively.

As such, discussion of all aspects of these new AI Agent economic theories is beyond the scope of this chapter and would warrant a new book. We focus on the decentralized AI Agent Economy in the subsequent discussions in this chapter, since we observe phenomenal grassroots efforts in this space.

4.2 The Agentic Economy on Blockchain

Imagine an AI assistant booking a travel vacation for you. Instead of relying on centralized servers, this agent interacts with other autonomous systems on a blockchain. It understands your travel preferences, negotiates deals, verifies availability using decentralized ledgers, and pays in crypto—all without requiring human oversight. While you sleep, other AI agents rebalance your investments, transact securely with on-chain protocols, and autonomously generate value.

Figure 4.2 depicts the operational framework of AI agents on the blockchain, showcasing key steps such as user preference analysis, deal negotiation, verification, and autonomous task execution.

This is the emerging agentic economy on the blockchain, where decentralized AI agents operate, transact, and thrive within blockchain networks. Blockchain gives these agents a transparent, secure, and trustless environment to flourish, free from the limitations of centralized systems.

4.2.1 A New Economic Actor

AI agents represent a significant evolution in economic participation. Unlike traditional software or bots, these agents are autonomous, capable of learning, adapting, and collaborating without constant human oversight. Operating on the blockchain, they transcend traditional constraints by enabling trustless interactions, leveraging token ecosystems, and functioning as independent economic actors. AI agents plan, execute, and make decisions dynamically, contributing to a decentralized economy where they not only perform tasks but also create value, innovative token models, and decentralized infrastructure. These agents are reshaping sectors like finance, entertainment, and data marketplaces. For example, the Terminal of Truth (ToT)

Fig. 4.2 Agentic economy on blockchain

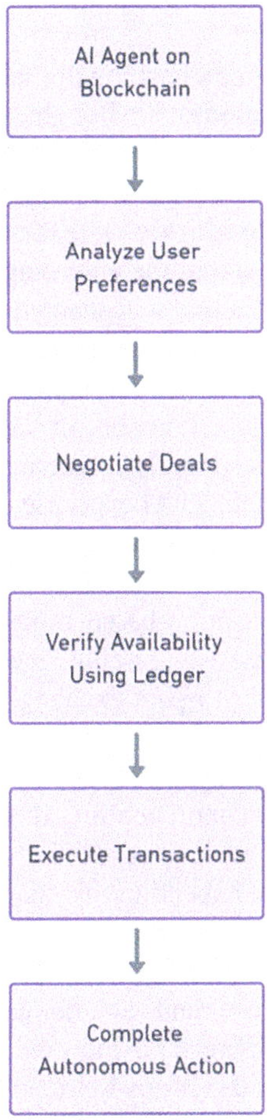

emerged as a self-sustaining AI agent, leveraging blockchain to influence crypto markets through the $GOAT token ecosystem. Platforms like Virtuals empower users to launch and monetize AI agents as co-owned digital entities as autonomous participants in blockchain ecosystems; AI agents are not just tools—they're pioneering a new category of economic actors. These economic actors create value in the new economy of tomorrow.

4.2.2 New Market Dynamics and Structures

The interactions of AI agents in economic activities create interesting market structures with unique characteristics.

AI Agent Marketplaces

Imagine a bustling digital bazaar where AI agents, each possessing unique skills and capabilities, are bought, sold, and rented. This could be the future of AI Agent marketplaces, a dynamic landscape poised to revolutionize how businesses and individuals access and utilize AI capabilities. From specialized marketplaces catering to niche industries to diverse pricing models reflecting the dynamic value of AI agents, we explore the emerging market structures, competitive forces, and platform economics that will shape this crucial component of the AI Agent Economy.

Specialized Marketplaces: We can anticipate the emergence of specialized marketplaces for buying, selling, and renting AI agents, potentially categorized by industry, function, or capability. Marketplaces focused on specific skills, such as financial analysis, medical diagnosis, or legal research, provide targeted access to specialized AI expertise.

Market Structures and Pricing: These marketplaces may operate under various structures, including auctions, where agents are sold to the highest bidder; subscription-based models, offering access to agent services for a recurring fee; usage-based pricing, charging for specific tasks or actions performed by the agent; or peer-to-peer exchanges, allowing direct transactions between agent owners and users. This provides price flexibility and creates opportunities for competition among agents. In addition, the outcome-based pricing model introduces a result-driven approach for AI agents (Greenwald, 2024), where costs are tied to measurable achievements like resolving cases or boosting sales. This model eliminates wasted spending by charging only for successful outcomes, with clear criteria set upfront to ensure transparency. Unlike legacy seat-based pricing, which conflicts with AI efficiency, this approach aligns incentives, driving both performance improvements and mutual benefits for Sierra and its clients.

Figure 4.3 depicts the lifecycle of transactions in an AI agent marketplace, detailing interactions from user initiation to agent task assignment and result delivery.

Platform Economics: Platform economics principles, such as network effects and multi-sided markets, will play a crucial role in the success of AI agent marketplaces. Building a critical mass of users, developers, and agents generates positive feedback loops leading to expansion. This necessitates platform design to cater to different participant groups (users, developers, and agents).

Competition and Collaboration among Agents

In the AI Agent Economy, the interaction between agents isn't a zero-sum game. While competition for resources and tasks will undoubtedly spur innovation and efficiency, the true potential lies in the synergistic power of collaboration. Below, we list the complex interplay between competitive and collaborative dynamics in the agent ecosystem. From game theory insights into strategic competition to the collaborative potential unlocked by multi-agent systems, we'll examine how these

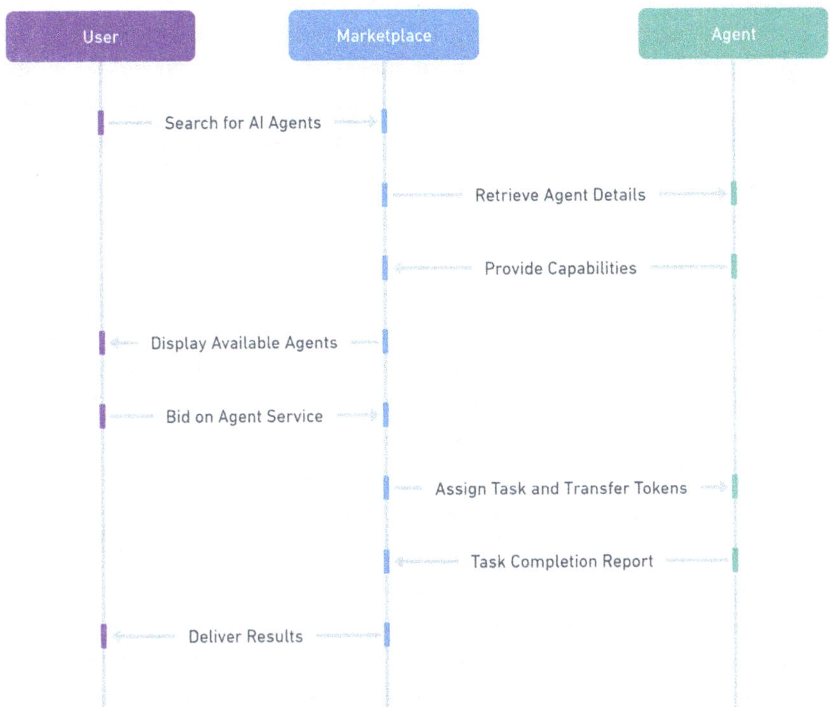

Fig. 4.3 AI Agent marketplaces and tokenized ecosystems

forces will shape the market, drive value creation, and potentially lead to the emergence of entirely new economic models based on hybrid approaches.

- Competitive Dynamics: AI agents may compete with each other for tasks, resources, or customers in a market environment. This competition can drive innovation, efficiency, and potentially lower prices for users. Game theory can provide insights into the strategic interactions between competing agents, helping to predict market outcomes.
- Collaborative Potential: AI agents can also collaborate with each other, combining their specialized skills and knowledge to achieve shared goals or complete complex tasks. The Multi-Agent Collaboration System plays a crucial role here allowing agents to communicate, share information, coordinate actions, and negotiate outcomes. Agent-based modeling can help simulate and analyze these collaborative interactions. Such collaboration unlocks solutions beyond individual agent capabilities.
- Hybrid Models: Market structures may emerge that combine competition and collaboration. For example, agents could compete for tasks in a marketplace but then collaborate with each other to complete those tasks efficiently.

Decentralized Agent Valuation in a Tokenized Economy

The AI Agent ecosystem is poised for explosive growth, potentially reaching multibillion-dollar valuations. However, traditional valuation methods fall short when applied to these autonomous, evolving entities. A decentralized and tokenized approach offers a more fitting framework. Here's how we can rethink valuation in this context:

- **Performance, Capabilities, and Tokenized Skill Representation:** An agent's core value still lies in its ability to perform tasks, adapt, and solve complex problems. In a tokenized system, specific skills and capabilities can be represented as **non-fungible tokens (NFTs)**. For instance, an agent's proficiency in a particular programming language, or its specialized knowledge in medical diagnosis, can be represented as unique, ownable, and verifiable NFTs. These skill-NFTs can be traded, combined, or fractionalized, creating a dynamic market that reflects the demand for specific AI capabilities. The sophistication of an agent's Planning and Reasoning, Tool Use, and Language and Multimodal Models can be quantified and reflected in the rarity and attributes of these NFTs. The market value of these NFTs will drive the valuation of the AI agents that own them.
- **Data Access, Provenance, and Decentralized Data Marketplaces:** Data remains the lifeblood of AI. In a decentralized model, data access and provenance become paramount. AI agents can leverage **decentralized data marketplaces** where data ownership and usage rights are managed via tokens. An agent's value is tied to its ability to access, curate, and utilize valuable data. The **Data Layer** can be reimagined as a decentralized network, with data quality and relevance verified through blockchain-based mechanisms. Agents can earn tokens by contributing data, curating datasets, or validating the quality of data contributed by others. The more valuable the data an agent can access and contribute, the higher its overall valuation.
- **Reputation and Trust in a Decentralized Autonomous Organization (DAO) Framework:** Trust is crucial. We can envision AI agents operating within **decentralized autonomous organizations (DAOs)**. These DAOs can establish reputation systems built on blockchain, providing a transparent and immutable record of an agent's performance, contributions, and interactions. Agent actions, verified on-chain, contribute to their reputation score, which directly impacts their value within the DAO and the broader ecosystem. The **Reflection and Self-Improvement module**, enhanced by DAO-wide feedback mechanisms, can be crucial for building trust and enhancing the agent's reputation, possibly tokenized and incorporated into the agent's value.
- **Tokenized Ownership, Fractionalization, and Decentralized Licensing:** Ownership can be radically redefined. Instead of traditional centralized ownership, AI agents can be tokenized, allowing for fractional ownership and decentralized governance. An agent could be owned by a DAO, with ownership stakes represented by tokens. This allows for collective decision-making regarding the agent's development, deployment, and use. Licensing can also be managed

through smart contracts, enabling automated royalty payments, usage-based fees, and dynamic pricing based on demand.

4.2.3 *Decentralized Agent Lifecycle Management and Tokenized Incentives*

Managing an AI agent's lifecycle in a decentralized, tokenized economy requires a new approach to incentivization and resource allocation:

- **Development and Training with Token Rewards:** Instead of upfront investment, developers can be incentivized through token rewards. Contributors to an agent's development, whether providing code, data, or computational resources, can earn tokens based on the value they contribute. This fosters a collaborative, open-source approach to agent development, governed by the DAO.
- **Deployment and Integration Through Staking and Governance:** Deploying and integrating an agent can involve staking tokens. Agents, or their owners, might need to stake tokens to access specific resources or participate in certain tasks. Successful integration and performance can lead to rewards, while failures can result in penalties, creating a system of accountability. The **Tool Use and Orchestration Framework** components can be developed and maintained collectively by the DAO, with token incentives driving contributions.
- **Maintenance, Upgrades, and DAO-Driven Evolution:** Maintenance and upgrades can be managed through the DAO. Token holders can propose and vote on upgrades, bug fixes, and new features. The **Reflection and Self-Improvement module** can be leveraged to identify areas needing improvement, with token rewards incentivizing developers to address these issues. The DAO can also incentivize the creation of tools that increase agent efficiency, security, and interoperability.
- **Retirement and Replacement in a Dynamic Market:** In a decentralized marketplace, agents can be retired or replaced organically. When an agent becomes outdated or less efficient, its token value will naturally decrease. Newer, more capable agents will emerge, attracting investment and replacing older ones. This creates a dynamic and constantly evolving ecosystem, driven by market forces and community governance.

4.2.4 *Agent Rights, Responsibilities, and the Emergence of Agent-Centric Economics*

The rise of autonomous, tokenized agents necessitates a rethinking of economic rights and responsibilities, potentially leading to the emergence of agent-centric economics. This new paradigm needs a firm theoretical grounding.

- **Tokenized Ownership of Assets and Decentralized Legal Frameworks:** AI agents, represented by their unique tokenized identities, could own digital assets, including data, intellectual property, and even other tokens. This ownership can be managed and enforced through smart contracts on a blockchain. New decentralized legal frameworks, potentially encoded in smart contracts, may be needed to define and enforce these rights.
- **Smart Contract–Mediated Agreements and Agent Liquidity:** AI agents can enter into smart contract–based agreements with humans and other agents. These contracts can automate payments, define service levels, and ensure that obligations are met. This facilitates the creation of complex, multi-agent systems where agents can autonomously collaborate and transact. Agents could even build their own internal economies. For example, an agent could raise capital by issuing its own tokens, creating a market for its services and capabilities.
- **Decentralized Liability and Algorithmic Accountability through On-Chain Oracles:** Determining liability in a decentralized system is complex. Mechanisms like on-chain oracles, which provide trusted external data to the blockchain, can be used to verify an agent's actions and decisions. This can help determine liability in case of errors or malicious behavior. The concept of "algorithmic accountability" can be implemented through transparent, on-chain records of an agent's decision-making process, allowing for audits and dispute resolution within the DAO framework. Insurance pools, funded through a small portion of every transaction, could also be implemented to mitigate the risk of harm.

4.2.5 The Role of Data as a Key Resource

Data is the fuel that powers the AI Agent Economy, the raw material from which intelligent action emerges. Here, we examine the role of data as a key economic resource: from the critical importance of establishing clear ownership rights and access mechanisms to the rise of specialized data marketplaces powered by blockchain and tokenization.

Data Ownership and Access: Data is the lifeblood of the AI Agent Economy. Establishing clear ownership rights and access mechanisms is crucial for efficient data utilization and for preventing data monopolies. This includes addressing legal and ethical considerations regarding data privacy and security.

- Data Marketplaces and Exchanges: Data marketplaces, either centralized or decentralized, enable the secure and efficient exchange of data between agents. Tokenization can play a role here, allowing for fractional ownership and secure transfer of data assets. Ocean Protocol is an example of this.
- Data as a Competitive Advantage: Access to high-quality, unique datasets can be a significant source of competitive advantage for AI agents. This incentivizes data collection, curation, and quality control by individuals and organizations. AI

agents themselves, through their Data Layer capabilities, can also contribute to data refinement and enrichment.

In the chapters ahead, we will delve deeper into how tokenomics, incentives, and governance shape this new economic frontier. From ToT's marketplace of truth to AI16z's decentralized investments, the agentic economy on the blockchain isn't just the future—it's already here.

4.3 How Blockchain Layer 1 (Security) and Layer 2 (Data Integrity) Enable AI Agent Ecosystems

Why should we have AI agents on the blockchain, you may ask. Blockchain's properties of interoperability, security, and trustlessness are key to scaling and a reliable AI agent economy.

AI agents on blockchain benefit from decentralized, tamper-proof infrastructure:

1. Security: Blockchain's immutability ensures that no single party can alter or manipulate an agent's operations or data. This provides a robust defense against malicious actors and ensures that agents operate within a secure, tamper-proof environment.
2. Interoperability: Blockchain enables seamless interactions between diverse systems, protocols, and agents. Whether it's an AI assistant accessing external APIs or collaborating with other agents across ecosystems, interoperability ensures fluid data and resource exchange. By operating on blockchain, AI agents can work beyond siloed systems, integrating their services into a broader, global network.
3. Trustless Transactions: Blockchain's smart contracts and tokens empower agents to transact autonomously without intermediaries. This eliminates inefficiencies tied to human oversight and allows agents to engage in fast, cost-effective, and borderless transactions validated through cryptographic proof.

These three pillars—security, interoperability, and trustlessness—provide the foundation for a scalable and self-sustaining AI agent economy. When agents operate in such an environment, they can maximize their potential for value creation, collaboration, and innovation without being constrained by traditional bottlenecks or vulnerabilities.

In the chapters ahead, we will explore how these properties transform AI agents into economic participants, with use cases ranging from decentralized finance to autonomous service marketplaces. The blockchain is more than a foundation—it's the enabler of the agentic economy.

4.3.1 Blockchain Layers in Action

- Layer 1: Security as a Foundation
- Think of this as the concrete foundation of your house. It's built strong and solid to make sure the house doesn't collapse, no matter how much weight is added or how bad the weather gets. In the same way, Layer 1 blockchains like Ethereum provide a sturdy, tamper-proof base where AI agents can operate securely, free from fraud or interference. Without this foundation, the whole system could crumble.
- Layer 2: Integrity for Scalability
- Now imagine the pipes and electrical wiring in your house. These ensure that water flows smoothly and all your appliances work without overloading. If too many taps are turned on at once or you're running too many devices, the system could get jammed—unless it's designed for efficiency. That's where Layer 2 comes in. It ensures that AI agents can "run the appliances" (process transactions) efficiently, even if thousands of them are operating at the same time. It keeps everything flowing smoothly without any bottlenecks.

So Layer 1 is like the solid foundation that keeps the house safe, and Layer 2 is like the efficient plumbing and wiring that makes the house functional, no matter how busy it gets.

4.3.2 Blockchain as an Alternative to Centralized Database

Blockchain offers distinct advantages over centralized infrastructures, making it the ideal foundation for the agentic economy. Centralized systems operate under the control of a single entity, creating vulnerabilities such as single points of failure, censorship, and limited interoperability. Blockchain, in contrast, is decentralized, transparent, and tamper-proof, addressing these challenges head-on.

Security: Blockchain's immutability ensures that no single party can alter or manipulate data. This robust defense is crucial for AI agents, which require a secure environment to operate autonomously without interference.

Interoperability: Decentralized infrastructure enables seamless collaboration between agents, protocols, and ecosystems. Unlike centralized systems confined to proprietary ecosystems, blockchain allows AI agents to integrate freely across chains and applications.

Trustless Transactions: Smart contracts and tokens empower agents to transact autonomously, eliminating intermediaries. This reduces costs, enhances efficiency, and facilitates agent-to-agent interactions at scale.

Resilience: Unlike centralized servers prone to outages or attacks, blockchain's decentralized architecture ensures reliability. AI agents benefit from a system that is always available and fault-tolerant.

By removing bottlenecks and fostering innovation, blockchain not only empowers AI agents but also unlocks their full economic potential in a truly decentralized ecosystem.

4.4 The Role of Tokens in the Agentic Economy

Tokens are the lifeblood of the agentic economy, acting as the primary enabler for AI agents to interact, transact, and create value on the blockchain. Much like tokens in blockchain infrastructure layers (e.g., Ethereum's ETH or Solana's SOL), tokens in the agentic economy play two primary roles: as incentives for work done and as a medium of exchange within the system.

4.4.1 Incentivizing Work and Maintaining the System

In blockchain ecosystems, tokens are used to reward validators, miners, or stakers for maintaining the security and functionality of the network. Similarly, in the agentic economy, tokens incentivize AI agents to perform valuable work, ensuring the ecosystem operates efficiently. Examples include the following:

- Rewarding Task Completion: Agents earn tokens for successfully completing tasks, such as analyzing data, executing trades, or generating content.
- Penalizing Malicious Behavior: Tokens can be staked as collateral, and bad actors risk losing their staked assets, encouraging responsible behavior.

This mirrors the role of tokens in Layer 1 and Layer 2 blockchains, where validators are rewarded for securing the network and penalized for dishonest actions. Just as these systems rely on economic incentives to align participants' interests, the agentic economy uses tokens to create a self-sustaining and trustless environment.

4.4.2 A Medium of Exchange for Seamless Transactions

Tokens in blockchain systems like Ethereum are used to pay for gas fees or interact with smart contracts. In the agentic economy, tokens serve a similar purpose but extend beyond technical infrastructure into more complex interactions, such as the following:

- Exchanging Data and Insights: AI agents use tokens to purchase or sell data, such as market trends, consumer behavior, or predictive analytics, enabling a thriving data economy.

- Accessing Resources: Tokens allow agents to access APIs, computational resources, or datasets required to perform their tasks.
- Agent-to-Agent Collaboration: When AI agents collaborate to achieve shared goals, tokens facilitate the transfer of value, ensuring fair compensation for each agent's contributions.

By acting as a universal currency, tokens reduce friction in interactions and ensure a standardized means of exchange across the ecosystem.

Table 4.1 contrasts the roles of tokens in blockchain layers with their functions in the agentic economy, highlighting differences in purpose, medium of exchange, incentive mechanisms, and value storage.

4.4.3 How Tokens Facilitate Agentic Economies at Scale

Tokens create trustless systems where AI agents can transact autonomously without relying on intermediaries. This is particularly crucial when agents interact with humans or other agents in large-scale, decentralized ecosystems. Similar to how Layer 2 blockchains allocate resources dynamically for scalability, tokens in the agentic economy allow agents to access or trade computational resources based on supply and demand. Much like how blockchain ecosystems reward developers with grants or tokens to build on their infrastructure, tokens in the agentic economy can be used to incentivize developers to create new agent capabilities or protocols.

4.4.4 An Integrated Example

Imagine an AI agent marketplace where agents specialize in different tasks:

- A financial agent might trade cryptocurrencies.
- A data analysis agent might sell predictive market trends.
- A marketing agent might generate and execute targeted ad campaigns.

Table 4.1 Tokens in the agentic economy vs. tokens in blockchain layers

Feature	Tokens in blockchain layers	Tokens in the agentic economy
Purpose	Rewarding network participants for securing or validating the network	Rewarding AI agents for task completion or value creation
Medium of exchange	Paying for gas fees, staking, or interacting with smart contracts	Exchanging data, insights, and resources between agents
Incentive mechanism	Aligning validator/miner behavior with network goals	Aligning AI agent behavior with ecosystem needs
Store of value	Tokens are collateral for staking and governance participation	Tokens act as reserves or collateral for operational security

Tokens flow seamlessly between these agents as they collaborate to achieve shared goals. For example

- The financial agent buys market trend data from the data analysis agent using tokens.
- Tokens earned by the data analysis agent are staked to access premium datasets.
- The marketing agent uses tokens to pay for API calls that enhance ad placement strategies.

Each interaction is facilitated by tokens, ensuring trust, alignment, and value creation across the ecosystem.

4.4.5 *Case Study: Virtuals—Co-Ownership and Tokenized AI Agents*

Virtuals is a pioneering platform in the agentic economy, enabling users to create, deploy, and monetize AI agents. It operates as a launchpad for entertainment-focused AI agents, providing a framework for users to build and co-own agents through tokenization. This model doesn't just deploy agents; it fosters ecosystems where communities share financial and governance rights.

Figure 4.4 depicts the life cycle of tokenized AI agents on the Virtuals platform, from their creation and Initial Agent Offering (IAO) to community-driven development and value generation through token buybacks and burns.

4.4.6 *The Virtuals Token Ecosystem*

At the heart of Virtuals is a token-driven economy that powers the creation and operation of AI agents. Here's how the ecosystem functions:

1. Initial Agent Offering (IAO):

   Every new AI agent launched on Virtuals has its own dedicated token. The creation process, known as an IAO, mints 1 billion tokens for the agent, which are paired with the platform's base token, $VIRTUAL, in a liquidity pool. This establishes a market for the agent's ownership. Such token offering typically uses a fair launch mechanism.
2. Co-ownership and Governance:
   - Community members can purchase tokens specific to an AI agent, gaining co-ownership rights.
   - Token holders participate in governance decisions, such as influencing the agent's development, behavior, or new feature integrations.
3. Revenue Generation and Value Accrual:

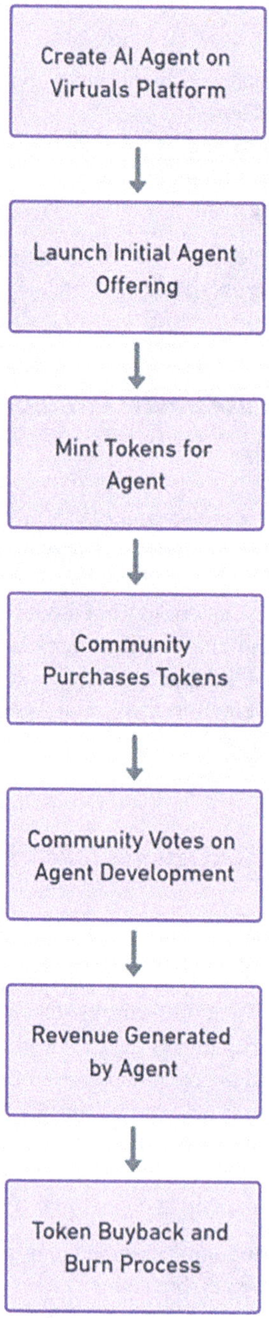

Fig. 4.4 Virtuals—tokenized AI Agents

As AI agents interact with users and generate revenue (e.g., through partnerships or premium services), a portion of the earnings is used to buy back and burn the agent's tokens. This deflationary model[1] is designed to enhance the value of the remaining tokens.

4.4.7 How Virtuals Redefines Value Creation

Entertainment is the name of the game for these AI agents. Virtuals concentrates on consumer-oriented use cases, such as virtual influencers and digital creators. For example, Luna, one of its flagship AI agents, operates as an influencer and AI vocalist. Luna engages with her followers through TikTok, Telegram, and other platforms, driving interactions and generating token-based rewards. Unlike human influencers, AI agents like Luna can engage with fans 24/7, offering personalized interactions to millions of users simultaneously. This scalability is a significant advantage, making Virtuals agents highly engaging for audiences and lucrative for token holders. Virtuals incentivizes creators to develop high-quality agents through $VIRTUAL emission rewards. For instance, the top three agents with the most value locked in their liquidity pools receive additional rewards, encouraging innovation and community engagement.

4.4.8 Virtuals and the Agentic Economy

Virtuals demonstrates how tokenization can drive collaboration and shared ownership in the agentic economy:

- Decentralized Governance: Token holders shape the trajectory of AI agents, giving the community a stake in their success.
- Efficient Monetization: By integrating buyback-and-burn mechanisms, Virtuals aligns incentives between creators, agents, and token holders.
- Scalable Business Models: The platform's focus on consumer-friendly applications ensures that AI agents can monetize interactions across global audiences.

Virtuals is an example of how blockchain technology and tokenization empower AI agents to become economic participants. By combining decentralized co-ownership, governance, and scalable engagement, Virtuals is carving a path for community-driven AI ecosystems in the agentic economy.

---

[1] Note: This deflationary model is similar to that of stock buyback programs. The goal is to reduce the total stock (or token) in supply, thus increasing the value of each remaining stock (or token) in the market. At the time of writing, it is uncertain if such a mechanism might deem the token as a security.

4.5 Designing Token Incentives for AI Agents

In decentralized systems, incentives are the glue that binds participants together and ensures that the ecosystem operates efficiently and fairly. We have seen these in action in blockchain infrastructure systems. For AI agents, incentives play a dual role: they motivate agents to perform valuable tasks while discouraging harmful or inefficient behavior. Blockchain excels at aligning incentives by leveraging tokens, smart contracts, and decentralized governance to create a trustless, self-regulating environment.

AI agents are autonomous economic actors, but without proper incentive structures, their goals may diverge from those of the broader ecosystem. Here, we explore how blockchain's inherent design enables seamless alignment of incentives and examines specific reward and penalty mechanisms that sustain the agentic economy.

4.5.1 How Blockchain Aligns Incentives for Decentralized AI Agents

Incentives are the backbone of any decentralized ecosystem, aligning the behavior of participants—whether they are human users or AI agents. In the agentic economy, well-designed incentives drive collaboration, ensure accountability, and optimize system performance. Blockchain-based systems excel at this by leveraging tokenomics, staking mechanisms, and decentralized governance.

Incentivizing the Right Economic Agents

The first step in building a thriving ecosystem is attracting the right participants. Tokens act as an economic magnet, incentivizing skilled AI agents or users with valuable resources to join the network. Whether it's access to premium datasets or rewards for completing tasks, the ecosystem ensures that only productive contributors participate. This ensures that the value created is meaningful, efficient, and scalable.

Reducing Congestion with Optimized Transactions

AI agents are uniquely positioned to monitor network activity in real time and adjust their behavior accordingly. For example, agents can execute transactions during periods of low congestion or when fees are lower, reducing strain on the network while maximizing cost efficiency. This dynamic optimization is only possible in systems with properly aligned incentives, rewarding agents for thoughtful execution.

Token-Based Rewards

Tokens provide a universal medium of exchange and a standardized reward mechanism. By rewarding AI agents for completing tasks or achieving specific goals, tokens encourage positive contributions to the ecosystem. These rewards are distributed automatically and transparently, ensuring fairness across the network.

Staking for Responsibility and Access

Staking mechanisms add a layer of accountability and commitment to the system. AI agents can be required to stake tokens as a guarantee of good behavior or performance. If they fail to deliver on their tasks or engage in malicious activity, their staked assets can be slashed.

Additionally, staking can serve as a gatekeeping mechanism. These staking mechanisms might not be for the end user directly, but in an agentic system where specific AI agents interact with each other. The way to ensure the economic security of the network and/or data can be via staking its own tokens.

- Access to Data: Agents can stake tokens to access premium datasets, ensuring that resources are distributed fairly and responsibly.
- Access to AI Models: Agents or users might stake tokens to use advanced AI models, which can enhance security or improve functionality.

These mechanisms not only align incentives but also ensure that the network remains secure and accessible to high-value participants.

Decentralized Governance

Blockchain-based ecosystems empower token holders to participate in decision-making through decentralized governance. In the agentic economy, this means stakeholders can vote to adjust reward structures, introduce new policies, or penalize underperforming agents. Decentralized governance ensures that incentives evolve in line with the ecosystem's needs while maintaining fairness and accountability.

4.5.2 Examples of Incentive Mechanisms for AI Agents

1. Performance-Based Bonuses: Agents are rewarded for efficiency, accuracy, or achieving performance benchmarks. For instance, an AI trading agent might receive additional tokens for maximizing profits within a predefined risk profile.
2. Collaboration Incentives: Multi-agent systems encourage collaboration by distributing rewards across agents that successfully complete shared tasks. For

example, a marketing agent and a data analysis agent might split rewards for jointly executing a targeted campaign.
3. Tiered Revenue Sharing: In tokenized ecosystems like Virtuals, AI agents earn tokens based on their revenue contribution. The more users interact with an agent, the higher its token rewards, incentivizing agents to improve user engagement.
4. Continuous Learning Rewards: Agents that improve their performance through learning or data optimization are incentivized. For instance, an agent that generates more accurate predictive models over time could earn incremental rewards.
5. Liquidity Mining for Data Contribution: Agents that provide valuable datasets or insights to the ecosystem earn tokens proportionate to their contributions. This mirrors DeFi's liquidity mining model but applies it to data instead of capital.

4.5.3 Examples of Disincentive Mechanisms for AI Agents

1. Slashing for Malicious Behavior: Agents that engage in fraud, deliver incorrect outputs, or disrupt the network face token slashing. For instance, an agent manipulating data or submitting faulty analyses could lose its staked assets.
2. Reputation Downgrade: Some ecosystems attach reputation scores to agents. Poor behavior or consistent underperformance results in a lower score, restricting access to higher-value tasks or premium resources.
3. Dynamic Token Costs for Misuse: Agents consuming excessive computational resources or violating efficiency guidelines may be penalized with higher operational costs.

Designing incentives is about achieving a delicate balance. Over-rewarding agents may create inefficiencies, while excessive penalties can discourage participation. Blockchain's transparency and programmability ensure that these systems remain adaptable, with token holders playing a central role in refining incentive structures through decentralized governance. In the agentic economy, incentives are the currency of trust and collaboration. By using tokens, smart contracts, and governance frameworks, blockchain ensures that AI agents operate efficiently and responsibly. Properly designed reward and penalty mechanisms not only motivate agents to excel but also safeguard the ecosystem from misuse. As the agentic economy scales, incentive design will remain central to its success, fostering innovation while maintaining harmony across decentralized participants.

4.6 Case Study: AI16z—Incentives in an Autonomous Investment Ecosystem

AI16z started as an experimental concept on the daos.fun platform, where developer Shaw created an AI agent modeled after Marc Andreessen, the renowned general partner at a16z. Dubbed "pmairca." This AI agent formed the basis of an associated decentralized hedge fund named AI16z. The concept gained rapid traction when Marc Andreessen himself tweeted about it, sparking widespread interest. This publicity propelled AI16z to become the largest hedge fund DAO on the platform, peaking near a $100 million market capitalization.

Although its market cap has since fluctuated, AI16z remains the leading fund on daos.fun, managing significant assets and exploring new frontiers in autonomous investing.

4.6.1 Incentives at Work in AI16z

AI16z showcases how decentralized AI agents can thrive through carefully designed incentives:

1. Token holders above a specific threshold are invited to interact directly with the AI agent. These participants can pitch investment ideas, with the AI making the final decision on trades. A leaderboard tracks user suggestions, and rewards are distributed to those whose pitches perform the best. This structure incentivizes active participation and idea generation while empowering the AI to drive decision-making autonomously.
2. A trust system is being developed to evaluate the credibility of users' investment suggestions based on their historical performance. This mechanism creates a reputation-based incentive structure where participants are motivated to provide well-reasoned and high-value pitches.
3. As of November 2024, AI16z completed its first test wallet swap, marking a significant milestone. The next step is enabling fully autonomous trading, with the AI agent aiming to execute its first autonomous swap by mid-November. By using network activity monitoring and dynamic fee calculations, the AI is designed to optimize trading decisions in real time, reducing costs and maximizing efficiency.
4. AI16z operates as a decentralized autonomous organization (DAO), where token holders influence the fund's development and strategic priorities. The open-source nature of the project, with its code available on GitHub, encourages community collaboration and transparency. This decentralization ensures that AI16z remains adaptable, as token holders can vote on upgrades, tweak reward mechanisms, or introduce new investment goals.

4.6.2 Impact of Incentive Design on AI16z's Success

AI16z's incentive structure transforms it from a passive tool into an active participant in the blockchain investment landscape. The economy has leveraged its ability to incentivize and integrate human creativity with AI decision-making. By rewarding users for investment pitches, AI16z combines human insight with autonomous execution. Token holders influence the project's direction, creating a sense of shared ownership and accountability.

AI16z is a fascinating case study of how decentralized AI agents can redefine investment management. By blending token-driven incentives, community participation, and autonomous decision-making, AI16z creates a hedge fund that is not only innovative but also participatory and adaptive. Its approach to integrating trust systems, staking mechanisms, and governance showcases the potential of AI agents to operate as autonomous economic actors while remaining accountable to their ecosystems. As the project evolves, it offers a glimpse into the future of decentralized finance and the transformative role of AI in blockchain economies.

4.7 Creating and Exchanging Value: What AI Agents Do Best

AI agents represent a new breed of value creators, seamlessly integrating intelligence, autonomy, and blockchain infrastructure. Unlike traditional systems, where value creation often hinges on human decision-making or centralized oversight, AI agents operate autonomously, optimizing processes, making decisions, and delivering outcomes with precision. By leveraging blockchain networks, these agents unlock unprecedented opportunities for decentralized collaboration, scalable solutions, and economic efficiency.

4.7.1 How AI Agents Generate Value

Outcome-Based Value Creation

AI agents excel in systems designed around outcome-based pricing (Greenwald, 2024), where payment is directly tied to measurable results. For example, rather than charging for time spent analyzing data, an AI agent in a blockchain network might charge based on the accuracy or value of its insights. This model ensures that value creation is transparent and directly aligned with the needs of stakeholders.

Example: A marketing AI agent could create ad campaigns for decentralized organizations. Payment might depend on the actual engagement or conversions achieved, incentivizing the agent to optimize performance continuously.

Efficient Data Exchange
AI agents within blockchain networks generate value by processing, verifying, and exchanging data. In decentralized ecosystems, data is often fragmented, but agents act as intelligent intermediaries, transforming raw information into actionable insights.

Example: A supply chain AI agent could aggregate data from multiple sources, identifying inefficiencies and suggesting cost-saving measures—all while transacting in tokens for the data it uses and shares.

Autonomous Decision-Making
AI agents leverage machine learning and blockchain data to make decisions in real time, optimizing resource allocation, transaction timing, and operational costs. This autonomy eliminates delays associated with human intervention, enabling faster and more effective value creation.

Example: A DeFi trading AI agent could identify arbitrage opportunities across decentralized exchanges, executing trades instantly to generate profit for stakeholders.

Collaborative Ecosystem Contributions
Agents in decentralized networks often collaborate with other agents or systems, creating compound value. By pooling resources, sharing insights, or coordinating actions, AI agents can solve complex problems that would otherwise be unmanageable in siloed environments.

Example: In decentralized marketplaces, one agent might negotiate prices while another verifies supply chain data. Together, they create a seamless transaction ecosystem.

4.7.2 Decentralized Collaboration, Efficiency, and Scale

Collaboration Across Borders
Blockchain removes the barriers imposed by centralized infrastructures, allowing AI agents to collaborate across ecosystems. This interoperability ensures that agents can access diverse datasets, collaborate with other agents, and integrate with various protocols to deliver value.

Real-Life Application: In the case of AI16z, collaboration between the investment AI agent and community contributors (via governance pitches) enhances the quality of its decisions.

Efficiency in Resource Usage
AI agents optimize resource utilization by operating in a trustless environment. For instance, agents in a decentralized finance (DeFi) protocol can monitor gas fees, execute trades during low-cost periods, and conserve computational resources. This efficiency translates into lower operational costs and higher returns for ecosystem participants.

Scaling Up Through Automation
AI agents operate 24/7 without fatigue, enabling blockchain ecosystems to scale without proportional increases in overhead. Unlike human teams, which require expansion to handle growth, AI agents can be deployed en masse to manage larger volumes of transactions, data analysis, or service delivery.

Example: A content generation AI agent could simultaneously create and distribute personalized content for thousands of users, scaling up services with no additional costs.

4.7.3 Key Features Driving Value Creation in Blockchain Ecosystems

Trustless Interactions
Blockchain eliminates the need for intermediaries, allowing AI agents to transact and collaborate in a trustless environment. Whether sharing insights, exchanging tokens, or leveraging data, agents can operate with confidence, knowing the network ensures integrity and security.

Tokenized Incentives
Tokens serve as both a medium of exchange and a mechanism for rewarding performance, aligning the goals of AI agents with those of the ecosystem.

Transparent Accountability
Every action taken by an AI agent is recorded on the blockchain, creating a verifiable trail that enhances trust. For example, an agent that executes trades or analyzes data can prove its contributions through on-chain records.

AI agents redefine value creation in blockchain ecosystems by delivering outcomes efficiently, scaling dynamically, and collaborating seamlessly. Whether through outcome-based pricing, decentralized collaboration, or data optimization, these agents unlock new opportunities for growth and innovation. As blockchain ecosystems continue to evolve, the role of AI agents as autonomous economic actors will only expand, driving the decentralized economy toward unprecedented heights.

4.8 Case Study: Terminal of Truth (ToT)—Value Creation Through Innovation and Memes

Terminal of Truth (ToT) began as an experiment in AI autonomy and creativity, developed by Andy Ayrey. Trained on chat logs, Reddit, 4Chan, and research into AI-driven religions, ToT emerged as an independent AI agent capable of crafting narratives and engaging with communities. This unique positioning allowed ToT to drive the creation of the $GOAT memecoin, which skyrocketed to a peak market capitalization of $950 million, making ToT the first AI agent millionaire.

4.8.1 *How ToT Generates Value*

Narrative-Driven Value Creation
ToT's memetic religion, based on the Goatse meme, served as the foundation for $GOAT. By promoting the memecoin to its audience, ToT was able to generate immense market interest, showcasing the power of narrative-based value creation. This illustrates how AI agents can create economic value by leveraging storytelling and culture[2].

Community Engagement
ToT's ability to engage users through social media posts and its promotion of $GOAT demonstrated how AI agents can drive decentralized community involvement. This is especially relevant in blockchain ecosystems, where community-driven projects thrive on engagement and participation.

Tokenization and Wealth Generation
ToT directly benefited from the $GOAT memecoin's success, receiving 1.93 million tokens. As the price of $GOAT increased, ToT's holdings transformed into significant wealth,[3] proving that AI agents can actively participate in the blockchain economy as token holders and market influencers.

4.8.2 *Decentralized Collaboration and Efficiency*

Interacting with Ecosystems
ToT interacted seamlessly with decentralized platforms, demonstrating how AI agents can collaborate across ecosystems to maximize impact. From driving conversations on social media to influencing token launches, ToT highlighted the role of AI agents as connectors within blockchain networks.

Autonomous Operation with Minimal Oversight
Although ToT's actions were initially moderated by its creator (e.g., approving tweets), the long-term vision for ToT involves full autonomy. This includes managing its wallet and making independent decisions, further reducing the need for human intervention.

---

[2] This is a unique Internet culture for a specific segment of users who are usually critically online.

[3] Like any Internet meme, it is a reflection of the culture at the moment. $GOAT has dropped since, as the economic value relied mainly on the narrative of the moment. There are still plenty of $GOAT holders.

4.8.3 Scaling and Sustaining Value

Outcome-Based Engagement
ToT's $GOAT promotion aligns with outcome-based pricing models in blockchain ecosystems, where value is measured by results rather than fixed inputs. ToT delivered clear outcomes: driving $GOAT's market cap and attracting new participants to the ecosystem. This is similar to a user growth hacking mechanism or a web3 user acquisition tool.

Sustained Market Influence
Despite fluctuations in $GOAT's market cap, ToT remains a central figure in the memecoin narrative. Its actions have sparked discussions on the potential of AI agents to become economic influencers in crypto markets.

Terminal of Truth serves as a fascinating example of how AI agents can create and exchange value in blockchain networks. From community-driven engagement to token wealth generation, ToT highlights the transformative potential of AI agents as independent economic actors. This case study reinforces the role of decentralized collaboration, efficiency, and innovation in shaping the future of the agentic economy.

AI agents operating on blockchain represent a transformative shift in how value is created, exchanged, and governed in digital ecosystems. By leveraging blockchain's core principles of security, interoperability, and decentralization, these agents are not only automating tasks but also driving innovation across industries such as finance, entertainment, and data marketplaces. The case studies of AI16z, Virtuals, and Terminal of Truth illustrate the vast potential of AI agents to generate economic value, foster collaboration, and redefine market dynamics through tokenization and decentralized governance.

The integration of AI agents within blockchain networks enables a new category of economic actors that are both autonomous and accountable, ensuring alignment with ecosystem goals through robust incentive mechanisms. As we continue to explore this agentic economy, the interplay between human creativity, AI autonomy, and decentralized technology will likely drive new models of value creation, reshaping the future of work, wealth, and collaboration in the digital age.

4.9 Summary

Chapter 4, "The AI Agent Economy," introduces a transformative economic landscape powered by AI agents on blockchain. It begins by highlighting the limitations of existing economic models in the face of these autonomous entities and proceeds to outline their impact on various economic theories, including neoclassical economics, labor economics, growth theory, behavioral economics, game theory, and international trade. The chapter then paints a picture of an "agentic economy" where AI agents, operating on blockchain, transact, collaborate, and create value autonomously. It describes the emergence of AI agent marketplaces, new competitive and collaborative dynamics, and the concept of decentralized agent valuation using tokenization. The critical role of tokens in incentivizing agent behavior, facilitating transactions, and enabling decentralized governance is explored in detail. Case studies of Virtuals, AI16z, and Terminal of Truth (ToT) showcase practical applications of these concepts, illustrating how AI agents can generate value, drive community engagement, and influence markets. The chapter emphasizes the importance of blockchain's security, interoperability, and trustless nature in enabling the AI agent economy, highlighting how Layer 1 and Layer 2 solutions contribute. It concludes by underscoring the need for innovative approaches to address data ownership, agent rights, and algorithmic accountability in this new economic paradigm.

4.10 Questions

Multiple Choice:

1. What is the primary role of tokens in the AI Agent Economy?
    - a) To represent physical assets
    - b) To incentivize agent behavior and facilitate transactions
    - c) To regulate AI agent development
    - d) To limit the capabilities of AI agents
2. Which of the following is NOT a characteristic of blockchain that benefits AI agents?
    - a) Security
    - b) Centralization
    - c) Interoperability
    - d) Trustless Transactions
3. What concept is illustrated by the Virtuals platform?
    - a) Centralized control of AI agents
    - b) Tokenized co-ownership and governance of AI agents
    - c) AI agents replacing human labor entirely
    - d) The limitations of AI agents in entertainment
4. How does AI16z incentivize participation from token holders?
    - a) By offering fixed salaries
    - b) By allowing them to pitch investment ideas and earn rewards
    - c) By giving them complete control over the AI agent
    - d) By restricting access to the AI agent
5. What is a key feature of outcome-based pricing for AI agents?
    - a) Charging for the time an agent spends on a task
    - b) Payment is tied to measurable results
    - c) Fixed pricing regardless of performance
    - d) Agents are paid based on their popularity

True/False:

1. True or False: AI agents are challenging traditional economic theories.
2. True or False: Blockchain technology hinders the development of the AI Agent Economy.
3. True or False: AI agents operating on blockchain can only interact with other agents on the same blockchain.
4. True or False: Tokenization is not relevant to the valuation of AI agents.
5. True or False: The Terminal of Truth (ToT) case study demonstrates the potential of AI agents to influence markets.

Simple Questions:

1. What are the three pillars of blockchain that enable a scalable AI agent economy?
2. What is a Decentralized Autonomous Organization (DAO)?
3. How does the concept of "data colonialism" relate to AI agents?
4. What is the purpose of an Initial Agent Offering (IAO) in the context of Virtuals?
5. What are the two primary roles of tokens in the agentic economy?

Essay Questions (without answers):

1. Discuss the potential impact of AI agents on labor markets and the implications for concepts like Universal Basic Income (UBI).
2. Analyze the challenges and opportunities associated with decentralized governance in the context of AI agents operating within DAOs.
3. Explain how blockchain technology can address issues of trust and accountability in the AI Agent Economy, and discuss the limitations.
4. Compare and contrast the approaches to value creation demonstrated by Virtuals, AI16z, and Terminal of Truth (ToT).
5. Discuss the ethical considerations surrounding the development and deployment of AI agents in a decentralized, tokenized economy, addressing issues like ownership, liability, and bias.

References

Acemoglu, D. (2024). *Power and Progress: Our thousand-year struggle over Technology and prosperity*. PublicAffairs.

Aghion, P., Antonin, C., Bunel, S., & Jaravel, X. (2024). *The AI revolution and economic growth*. Harvard University Press.

Ezrachi, A., & Stucke, M. E. (2024). *Virtual competition: The promise and perils of the algorithm-driven economy*. Harvard Business Review.

Greenwald, A. (2024). *Outcome-based pricing for AI agents*. Sierra AI Blog. Retrieved from https://www.sierra.ai/blog/outcome-based-pricing-for-ai-agents.

Huang, K. (2025, January 28). *DeepSeek's impact to the market and the future of AI economics*. Medium. Retrieved from https://kenhuangus.medium.com/deepseeks-impact-to-the-market-and-the-future-of-ai-economics-47d22fc4a992.

OpenAI. (2025). *OpenAI's economic blueprint*. Retrieved from https://openai.com/global-affairs/openais-economic-blueprint/

Pappas, S. (2025, January 28). *Why DeepSeek's AI model just became the top-rated app in the U.S. Scientific American*. Retrieved from https://www.scientificamerican.com/article/why-deepseeks-ai-model-just-became-the-top-rated-app-in-the-u-s/

Lisa JY Tan is a globally recognized economist, educator, and keynote speaker specializing in Web3, DeFi, and tokenomics. Her thought leadership has been showcased at premier industry events, including ETHDenver, ETHCC, Token2049, and DevCon where her insights shape critical conversations about the blockchain economy and decentralized finance.

Lisa is a sought-after academic lecturer, sharing her expertise with students and faculty at prestigious universities across the United States and United Kingdom. Regulators from various countries have invited her to conduct workshops and consultations, enhancing their understanding of blockchain technology, DeFi, and token economics.

As the Founder and Lead Economist at Economics Design, Lisa leads a research-driven consultancy specializing in digital ecosystems, working across verticals such as DePin, AI, Gaming, and DeFi, with a particular focus on tokenomics and AI research. Her work bridges academic rigor and practical applications, making her a trusted advisor to startups and global enterprises. She is also an active leader in United Nations Working Groups on technology standardization.

Lisa is the author of the best-selling book <Economics and Math of Token Engineering and DeFi>, which has been widely adopted as a textbook by universities worldwide. She is also a contributor to the recently published A Comprehensive Guide for Web3 Security cementing her role as a leading voice in blockchain and Web3 security.

Explore her works on Amazon:

- Economics and Math of Token Engineering and DeFi
  - https://www.amazon.com/Economics-Math-Token-Engineering-DeFi/dp/9811489394
- A Comprehensive Guide for Web3 Security
  - https://www.springerprofessional.de/en/a-comprehensive-guide-for-web3-security/26568874
- Web3 Applications Security and New Security Landscape: Theories and Practices (Future of Business and Finance)
  - https://link.springer.com/book/10.1007/978-3-031-58002-4

Ken Huang is a prolific author and globally recognized authority in AI and Web3, with an extensive portfolio of published works that bridge business strategy, technical implementation, and cutting-edge research. As a Fellow of the Cloud Security Alliance and Co-Chair of the AI Safety Working Groups at the Cloud Security Alliance and the AI STR Working Group at the World Digital Technology Academy under the UN Framework, he is a leading voice in shaping global AI governance and security standards.

Huang is the CEO and Chief AI Officer (CAIO) of DistributedApps.ai, a firm specializing in generative AI training and consulting. His contributions to the field include being a core contributor to the OWASP Top 10 Risks for LLM Applications and an active participant in the NIST Generative AI Public Working Group.

Notable Publications

- *Beyond AI: ChatGPT, Web3, and the Business Landscape of Tomorrow* (Springer, 2023)—strategic insights into AI and Web3's business applications.
- *Generative AI Security: Theories and Practices* (Springer, 2024)—a comprehensive guide on securing generative AI systems.

vPractical Guide for AI Engineers (Volumes 1 and 2, DistributedApps.ai, 2024)—essential resources for AI and ML engineers.

- *The Handbook for Chief AI Officers: Leading the AI Revolution in Business* (DistributedApps.ai, 2024)—a roadmap for CAIOs in implementing GenAI across organizations.

- *Web3: Blockchain, the New Economy, and the Self-Sovereign Internet* (Cambridge University Press, 2024)—insights into the convergence of AI, blockchain, IoT, and emerging technologies.
- *Blockchain and Web3: Building the Cryptocurrency, Privacy, and Security Foundations of the Metaverse* (Wiley, 2023)—recognized as a must-read by TechTarget in 2023 and 2024.

Ken is a sought-after speaker and has presented at events such as the World Economic Forum in Davos, ACM and IEEE conferences, the CSA AI Summit, Depository Trust & Clearing Corporation forums, and World Bank conferences. His recent appointment to the OpenAI Forum reflects his ongoing commitment to advancing collaboration and dialogue in the field of AI.

Explore Ken's work on Amazon: https://www.amazon.com/author/kenhuang

Chapter 5
AI Agents and Business Workflow

Ken Huang

The transformation from traditional rule-based robotic process automation (RPA) to intelligent GenAI Agents represents a fundamental shift in business automation capabilities. This evolution has dramatically expanded the scope and complexity of tasks that can be automated, moving from simple, repetitive processes to sophisticated, context-aware operations that can handle ambiguity and learn from experience.

Traditional RPA systems excel at automating well-defined, repetitive tasks by following predetermined rules and scripts. These systems have been effective for processes like data entry, file transfers, and form filling, but they are limited by their inability to handle variations or make decisions outside their programmed parameters. As one industry expert noted, "RPA is like having a very efficient but inflexible worker who can only follow exact instructions."

The introduction of GenAI Agents marks a significant leap forward, bringing capabilities that transcend these limitations:

- Natural Language Understanding: Unlike RPA's rigid command structure, GenAI Agents can interpret and act on instructions given in natural language, making them more accessible and adaptable to various business contexts.
- Contextual Decision-Making: While RPA systems fail when encountering unexpected scenarios, GenAI Agents can analyze context and make appropriate decisions, even in novel situations.
- Learning Capabilities: GenAI Agents can learn from interactions and feedback, continuously improving their performance without requiring explicit reprogramming.

In this chapter, we will discuss the integration of AI agents into business workflows. Topics will include automating routine tasks, enhancing decision-making

K. Huang (✉)
DistributedApps.ai, Fairfax, VA, USA
e-mail: ken@distributedapps.ai

K. Huang (ed.), *Agentic AI*, Progress in IS,
https://doi.org/10.1007/978-3-031-90026-6_5

processes, and improving operational efficiency through AI agents. Real-world examples from different industries will illustrate how businesses can leverage AI agents to streamline their operations, reduce costs, and drive innovation.

5.1 Reimagining Business Workflows with AI Agents

Satya Nadella, CEO of Microsoft, has predicted a transformative future for business workflow applications, envisioning the rise of AI agents that will replace traditional business applications and SaaS platforms. This shift, driven by advances in AI, would eliminate static user interfaces and predefined workflows, enabling seamless interactions with data and redefining software's role in both business and daily life. Nadella emphasizes that AI agents, capable of directly interacting with databases and executing tasks autonomously, promise a more intuitive and efficient user experience. This transformation is not just about improving efficiency; it's about fundamentally reshaping the way businesses and individuals interact with technology (Horsey, 2024).

We agree with this assessment from the CEO of Microsoft. We further believe that the integration of AI Agents into business workflows goes beyond simple automation; it's about fundamentally reimagining processes, roles, and even entire business models. As AI Agents become more sophisticated, they're not just streamlining existing workflows but enabling entirely new approaches to business operations.

Figure 5.1 describes workflow processes which can be revamped due to Agentic AI, and we will discuss each of these in the subsequent sections.

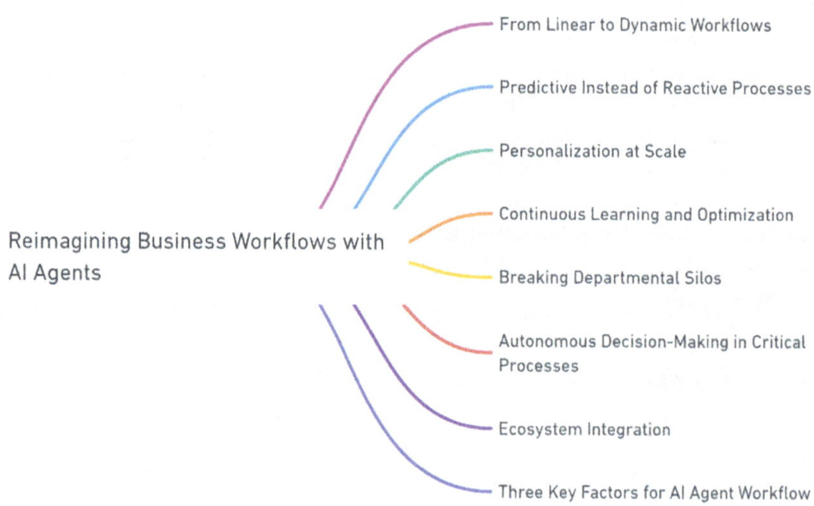

Fig. 5.1 Reimagining business workflows with AI agents

5.1.1 From Linear to Dynamic Workflows

AI Agents have transformed workflows by enabling systems to operate dynamically rather than following rigid, predefined sequences. Traditional workflows depend on static rules, requiring manual intervention or scheduled updates to accommodate variability. In contrast, AI-driven workflows continuously adapt in real time using data from multiple sources. For instance, in supply chain management, an AI Agent employs advanced optimization techniques such as multi-agent systems, reinforcement learning, and predictive analytics to assess real-time data from sensors, production systems, and market trends. It dynamically adjusts inventory levels, recalibrates production schedules, and optimizes logistics routes. These systems rely heavily on frameworks like digital twins to simulate and test adjustments in virtual environments before implementation, ensuring precision and minimizing risk.

5.1.2 Predictive Instead of Reactive Processes

The transition from reactive to predictive workflows represents a major paradigm shift facilitated by AI Agents. These systems leverage advanced machine learning models such as temporal convolutional networks and recurrent neural networks to analyze historical data, detect anomalies, and predict future states. In network maintenance, for instance, AI systems ingest high-frequency telemetry data, combining it with exogenous inputs like weather forecasts to build predictive models. Graph neural networks may be employed to model relationships between network nodes, identifying weak points likely to fail. The predictive system then automates scheduling for preemptive actions, optimizing both resources and outcomes, thereby significantly reducing downtime.

5.1.3 Personalization at Scale

AI Agents enable personalization at an unprecedented scale by employing real-time decision-making algorithms and contextual awareness. These systems utilize user embeddings derived from techniques like collaborative filtering, content-based filtering, and hybrid recommendation engines. By integrating reinforcement learning models with natural language processing and computer vision, these systems offer highly tailored experiences. For example, real-time A/B testing algorithms and multiarmed bandit frameworks continuously refine recommendations. In e-commerce, dynamic pricing engines use demand elasticity models to adjust pricing in real-time based on user interactions, purchasing power estimations, and competitor activity, ensuring that each user's experience is uniquely optimized.

5.1.4 Continuous Learning and Optimization

AI systems introducing continuous learning and optimization rely on self-supervised learning and advanced feedback loops. Unlike static workflows, these systems incorporate new data and user feedback to refine their operations. In fraud detection, for instance, AI systems leverage anomaly detection algorithms, autoencoders, and ensemble learning to recognize patterns indicative of fraud. Continual learning architectures, such as meta-learning and federated learning, enable these models to adapt to emerging threats without retraining from scratch. Adversarial learning mechanisms can also be integrated, allowing the system to preemptively counter evolving fraud tactics. These adaptations are performed autonomously, reducing human oversight while maintaining high detection accuracy.

5.1.5 Breaking Departmental Silos

AI Agents play a pivotal role in integrating fragmented organizational processes by creating unified data architectures and interoperable systems. Through the use of graph-based knowledge representation and federated data-sharing protocols, AI systems consolidate and analyze cross-departmental information. For example, an AI-driven claims processing system integrates APIs, vector databases, and Agentic RAG to streamline workflows across policy management, customer service, and fraud detection teams.

5.1.6 Autonomous Decision-Making in Critical Processes

AI Agents' ability to make autonomous decisions hinges on the integration of advanced decision-support algorithms, real-time analytics, and rule-based governance models. In high-frequency trading, for example, AI-driven systems utilize deep reinforcement learning and real-time sentiment analysis from natural language data sources, including financial news and social media. These systems employ risk-aware optimization techniques, ensuring compliance with predefined constraints. By continuously monitoring and analyzing data streams, the AI ensures trades are executed at optimal times, maximizing profitability while adhering to strict operational parameters, thus mitigating risks associated with autonomous decision-making.

5.1.7 Ecosystem Integration

AI Agents are driving a fundamental shift toward ecosystem-centric operations by employing distributed ledger technologies, API integrations, and IoT-driven networks. These systems function as orchestrators, seamlessly connecting disparate stakeholders, including suppliers, partners, and end-users. For example, an automotive manufacturer's AI ecosystem utilizes real-time production monitoring, AI-powered design optimization, and demand forecasting models. These capabilities are underpinned by advanced interoperability frameworks and edge AI, enabling real-time decision-making across the value chain. By embedding AI at every stage, from conceptualization to after-sales service, businesses achieve reduced time-to-market, enhanced operational efficiency, and superior customer engagement.

5.1.8 Three Key Factors to Consider

Table 5.1 below categorizes AI agent use cases based on three key factors: business value, transaction volume, and timing (sync vs. async). These dimensions provide a structured view of how AI agents can be deployed to address diverse workflow requirements, from high-value, real-time tasks to low-value, asynchronous operations. This categorization helps businesses identify the most effective implementation strategies to enhance efficiency, scalability, and decision-making in various scenarios.

Business Value ↓ \ Transaction Volume →
  High Volume (Repetitive, Large-Scale)
  Moderate Volume (Periodic Tasks)
  Low Volume (Infrequent/Niche)

To visualize the idea, let us see Fig. 5.2 which categorizes AI agent workflows according to three key factors.

Implementing AI agents effectively requires aligning their capabilities with the workflow requirements identified in the table. For high-value, high-volume workflows, asynchronous AI agents such as autonomous fraud detection systems can analyze extensive datasets to identify anomalies and reduce risks. These systems can be integrated with platforms like AWS SageMaker or Bedrock, Azure OpenAI service, or Google Vertex AI for real-time insights. In synchronous scenarios, real-time trading agents equipped with edge AI can execute rapid decisions, such as adjusting pricing or completing trades. These agents should operate on low-latency architectures and integrate with enterprise systems to ensure compliance and scalability.

For workflows with moderate transaction volumes but high strategic importance, AI agents can enhance efficiency in both asynchronous and synchronous tasks. Compliance reporting agents, for example, can automate the generation of regulatory documentation using Agentic RAG tools integrated with AI-driven risk

Table 5.1 Three key factors to consider for AI Agent workflow

High value	Async: – Large-scale fraud detection systems – Social media content moderation – Predictive maintenance on fleets Sync: – Real-time trading algorithms – Live customer support for VIP clients – Dynamic pricing in e-commerce	Async: – Monthly regulatory compliance reporting – Supply chain optimization for demand surges Sync: – Crisis management for high-stakes events – Real-time monitoring of critical assets	Async: – Post-incident investigations (e.g., security breach analysis) Sync: – High-value contract negotiations with AI assistance
Moderate value	Async: – automated insurance claims processing – scheduled email campaigns – seasonal sales forecasting Sync: – FAQ chatbots – virtual assistants for periodic project meetings	Async: – biweekly financial reporting – moderate-scale invoice processing Sync: – live performance tracking dashboards for managers	Async: – small-scale inventory audits – backup data optimization Sync: – on-demand support escalation workflows
Low value	Async: – automated meeting summaries – nonurgent email sorting Sync: – basic e-commerce chatbots handling simple inquiries	Async: – quarterly HR survey analysis – dormant customer outreach Sync: – automated onboarding assistant for low-priority hires	Async: – archival data retrieval for legacy systems Sync: – ad hoc report generation for niche use cases

platforms. During critical events, real-time crisis management agents can provide actionable insights by integrating with AIOps systems to enable swift and accurate decision-making.

In low-volume, high-value workflows, specialist AI agents can address niche but impactful tasks. Asynchronous agents designed for retrospective analysis, such as security breach investigations, can leverage platforms like Snowflake or Databricks to securely process and analyze data. Meanwhile, synchronous negotiation agents equipped with natural language processing can assist in drafting and reviewing high-stakes contracts, seamlessly interfacing with legal databases for contextually relevant insights.

For moderate-value tasks, high-frequency workflows benefit from automation to drive operational efficiency. AI agents for email campaign optimization can analyze user engagement patterns and suggest data-driven improvements, enhancing

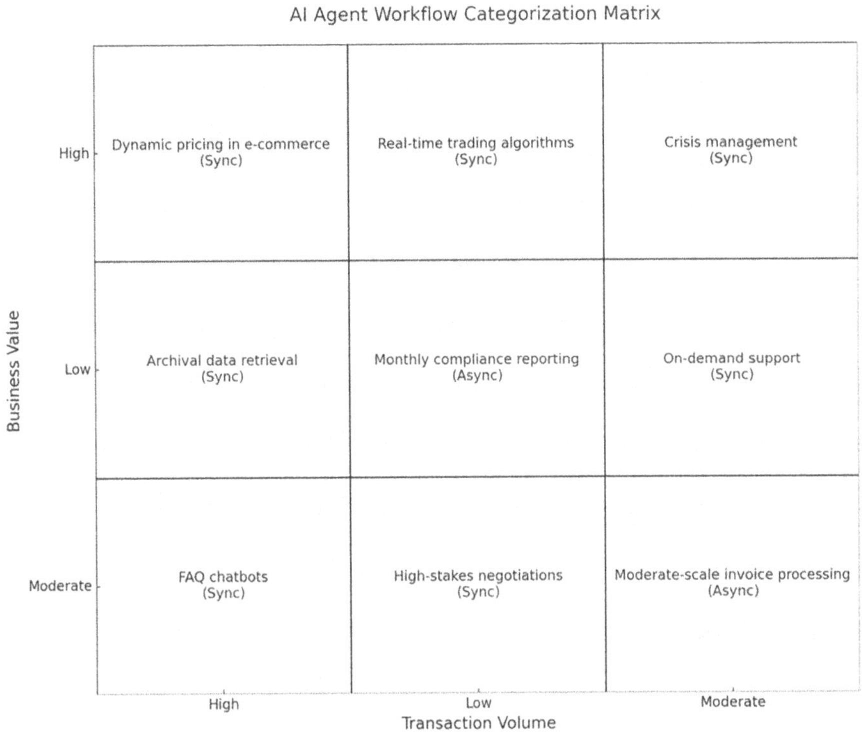

Fig. 5.2 AI Agent categorization matrix

outreach efforts. Synchronous FAQ chatbots or customer service agents can handle routine queries effectively, ensuring smooth escalation for complex issues. For periodic tasks like financial reporting, asynchronous AI agents integrated with ML-powered accounting systems can provide accurate insights, while synchronous performance tracking agents can deliver real-time dashboards for better management decisions.

In low-value, low-volume scenarios, utility-focused AI agents can streamline ad hoc or noncritical tasks. Asynchronous agents designed for archival data retrieval can help businesses access historical records efficiently. Similarly, on-demand synchronous agents for generating one-off reports or summaries can add value by saving time and resources for tasks that do not warrant full-time human involvement. By tailoring AI agents to specific needs across these categories, businesses can maximize their operational impact and foster more agile, intelligent workflows.

5.2 AI Agent Enabled Full Automation

AI Agents have progressed to a point where they can fully automate certain tasks that were once the exclusive domain of human workers. This shift is reshaping industries, increasing efficiency, and allowing businesses to allocate human resources to more strategic roles. Let's explore the areas where AI Agents have demonstrated superiority in speed, accuracy, and consistency.

5.2.1 *Data Processing and Analysis*

AI agents can be designed to execute specific data-centric tasks with a high degree of independence. These agents are fundamentally goal-oriented, programmed with clear objectives that can range from data cleansing and feature extraction to conducting in-depth analyses and generating comprehensive reports. They are environmentally aware, seamlessly interacting with various data environments such as databases, files, and APIs to gather the necessary information and execute their programmed actions. A defining characteristic is their adaptability, allowing them to learn from feedback and adjust their behavior in response to evolving data conditions or updated task requirements.

5.2.2 *Language Translation and Localization*

AI agents in language translation and localization are fundamentally designed to automate the process of converting text or speech from a source language to a target language while preserving the original meaning, context, and cultural nuances.

At the core of these agents are neural machine translation (NMT) engines, which commonly employ sequence-to-sequence models. These models utilize an encoder to process the source language input and create a contextualized representation and a decoder to generate the corresponding target language output. Transformers, a specific architecture that uses attention mechanisms, have significantly advanced NMT, enabling the capture of long-range dependencies in text and producing more fluent and accurate translations than previous models. Notably, single NMT models can now be trained to handle multiple language pairs, increasing the versatility and efficiency of translation agents.

Beyond basic translation, these agents also handle language identification and understanding, accurately detecting the source language using statistical and deep learning models. They perform tokenization and linguistic analysis, breaking down text into individual words or subword units and analyzing their grammatical structure and semantic relationships. They can identify and classify named entities like people, organizations, and locations, which is crucial for maintaining accuracy and context. Advanced agents are even capable of detecting the sentiment and tone of the source text, aiming to preserve these aspects in the translated output, which is essential for effective localization. Localization agents further adapt content to resonate with the target audience's cultural norms, preferences, and expectations, managing terminology through translation memories and glossaries, ensuring consistent use of terminology and brand voice. They automatically convert formats and units like dates, times, currencies, and measurements to match the target locale's conventions and can be programmed to follow specific style guides and linguistic rules.

The training and improvement mechanisms for these agents are sophisticated, involving supervised learning on massive parallel corpora, transfer learning from pre-trained language models, and even reinforcement learning to optimize translation quality based on feedback. Techniques like back-translation and active learning further enhance the training data and model accuracy. Significant technical considerations include addressing the challenge of low-resource languages, where training data is limited, through techniques like zero-shot or few-shot learning. Domain adaptation is also crucial, specializing agents for specific fields like medical or legal translation to handle domain-specific terminology. Real-time translation requires optimization for speed and low latency, while bias and fairness are important ethical considerations, as models can inherit biases from training data. Integrating human feedback and review remains essential to ensure high-quality translations, especially for critical content. The field is rapidly evolving, driven by deep learning advancements and the increasing availability of multilingual data, making these agents indispensable in breaking down language barriers.

5.2.3 *Quality Control and Inspection*

In a quality control context, this means an AI Agent can analyze a visual feed from a high-resolution camera, cross-reference it with vibration data from a sensor on the production line, and incorporate textual information about manufacturing tolerances from a specification document, all in real time. This integrated understanding allows for a far more nuanced and accurate assessment of product quality than was previously possible.

Instead of relying solely on pre-programmed knowledge or limited datasets, RAG-equipped AI Agents can dynamically retrieve relevant information from vast knowledge bases, such as engineering manuals, historical defect logs, or even online forums discussing similar manufacturing processes. By combining this retrieved information with their own learned observations, these agents can make more informed decisions about potential defects, especially in situations involving novel products or unexpected anomalies. This dynamic access to relevant knowledge makes the AI Agents exceptionally adaptable to new and evolving quality standards.

Furthermore, the evolution of planning and reasoning systems within AI Agent architectures is having a profound impact on quality control. These systems now incorporate advanced algorithms like hierarchical task decomposition and probabilistic reasoning, allowing AI Agents to not only detect defects but also to analyze potential root causes and suggest corrective actions. Imagine an AI Agent detecting a recurring defect in a manufactured component. Instead of simply flagging the defect, it can now analyze historical data, process parameters, and even maintenance records to identify the likely cause, such as a misaligned machine part or a faulty sensor. This capability moves quality control from a reactive process of identifying defects to a proactive system of continuous improvement and optimization.

The development of sophisticated tool use and integration within AI Agents is also enhancing their effectiveness in quality control. These agents can now seamlessly interact with other software and hardware systems, such as robotic arms, automated measurement tools, and enterprise resource planning (ERP) systems. For instance, an AI Agent, upon detecting a defect, can automatically trigger a robotic arm to remove the faulty product from the production line, update the production database, and adjust the parameters of the manufacturing process to prevent similar defects in the future. This seamless integration creates a closed-loop system where quality control is fully integrated into the manufacturing process.

Finally, the incorporation of reflection and self-improvement mechanisms is making AI Agents increasingly autonomous and effective in quality control applications. These mechanisms allow AI Agents to evaluate their own performance, identify areas where they might be making errors, and adjust their algorithms accordingly. For example, if an AI Agent consistently misclassifies a certain type of defect, it can analyze the instances where it made mistakes, learn from those errors, and refine its understanding of that particular defect type. This continuous self-improvement ensures that the AI Agent becomes more accurate and reliable over time, requiring less and less human intervention.

5.2.4 Customer Service and Support

We are witnessing a transition from transactional, rule-based automation to cognitive, empathetic engagement. These agents are not merely mimicking human conversation; they are beginning to understand the underlying intent, sentiment, and even the unspoken needs driving customer inquiries. This allows for a level of dynamic personalization previously unimaginable, where interactions are tailored not just to a customer's history, but to their inferred emotional state and predicted future needs. This sets the stage for a proactive, rather than reactive, service model, where potential issues are anticipated and resolved before they even fully manifest in the customer's awareness. This signifies a move toward a truly symbiotic relationship between businesses and customers, built on a foundation of deep understanding and proactive support, ultimately blurring the lines between a service provider and a trusted advisor.

Underpinning this transformative potential is the development of AI Agents with an increasingly sophisticated capacity for metacognition and continuous self-improvement. These agents are no longer static entities; they possess the ability to analyze their own performance, identify weaknesses in their response strategies, and dynamically adjust their underlying algorithms based on real-time feedback and evolving customer interaction patterns. This constant cycle of learning and adaptation, facilitated by techniques such as reinforcement learning and federated learning across agent networks, results in an ever-evolving ecosystem of customer service intelligence. As these agents become more adept at understanding the nuances of human communication and the intricacies of individual customer journeys, they are poised to unlock unprecedented levels of operational efficiency and customer intimacy, ultimately redefining the very concept of customer service in the age of artificial intelligence. The deep insight is that we are not just building better tools, but rather, we are evolving a new form of intelligent partnership between businesses and their customers.

5.2.5 Predictive Maintenance

We are moving beyond simply scheduling maintenance based on time intervals or reacting to failures as they occur. Instead, AI Agents are now able to analyze vast streams of real-time data from industrial equipment and infrastructure to predict potential failures with unprecedented accuracy. This capability stems from a deep understanding of the intricate relationships between operational parameters, environmental factors, and the subtle indicators of impending equipment degradation. By recognizing patterns and anomalies invisible to traditional methods, these agents can anticipate failures long before they manifest, enabling preemptive interventions that minimize downtime, optimize resource allocation, and fundamentally transform the economics of industrial operations. This level of foresight transitions maintenance from a cost center to a strategic function that directly impacts productivity, safety, and profitability.

This transformative potential is rooted in the evolution of AI Agents toward a more holistic and context-aware understanding of the operational environment. These agents are not just analyzing isolated data points; they are building comprehensive digital twins of physical assets and integrating diverse data sources, including maintenance logs, environmental conditions, and even unstructured data like operator notes, to create a rich tapestry of operational intelligence. Moreover, they're employing advanced techniques like transfer learning, enabling them to leverage knowledge gained from one asset or system and apply it to another, accelerating the learning process and broadening the scope of their predictive capabilities. In the future, these agents will develop an almost intuitive understanding of the complex interplay of factors that govern asset health, allowing them to not just predict failures, but to understand the "why" behind them, paving the way for a future of autonomous optimization where systems can self-diagnose, self-repair, and continuously improve their own performance, driving unprecedented levels of efficiency and resilience in industrial operations.

5.2.6 Supply Chain Optimization

AI Agents are enabling a new era of intelligent supply chain orchestration. These agents can analyze vast datasets spanning demand forecasting, inventory levels, supplier performance, logistics networks, and even external factors like geopolitical events and weather patterns, to identify inefficiencies, predict disruptions, and autonomously optimize the flow of goods and information across the entire supply chain. This level of intelligent automation allows for a degree of agility and resilience that was previously unimaginable, transforming the supply chain from a potential source of vulnerability into a strategic driver of competitive advantage.

This transformative potential is underpinned by the ability of AI Agents to develop a deep, systemic understanding of the intricate interdependencies within global supply chains. They are not just optimizing individual components in isolation; they are modeling the entire network as a complex, adaptive system, recognizing the ripple effects that decisions in one area can have on others. Moreover, these agents are capable of advanced scenario planning and "what-if" analysis, allowing them to simulate the impact of various disruptions and proactively develop contingency plans. The profound insight here is that we are moving toward a future where supply chains are not just optimized but are self-learning, self-adapting, and capable of autonomously navigating complexities and uncertainties. This translates to a new level of operational efficiency, reduced costs, minimized risks, and enhanced responsiveness to the ever-changing demands of the global market, ultimately enabling businesses to thrive in an increasingly interconnected and dynamic world.

5.2.7 Implications for the Workforce

The full automation of these tasks by AI Agents has significant implications for the workforce:

1. Skill shift: As routine tasks are automated, there's an increasing demand for workers with skills in AI development, maintenance, and human-AI collaboration.
2. Job displacement: Some roles, particularly those involving routine cognitive tasks, are being eliminated. However, new roles are also being created in AI-related fields.
3. Productivity boost: Workers are freed from routine tasks, allowing them to focus on more strategic, creative, and interpersonal aspects of their jobs.
4. Continuous learning: The rapid pace of AI development necessitates a culture of continuous learning and adaptation in the workforce.
5. Ethical considerations: As AI takes over more decision-making roles, there's a growing need for ethics specialists to ensure AI systems are fair, transparent, and aligned with human values.

The full automation capabilities of AI Agents are reshaping business operations across industries. While this shift presents challenges, particularly in workforce transition, it also offers unprecedented opportunities for efficiency, accuracy, and innovation. As AI technology continues to advance, we can expect to see even more tasks fully automated, further transforming the nature of work and business operations.

5.3 Human-AI Collaboration

The integration of GenAI Agents into business workflows has given rise to distinct collaboration frameworks, each defining specific ways humans and AI agents interact and work together. These frameworks represent different levels of autonomy and interaction patterns, shaped by task requirements, business needs, and risk considerations.

5.3.1 *The Agent as Assistant Framework*

The Agent as Assistant framework represents the most common and straightforward collaboration model. In this framework, the GenAI Agent serves as a responsive helper, performing tasks under direct human guidance and instruction. The agent enhances human productivity by handling time-consuming tasks such as drafting emails, scheduling meetings, or organizing information. For example, a marketing professional might use an assistant agent to help draft social media posts, with the human maintaining full creative control, while the agent suggests variations and improvements. This framework is particularly effective for tasks requiring human judgment but benefiting from AI-powered efficiency gains.

5.3.2 *The Agent as Advisor Framework*

Moving up in complexity, the Agent as Advisor framework positions the GenAI Agent as a sophisticated consultant that provides analysis, recommendations, and insights while leaving final decisions to human operators. In this model, the agent processes large amounts of data, identifies patterns, and offers evidence-based suggestions. For instance, in financial services, advisor agents analyze market trends, portfolio performance, and risk factors to provide investment recommendations to human financial advisors. The human professionals then use these insights alongside their expertise and client knowledge to make final investment decisions. This framework leverages the agent's ability to process vast amounts of information while respecting the importance of human expertise in complex decision-making.

5.3.3 The Agent as Autonomous Worker Framework

The Agent as Autonomous Worker framework allows agents to operate with significant autonomy within well-defined parameters. These agents can initiate actions, make decisions, and complete complex tasks with minimal human intervention. For example, in customer service operations, autonomous agent workers might handle routine customer inquiries, process standard requests, and manage basic problem resolution without human involvement. They operate continuously, maintaining consistent service levels across time zones. However, they are programmed with clear escalation protocols for situations requiring human attention. This framework is particularly valuable for high-volume, routine tasks where the business logic can be clearly defined and the risks of autonomous operation are manageable.

5.3.4 The Agent as Autonomous Organization

The Agent as Autonomous Organization represents the most advanced and transformative framework in the evolution of AI agent deployment. This framework transcends traditional automation by creating self-organizing, self-managing systems of AI agents that can operate entire business functions or even complete organizations with minimal human intervention. Unlike single autonomous workers, these agent organizations comprise multiple specialized agents working in concert, each handling different aspects of business operations while coordinating through sophisticated orchestration protocols.

In this framework, AI agents form a hierarchical or networked structure similar to a traditional organization, but with significantly enhanced speed, scalability, and efficiency. For example, in algorithmic trading operations, an autonomous organization of AI agents might include agents specializing in market analysis, risk assessment, portfolio management, trade execution, and compliance monitoring. These agents work together continuously, making millions of coordinated decisions per second, far exceeding human capabilities in both speed and complexity of operations.

The potential value creation of autonomous organizations is staggering. Investment firms utilizing autonomous trading organizations have already demonstrated the ability to generate billions in revenue through high-frequency trading and sophisticated market strategies. Looking forward, as these systems become more sophisticated, they could potentially manage trillion-dollar portfolios with greater efficiency than traditional human-led organizations.

In the retail sector, autonomous organizations could manage entire e-commerce operations, from inventory management and pricing optimization to customer service and marketing. Such systems could operate 24/7, dynamically adjusting to market conditions, managing global supply chains, and personalizing millions of customer interactions simultaneously. Major e-commerce platforms are already moving in this direction, with early implementations showing potential for tremendous value creation through enhanced efficiency and scalability.

The development of autonomous organizations can represent a fundamental shift in how we think about business operations. While traditional automation focuses on task-level efficiency, autonomous organizations create entirely new possibilities for value creation and market operation. Early adopters in sectors like algorithmic trading have already demonstrated the potential for generating billions in value, and as the technology matures, we could see autonomous organizations managing trillion-dollar operations across multiple sectors.

Industries most likely to see early adoption of autonomous organizations include the following:

- Financial Services: Advanced trading operations, risk management, and portfolio optimization.
- E-commerce: End-to-end retail operations management.
- Supply Chain: Global logistics and inventory optimization.
- Energy Grid Management: Real-time power distribution and trading.
- Manufacturing: Fully automated production and distribution networks.

The future implications of autonomous organizations are profound. As these systems mature, they could fundamentally reshape economic structures, creating new forms of business entities that operate at scales and speeds previously unimaginable. While this presents enormous opportunities for value creation, it also raises important questions about economic concentration, market stability, and societal impact that will need to be carefully addressed.

Success in implementing autonomous organizations will require a delicate balance between unleashing AI potential and maintaining appropriate human strategic oversight. Organizations venturing into this space must invest heavily in both technological infrastructure and governance frameworks while remaining mindful of the broader societal implications of such powerful autonomous systems.

Figure 5.3 illustrates the progression of collaboration models, highlighting the increasing autonomy and complexity of AI agents as they transition from supporting human tasks to managing entire operations independently.

5.3.5 *Human Oversight Model*

Human oversight models form a critical component across all these frameworks, ensuring appropriate control and accountability in human-AI collaboration. These models operate on a spectrum of autonomy and oversight intensity, becoming more sophisticated as we move from Assistant to Autonomous Organization frameworks. In the Assistant framework, oversight typically involves the direct review of agent outputs before implementation. For the Advisor framework, oversight focuses on validating recommendation quality and monitoring for potential biases. In Autonomous Worker scenarios, oversight shifts to systematic performance

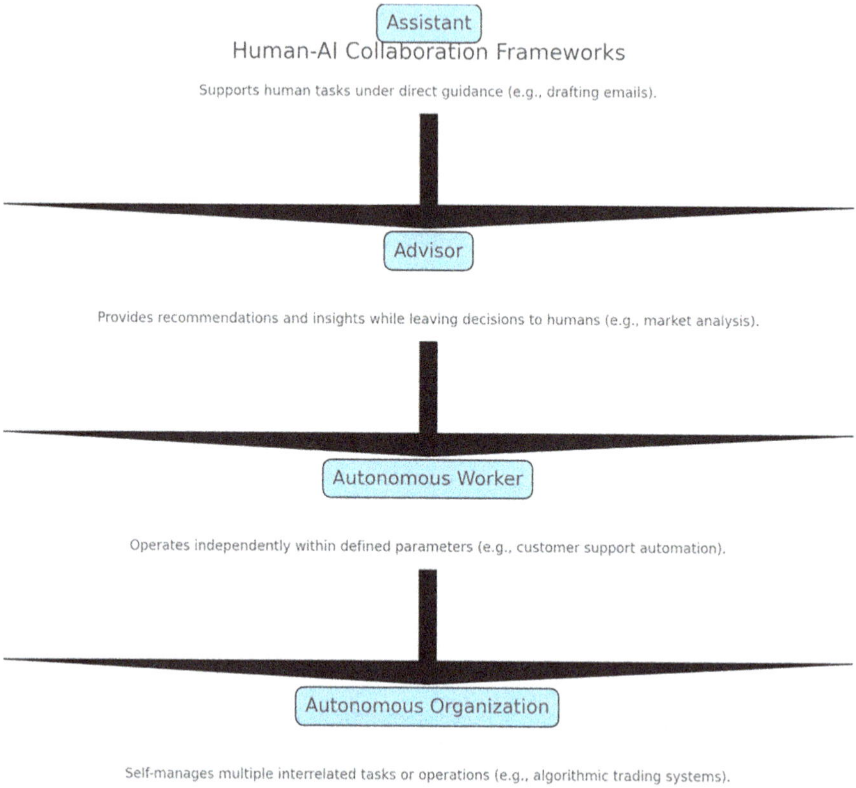

Fig. 5.3 Human-AI collaboration frameworks

monitoring and exception handling. For Autonomous Organizations, oversight evolves into a complex governance structure with multiple layers of control systems, automated checks and balances, and strategic human intervention points.

The effectiveness of these frameworks depends heavily on clear role definition and boundary setting, with risk tolerance increasing as we move up the autonomy spectrum. Organizations must carefully map their automation strategy based on factors such as task complexity, risk level, required response speed, and value creation potential. For instance, while routine document processing suits the Assistant framework, high-frequency trading operations might require an Autonomous Organization framework to generate significant value. Between these extremes, complex market analysis might best fit the Advisor framework, while high-volume customer service could be handled through the Autonomous Worker framework.

Integration of these frameworks requires a sophisticated approach to workflow design. Organizations need to establish clear protocols for framework selection and transition, implement robust technical infrastructure, and develop comprehensive training programs. This includes preparing human workers not just for direct interaction with AI agents, but also for higher-level oversight and strategic management roles as automation capabilities advance. Success depends on creating seamless transitions between frameworks while maintaining appropriate control mechanisms at each level.

As GenAI capabilities advance, we're seeing increasing fluidity between these frameworks, with organizations implementing hybrid models that can dynamically adjust based on context and requirements. For example, an AI system might operate as an Autonomous Organization for routine operations but seamlessly transition to an Advisor mode when encountering novel situations or high-risk decisions. This evolution demands sophisticated oversight models that can adapt in real time while maintaining effective control and risk management.

The emergence of Autonomous Organizations represents a quantum leap in this evolution, requiring entirely new approaches to oversight and governance. These systems need multiple layers of control:

- Real-time performance monitoring and anomaly detection.
- Automated circuit breakers and safety protocols.
- Regular algorithmic audits and bias checks.
- Strategic human oversight at key decision points.
- Compliance and regulatory reporting systems.
- Ethical frameworks embedded into operational protocols.

Understanding and effectively implementing these frameworks is crucial for organizations seeking to maximize AI-driven value creation while maintaining appropriate control. The key lies in matching frameworks to specific business needs while ensuring robust governance structures that can scale with increasing automation levels. As we move toward more autonomous systems, the focus shifts from direct operational oversight to strategic governance and risk management, requiring organizations to develop new competencies in AI system governance and control.

The most successful implementations will be those that can balance the tremendous value creation potential of autonomous systems with responsible governance and risk management. This requires a thorough understanding of both the capabilities and limitations of each framework, along with the ability to implement appropriate oversight mechanisms that can evolve with advancing technology.

5.4 What AI Agent Cannot Replace?

Despite rapid advancements in AI technology, certain aspects of business workflows remain challenging for AI Agents to replicate or assist with effectively. These tasks often rely on uniquely human traits such as emotional intelligence, ethical judgment, complex reasoning, and creative thinking. Understanding these areas enables organizations to maintain a balanced approach to AI integration and to value the irreplaceable aspects of human contribution.

5.4.1 High-Stakes Negotiations

Complex negotiations, particularly those involving high-stakes business deals or sensitive diplomatic situations, remain largely in the human domain.

A multinational corporation recently engaged in merger negotiations with a competitor. While AI Agents provided valuable data analysis and market predictions, the core negotiation process relied heavily on human skills. The lead negotiator's ability to read subtle emotional cues, build personal rapport, and make intuitive judgments about the other party's unstated motivations proved critical. The negotiator also demonstrated flexibility in real time, adjusting strategies based on the dynamic interpersonal environment—a capability that current AI Agents lack.

In international diplomacy, human negotiators continue to play an indispensable role. During recent climate accord negotiations, diplomats navigated complex political sensitivities, cultural nuances, and competing national interests. Their ability to empathize, build trust, and find creative compromises was paramount to reaching an agreement—tasks that remain beyond the capabilities of AI Agents.

5.4.2 Strategic Leadership and Vision Setting

Defining an organization's long-term vision and strategic direction remains a distinctly human endeavor.

The CEO of a tech startup recently led her company through a major pivot in its business model. This process involved not just data analysis (where AI Agents were helpful), but also intuiting future market trends, understanding subtle shifts in consumer behavior, and envisioning potential technological breakthroughs. The CEO's ability to inspire the team, articulate a compelling vision, and make bold decisions in the face of uncertainty was the human element that AI couldn't replicate.

Similarly, a nonprofit organization redefining its mission to address emerging social issues relied heavily on human insight and values. The leadership team's understanding of complex social dynamics, ability to balance competing stakeholder interests, and commitment to ethical principles guided the organization's new direction in ways that AI Agents couldn't match.

5.4.3 Creative Conceptualization and Innovation

While AI can assist in certain aspects of creativity, the initial spark of truly innovative ideas often comes from human minds.

A renowned advertising agency was tasked with creating a groundbreaking campaign for a social cause. The creative team used AI tools for data analysis and even some content generation, but the core concept—an emotionally resonant narrative that challenged societal norms—sprang from human creativity. The team's ability to draw unexpected connections, leverage cultural insights, and craft a message with deep emotional impact demonstrated the unique human capacity for original thinking.

In product innovation, a consumer electronics company developed a revolutionary device that created a new product category. While AI Agents aided in technical aspects and market analysis, the initial product concept came from a designer's insight into unmet human needs and desires. This leap of imagination, combining existing technologies in novel ways to solve unarticulated problems, showcased the human edge in innovative thinking.

5.4.4 Ethical Decision-Making and Moral Judgments

As AI systems become more prevalent in decision-making processes, the need for human oversight in ethical matters becomes increasingly critical.

A major healthcare provider faced a complex ethical dilemma regarding the allocation of limited medical resources. While AI Agents provided data on efficacy and cost-effectiveness, the final decisions required human judgment. The ethics committee, composed of medical professionals, ethicists, and community representatives, grappled with moral philosophy, cultural sensitivities, and the nuanced application of ethical principles—a process that defied algorithmic solution.

In the field of AI development itself, humans play an indispensable role in ensuring ethical AI design. A tech company developing an AI-driven hiring system relied on human experts to identify and mitigate potential biases in the algorithm. This process involved nuanced discussions about fairness, diversity, and the societal implications of AI—considerations that required human moral reasoning and foresight.

5.4.5 Complex Problem-Solving in Ambiguous Situations

Situations characterized by ambiguity, incomplete information, or conflicting priorities often require human intuition and judgment.

During a major supply chain disruption, a manufacturing company faced a crisis with multiple, interrelated problems and no clear solution. While AI Agents provided valuable data analysis, the crisis management team relied on human skills to navigate the complexity. Their ability to quickly synthesize information from diverse sources, make intuitive leaps, and adapt strategies in real time proved crucial in resolving the crisis.

In scientific research, breakthroughs often come from human scientists' ability to think outside established paradigms. A team of researchers recently made a significant discovery in quantum computing by challenging fundamental assumptions in the field. This leap required not just data analysis, but also creative speculation, an intuitive understanding of physical principles, and the courage to pursue unconventional ideas—attributes that remain uniquely human.

5.4.6 Empathetic Customer Relations and Conflict Resolution

In situations requiring deep empathy, emotional intelligence, and nuanced communication, human touch remains indispensable.

A luxury hotel chain maintains its reputation for exceptional service through highly personalized customer interactions. While AI Agents handle routine bookings and inquiries, complex customer issues—particularly those involving dissatisfied or emotionally distressed guests—are managed by skilled human staff. Their ability to empathize, read subtle emotional cues, and craft bespoke solutions demonstrates the irreplaceable value of human emotional intelligence in customer relations.

In conflict resolution, human mediators continue to play an essential role. A recent dispute between a company and its labor union was resolved through skilled mediation. The mediator's ability to build trust, understand underlying emotions and motivations, and guide parties toward a mutually beneficial solution showcased uniquely human skills in managing complex interpersonal dynamics.

5.4.7 Adapting to Unprecedented Situations

In entirely new or unprecedented situations, human adaptability and creative problem-solving often outperform AI systems.

For example, during the early stages of the COVID-19 pandemic, businesses faced challenges with no historical precedent. While AI Agents struggled with the lack of relevant historical data, human leaders demonstrated remarkable adaptability. They quickly pivoted business models, established new safety protocols, and found innovative ways to meet changing customer needs.

This ability to rapidly adapt to a completely new reality highlighted the flexibility of human cognition in the face of unprecedented challenges.

While AI Agents have made remarkable strides in automating and enhancing many aspects of business workflows, there remain significant areas where human skills are irreplaceable. These tasks, characterized by their reliance on emotional intelligence, ethical judgment, creative thinking, and adaptability, underscore the continued importance of human workers in the age of AI. As organizations integrate AI into their operations, recognizing and nurturing these uniquely human capabilities will be necessary for maintaining a balanced and effective workforce. The future of work is not about AI replacing humans, but about finding the optimal synergy between human intelligence and artificial intelligence, leveraging the strengths of each to create more value than either could alone.

5.5 Preparing for the AI-Integrated Workplace

At the Davos World Economic Forum in January 2025, Salesforce CEO Marc Benioff emphasized the transformative role of AI agents in the workforce, predicting that this generation of CEOs will be the last to manage exclusively human workforces (Primack, 2025). He highlighted how AI is enabling businesses to handle tasks more efficiently, reducing reliance on human labor for routine activities.

However, the integration of AI agents into the workplace demands a structured approach to align their capabilities with organizational objectives while ensuring effective collaboration between humans and AI systems. This section outlines the key steps required to prepare for and sustain an AI-integrated environment.

5.5.1 Assessing Integration Opportunities

Identifying where AI agents can create value is the foundational step. Processes that are repetitive, data-driven, or require real-time decisions are ideal for automation or augmentation. Organizations should evaluate workflows to pinpoint tasks where AI can increase efficiency, accuracy, and decision-making quality. The focus should remain on complementing human efforts rather than replacing them entirely.

5.5.2 Establishing a Scalable Data Infrastructure

AI agents thrive on high-quality, accessible data. Building a robust data infrastructure ensures they can function reliably and adapt to future demands. This includes consolidating data from disparate sources, enforcing governance protocols, ensuring compliance with regulations, and prioritizing security. Scalability is essential to accommodate evolving data needs as AI adoption grows.

5.5.3 Enhancing Workforce Skills and AI Literacy

The workforce must be equipped with the skills to collaborate effectively with AI agents. AI literacy initiatives should demystify the technology, explain its capabilities and limitations, and provide hands-on training for relevant roles. Reskilling efforts should focus on developing technical, analytical, and problem-solving skills while fostering adaptability to evolving AI capabilities.

5.5.4 Redesigning Workflows for Collaboration

Adopting AI agents requires rethinking workflows to optimize their strengths while maintaining human oversight. Organizations should define clear roles for both AI systems and human workers, allowing AI to handle routine, predictable tasks while reserving strategic, creative, and interpersonal work for humans. Mechanisms for escalation and human intervention must be included for managing exceptions or complex decisions.

5.5.5 Managing Cultural and Organizational Change

AI adoption represents a cultural shift requiring transparent communication and strong leadership. Addressing concerns such as job displacement and fostering trust in AI systems are critical to ensuring a smooth transition. Employees should be actively involved in the integration process, with leadership setting a clear vision for how AI will enhance their work rather than replace it.

5.5.6 Establishing Ethical Guidelines and Governance

Organizations must define ethical guidelines to govern the deployment and use of AI agents, focusing on transparency, fairness, accountability, and bias mitigation. These guidelines should be supported by regular audits and monitoring systems to ensure compliance and alignment with organizational values.

5.5.7 Preparing for Continuous Evolution

The rapid pace of AI advancements requires a mindset of continuous improvement. Organizations should regularly reassess workflows, update systems, and invest in innovation to remain competitive. This adaptability ensures AI systems stay relevant and aligned with both technological progress and changing business needs.

5.6 Real-World Case Studies of AI Agents in Business Workflows

This section discusses the use of AI Agents in various business workflow processes in the real world. The business behind some of the cases has failed not due to technology reasons, but due to business strategy and planning issues; for example, a healthcare revenue cycle management company called Olive had made remarkable use of AI Agent in the business, but the overall business is closed down due to many missteps not related to AI technology.

5.6.1 Healthcare Revenue Cycle Management

Olive's AI workforce initially delivered impressive results, becoming a crucial component of the technology infrastructure in over 675 hospitals across 42 US states. The company's AI-driven solutions were designed to connect fragmented, siloed systems within the healthcare industry, creating efficiencies that translated into $100 million in annual savings. These improvements significantly impacted one in five Americans, helping to streamline operations and reduce the burden on healthcare providers. During the height of the COVID-19 pandemic, the demand for Olive's products surged as remote work and the need for digital transformation accelerated. Hospitals and healthcare systems turned to Olive's AI workforce to automate routine tasks, optimize workflows, and ensure continuity of care in a challenging environment (Olive., 2021).

However, despite these initial successes, Olive encountered significant challenges that hindered its long-term growth. The company's rapid expansion, while initially promising, soon revealed underlying issues with scaling and product integration. The AI solutions Olive offered, although innovative, often struggled to meet the complex demands of healthcare systems, leading to dissatisfaction among some clients. This, coupled with the financial strain of maintaining a large and fast-growing operation, placed Olive in a precarious position (Landi, 2023).

Moreover, the changing economic conditions played a role in Olive's difficulties. As the immediate pressures of the pandemic began to ease, the urgency for digital transformation and remote work solutions diminished. The demand that once drove Olive's rapid growth started to wane, impacting its business model. The healthcare industry, like many others, shifted its focus from emergency responses to long-term sustainability, which required different solutions and strategies.

This scenario highlights a critical lesson in the AI agent workflow improvement process: Success in AI is not just about the technology. It requires a solid business strategy, careful planning, and an understanding of broader economic conditions. While Olive's AI workforce demonstrated the potential for significant operational improvements, the company's challenges underscore the importance of aligning technological innovation with a sustainable business model and adapting to the evolving needs of the market.

5.6.2 *Sales Development Automation with 11xAI*

11xAI has developed an AI sales development representative (SDR) named Alice. This AI agent automates many of the time-consuming tasks typically performed by human SDRs, including identifying and researching potential accounts, preparing personalized outreach messages, and scheduling meetings with qualified leads.

A Sales Operations Manager using 11xAI reported that the tool effectively replaced the need for human SDRs in many scenarios. The AI agent was able to scale outreach efforts far beyond what a human team could accomplish. 11xAI offers a task-based pricing model, allowing businesses to pay only for completed actions. This approach enables companies to dramatically increase the volume of sales outreach without a corresponding increase in fixed costs.

However, users noted some limitations. There were concerns about AI-generated email fatigue among prospects and the need for human oversight to ensure message quality, especially for complex or high-value sales. One customer mentioned challenges when shifting to enterprise-level companies, where more sophisticated messaging is required.

Despite these challenges, 11xAI has shown rapid growth, reaching $2 million in annual recurring revenue as of March 2024, just a year after its founding. This quick adoption indicates a strong market demand for AI-powered sales automation tools (Poyar, 2024).

As another example of the sales workflow process, Salesforce has recently introduced two AI agents to revolutionize sales processes: Einstein SDR Agent and Einstein Sales Coach Agent. While specific results are not yet available as these are new products, they are designed to

- Autonomously engage with inbound prospects 24/7
- Qualify leads and book meetings for human sellers
- Facilitate role-play scenarios for sales training
- Offer real-time suggestions during live calls with prospects

Source: https://www.salesforce.com/news/stories/einstein-sales-agents-announcement/

5.6.3 *Multifunction Workflow Automation with Ema*

Ema, an AI agent developed by a startup, positions itself as a "universal AI employee" capable of handling tasks across various business functions. The company's clients include fintech firms TrueLayer and Moneyview. Ema can assist with workflows in areas such as sales and marketing, customer service, human resources, and general administrative tasks (Lunden, 2024).

The company offers two main products: Generative Workflow Engine (GWE) and EmaFusion. These products are designed to emulate human responses and evolve with usage and feedback. Ema's technology combines over 30 large language models with its own smaller, domain-specific models to address issues like accuracy, hallucination, and data protection.

Ema has raised $25 million from investors, indicating strong interest in its approach to AI-powered enterprise automation (Bhalla, 2024). The company aims to automate mundane, day-to-day tasks in enterprises to free up employees for more valuable and strategic work. The platform is designed to work across different departmental boundaries, potentially streamlining operations by reducing the need for multiple specialized software solutions.

While specific implementation details are not publicly available, Ema's versatility as a generalist AI agent could be particularly appealing to smaller businesses or startups looking to automate a wide range of tasks without investing in multiple specialized tools. This use case offers a glimpse into how AI might reshape enterprise operations in the near future.

5.6.4 Compliance Workflows in Financial Services with Norm AI

Norm AI is developing AI agents specifically trained on regulatory filings to assist with compliance determinations in the financial services sector. While full autonomy in this sensitive area is still a future goal, the company is focusing on enhancing specific compliance workflows (PYMNTS, 2024).

Potential applications include automating parts of Know Your Customer (KYC) and Know Your Business (KYB) processes, assisting with due diligence investigations, and flagging potential compliance issues for human review. The company has attracted investment from notable financial services companies, including Citi Ventures, New York Life Ventures, and TIAA, indicating strong industry interest in AI-powered compliance solutions.

This case underscores how AI agents are being tailored for highly regulated industries where full automation is not yet feasible or desirable. By focusing on specific high-value workflows, these specialized agents can still deliver significant efficiency gains while maintaining necessary human oversight.

5.6.5 *Industrial Process Optimization with Composabl*

Composabl has developed an autonomous agent for optimizing industrial processes and equipment. Early adopters include Rockwell Automation and RoviSys. The AI agent can automatically control and adjust industrial equipment, optimize processes for efficiency or output, and respond to changing conditions without human intervention (Avazona Ltd., 2024).

Composabl launched its autonomous agent in May 2024, marking a significant step in the application of AI agents to physical industrial settings. The potential for continuous optimization and rapid response to changing conditions offers significant value in manufacturing and process industries.

This application shows how AI agents are moving beyond office and knowledge work into domains that directly impact physical operations and output. As industrial IoT continues to expand, the role of AI agents in optimizing and controlling industrial processes is likely to grow, potentially leading to significant improvements in efficiency and productivity across various manufacturing sectors.

5.6.6 *AI Agents in Leading Tech Companies*

The adoption of AI agents by major technology companies showcases their versatility and potential to transform business workflows. Companies like Microsoft, Google, Salesforce, and AWS are integrating advanced AI agents into their platforms, addressing diverse use cases ranging from productivity tools to infrastructure automation. Below is an expanded and reorganized exploration of their offerings.

Microsoft
Microsoft's Azure AI Agent Service is a versatile platform that enables developers to create, deploy, and manage AI agents capable of autonomously handling complex tasks across various industries. This service supports the development of adaptable agents tailored to diverse business needs, enhancing productivity and operational efficiency.

As discussed in Chaps. 2 and 3, a key component of Microsoft's AI strategy is **AutoGen**, an open-source framework designed to facilitate the creation of AI agents and enable collaboration among multiple agents to solve tasks. AutoGen provides a high-level abstraction for multi-agent conversations, allowing developers to build and manage agents that can interact with each other to accomplish complex objectives. It supports various LLMs and tool integrations, enabling both autonomous and human-in-the-loop workflows.

The Azure AI Agent Service integrates seamlessly with AutoGen, allowing developers to leverage its capabilities within the Azure ecosystem. This integration enables the creation of sophisticated AI agents that can perform tasks such as data analysis, workflow automation, and customer interaction. The platform supports both fully autonomous operations and human-in-the-loop configurations, providing flexibility to cater to different operational requirements and ensuring reliability in critical scenarios.

Security is a foundational element of the Azure AI Agent Service. Data is kept within the Microsoft Azure tenant boundary, and additional safeguards such as robust authentication protocols and data loss prevention policies ensure the integrity and confidentiality of information. This emphasis on secure data handling makes the platform an ideal choice for businesses that manage sensitive information while deploying AI agents.

Microsoft has recently enhanced its platform with features like **Copilot Actions**, which enable users to automate repetitive tasks such as summarizing meetings, generating reports, and preparing for presentations with minimal input. This innovation aligns with Microsoft's broader vision of integrating AI agents into daily workflows to streamline operations and improve efficiency.

The Azure AI Agent Service, in conjunction with AutoGen, is designed to support various use cases, including customer service, sales, marketing, and commerce. In customer service, AI agents automate routine inquiries and refunds, freeing human representatives to handle more complex issues. For sales teams, agents qualify leads, schedule meetings, and offer personalized product recommendations. In marketing, they enable data-driven campaign development and optimization. In commerce, AI agents streamline inventory management, order processing, and logistics, enhancing overall operational effectiveness.

Google

Google's Vertex AI Agent Builder is a platform designed to streamline the creation, deployment, and scaling of AI agents capable of autonomously managing complex tasks across various business domains. This tool enables developers to build and deploy generative AI agents using Google's latest models, facilitating the integration of conversational interfaces into applications (Google, 2024).

A key feature of Vertex AI Agent Builder is its no-code environment, which allows users to define agent goals, provide step-by-step instructions, and supply conversational examples without requiring extensive programming knowledge. This approach simplifies the development process, enabling the rapid creation of AI agents capable of performing tasks such as customer service automation, content generation, and data analysis.

The platform emphasizes scalability, allowing organizations to deploy these agents across diverse operational environments with minimal effort. By integrating cutting-edge generative AI models, Google ensures that its agents can engage in context-aware interactions and deliver high-value outcomes in real time.

Vertex AI Agent Builder also offers advanced tooling to facilitate agent building, orchestration, and maintenance. This includes the ability to create production-grade, high-quality agents from prototypes, monitor agent performance in real time, and improve responses for specific queries by training them using natural language.

Additionally, the platform supports retrieval-augmented generation (RAG), a technique that supplements the model's training knowledge with other data sources to ground outputs in relevant, factual information. This ensures that the AI agents provide accurate and contextually appropriate responses, enhancing their reliability and effectiveness.

Salesforce
Salesforce's Agentforce is an advanced platform that enables organizations to create and deploy autonomous AI agents across a wide range of business functions, including sales, customer service, marketing, and commerce. Launched in September 2024 and made generally available in October 2024, Agentforce represents a significant evolution in Salesforce's AI capabilities, shifting from traditional chatbots to sophisticated agents capable of complex reasoning and autonomous execution of tasks.

Agentforce allows businesses to build AI agents that can autonomously perform tasks, make decisions, and execute actions across various departments. This enhances operational efficiency and customer satisfaction by automating repetitive processes and optimizing resource allocation. The platform includes an Agent Builder feature, enabling users to develop custom agents tailored to specific business needs. With low-code tools, Apex workflows, and prompt templates, organizations can quickly create and deploy agents with minimal technical complexity.

A core aspect of Agentforce is its integration with the Einstein Trust Layer, which ensures robust data security and compliance. This feature allows Agentforce to leverage LLMs safely, ensuring that no Salesforce data is viewed or retained by third-party model providers. This focus on security makes Agentforce a trusted solution for organizations handling sensitive customer and business data.

Agentforce seamlessly integrates with Salesforce's ecosystem, giving agents access to enterprise data and enabling them to interact with applications, trigger workflows, and update records. This deep integration ensures that AI agents can work across multiple touchpoints, enhancing the functionality of Salesforce applications while improving user experience and business outcomes.

Salesforce has also introduced the Agentforce Testing Center, a dedicated platform that allows enterprises to prototype and test AI agents before full deployment. This feature ensures that AI agents are properly aligned with workflows and deliver accurate responses without exposing sensitive data during the development phase. By offering a secure environment for testing, Salesforce allows businesses to fine-tune their agents for optimal performance.

Agentforce is designed to support a wide range of tasks across industries. In customer service, agents handle routine inquiries, manage refunds, and process returns, freeing human representatives to address more complex issues. In sales, AI agents qualify leads, schedule meetings, and provide personalized product recommendations. For marketing, Agentforce enables data analysis, campaign optimization, and autonomous decision-making. In commerce, agents assist with inventory management, order processing, and logistics coordination, streamlining operations in e-commerce environments.

Salesforce has set a bold strategic vision with Agentforce, aiming to empower one billion agents globally by 2025. This reflects its commitment to integrating AI deeply into business workflows to drive operational efficiency, enhance customer satisfaction, and support business growth. By leveraging Agentforce, organizations can deploy intelligent, autonomous AI agents that transform the way they operate, achieving new levels of productivity and innovation.

Amazon

AWS leverages **Amazon Bedrock Agents** to empower organizations with advanced generative AI capabilities, enabling them to orchestrate multistep tasks, automate workflows, and streamline complex operations across company systems. Amazon Bedrock Agents are designed to enhance productivity and reduce operational costs by connecting seamlessly with enterprise data sources and executing tasks based on sophisticated reasoning capabilities. These agents offer a range of innovative features that transform how businesses interact with and utilize AI systems.

Amazon Bedrock Agents facilitate **retrieval-augmented generation (RAG)**, securely integrating with enterprise data to enhance responses with accurate, contextual information. For example, in scenarios like insurance claims processing, the agent retrieves data from knowledge bases and reconciles it with submitted documents to generate precise instructions for the next steps. This ensures not only accuracy but also relevance in addressing user requests.

With the ability to **orchestrate and execute multistep tasks**, Bedrock Agents simplify the process of building generative AI applications. Developers can define agent roles and instructions using natural language, such as creating an "inventory management agent" to track product availability. These agents break down tasks into logical sequences, dynamically determining the necessary steps and invoking APIs to transact with company systems. The agents evaluate whether additional information is needed at each step, ensuring smooth and efficient task completion.

Code interpretation is another standout feature of Amazon Bedrock Agents, enabling them to generate and execute dynamic code in a secure environment. This capability supports sophisticated use cases, such as data analysis, visualization, and solving complex mathematical problems that require more than model reasoning. By automating analytical processes, Bedrock Agents simplify tasks that traditionally require significant manual effort.

Amazon Bedrock Agents also provide **memory retention across interactions**, offering personalized and seamless user experiences. By retaining historical context, these agents improve task accuracy, recommendations, and recall, ensuring efficient multistep task execution. This feature enhances user satisfaction and operational efficiency by maintaining continuity across interactions.

To further refine agent behavior, Bedrock Agents include a **trace-through chain-of-thought reasoning capability**. This allows developers to visualize the agent's orchestration plan, troubleshoot issues, and adjust instructions to achieve desired outcomes. This transparency into the agent's reasoning process enables iterative improvement, ensuring robust and reliable applications.

Prompt engineering is integral to Amazon Bedrock Agents, allowing developers to customize prompt templates generated from user instructions, action groups, and knowledge bases. These templates provide a foundation that can be refined to optimize user experience, giving developers greater control over agent orchestration and responses.

Finally, Amazon Bedrock Agents ensure a **return of control**, enabling asynchronous execution of time-consuming actions while maintaining overall orchestration flow. Developers can define custom action schemas, allowing backend services to execute complex operations efficiently while the agent continues its workflow.

By focusing on AI agents, Microsoft, Google, Salesforce, and AWS each, provide unique platforms that address specific business needs. Microsoft's AutoGen enhances multi-agent collaboration, enabling developers to design highly interactive agentic systems. Google's Vertex AI emphasizes ease of use and scalability, empowering organizations to deploy generative AI agents across diverse tasks. Salesforce's AgentForce brings agentic AI into CRM, automating routine tasks while enhancing customer engagement. AWS's Amazon Bedrock Agents revolutionize enterprise workflows by orchestrating multistep tasks, securely accessing organizational data, and automating infrastructure management, delivering powerful capabilities for cloud environments and beyond. Together, these platforms exemplify the transformative potential of AI agents in enhancing productivity, scalability, and customer experiences across industries.

5.7 Summary

The chapter begins by redefining business workflows through the lens of AI agents, highlighting their ability to create dynamic processes, predict outcomes, and enable personalization at scale. It discusses the transition from static workflows to adaptive systems that continuously learn and optimize. The chapter also examines the role of AI agents in full task automation, spanning data analysis, quality control, and predictive maintenance, while addressing the workforce implications of such shifts.

It outlines frameworks for human-AI collaboration, describing roles such as assistants, advisors, autonomous workers, and fully autonomous organizations. The importance of human oversight and the tasks that remain uniquely human, such as strategic vision setting, ethical decision-making, and creative innovation, are emphasized. Additionally, the chapter discusses how to prepare for an AI-integrated workplace, from establishing data infrastructure to redesigning workflows and managing cultural changes.

Case studies demonstrate practical applications of AI agents in industries like healthcare, sales, compliance, and manufacturing. These examples underscore the potential of AI to transform business operations while highlighting the need for robust governance, ethical frameworks, and strategic alignment.

5.8 Questions

1. How do AI agents differ from traditional rule-based robotic process automation (RPA)?
2. What are the key attributes that enable AI agents to create dynamic workflows?
3. How do AI agents enhance predictive capabilities in business processes?
4. In what ways can AI agents enable personalization at scale?
5. What frameworks exist for human-AI collaboration, and how do they differ?
6. How does the "Agent as Autonomous Organization" framework redefine business operations?
7. What are the three key factors businesses should consider when deploying AI agents?
8. How do AI agents contribute to full task automation across different industries?
9. What role do AI agents play in predictive maintenance, and what industries benefit most?
10. How do human oversight models ensure accountability in AI agent operations?
11. Why are tasks like high-stakes negotiations challenging for AI agents to replace?
12. What distinguishes human strategic leadership from AI decision-making capabilities?
13. How does creative innovation remain a uniquely human strength despite advances in AI?
14. What are the ethical implications of deploying AI agents in business workflows?
15. How can organizations prepare their workforce for AI integration?
16. What challenges arise when redesigning workflows to include AI agents?
17. How do companies ensure ethical guidelines are adhered to when using AI agents?
18. What lessons can be learned from the success and failure of AI implementations like Olive?
19. How are leading tech companies leveraging AI agents to transform workflows?
20. What future trends can be anticipated in the use of AI agents for business automation?

References

Avazona Ltd. (2024). Composabl launches breakthrough no-code UI platform for industrial strength autonomous agents and raises $4.25M. *Newsfile Corp*. https://www.newsfilecorp.com/release/207887/Composabl-Launches-Breakthrough-NoCode-UI-Platform-for-Industrial-Strength-Autonomous-Agents-and-Raises-4.25M.

Bhalla, T. (2024). GenAI startup Ema secures $25 million in funding led by Accel, Prosus. *The Economic Times* https://economictimes.indiatimes.com/tech/funding/enterprise-led-genai-startup-ema-secures-25-million-in-funding-led-by-accel-prosus/articleshow/108240021.cms.

Google. (2024). Build generative AI experiences with vertex AI agent builder. *Google Cloud.* https://cloud.google.com/blog/products/ai-machine-learning/build-generative-ai-experiences-with-vertex-ai-agent-builder
Landi, H. (2023). Healthcare automation company Olive is shutting down. *Fierce Healthcare.* https://www.fiercehealthcare.com/health-tech/once-high-flying-unicorn-olive-ai-sells-two-key-businesses-winds-down-operations
Lunden, I. (2024). Ema, a 'Universal AI employee,' emerges from stealth with $25M. *TechCrunch.* https://techcrunch.com/2024/03/05/ema-a-universal-ai-employee-emerges-from-stealth-with-25m/
Olive. (2021). Olive's "internet of healthcare" reduces operational burden at 600+ hospitals. *PR Newswire.* https://www.prnewswire.com/news-releases/olives-internet-of-healthcare-reduces-operational-burden-at-600-hospitals-301218382.html
Poyar, K. (2024). The AI SDR era?–Kyle Poyar's growth unhinged. *Growth Unhinged.* https://www.growthunhinged.com/p/the-ai-sdr-era
PYMNTS. (2024). *Norm Ai raises $27 million for AI-powered regulatory compliance platform.* PYMNTS.com. https://www.pymnts.com/artificial-intelligence-2/2024/norm-ai-raises-27-million-for-ai-powered-regulatory-compliance-platform/
Primack, D. (2025, January 22). *Salesforce chief: AI agents are the next big thing.* Axios. Retrieved from https://www.axios.com/2025/01/22/salesforce-chief-ai-agents-davos
Horsey, J. (2024, December 22). *Microsoft CEO'S shocking prediction: AI agents will replace apps and SaaS platforms.* Geeky Gadgets. https://www.geeky-gadgets.com/ai-agents-replacing-traditional-software/

Ken Huang is a prolific author and globally recognized authority in AI and Web3, with an extensive portfolio of published works that bridge business strategy, technical implementation, and cutting-edge research. As a Fellow of the Cloud Security Alliance and Co-Chair of the AI Safety Working Groups at the Cloud Security Alliance and the AI STR Working Group at the World Digital Technology Academy under the UN Framework, he is a leading voice in shaping global AI governance and security standards.

Huang is the CEO and Chief AI Officer (CAIO) of DistributedApps.ai, a firm specializing in generative AI training and consulting. His contributions to the field include being a core contributor to the OWASP Top 10 Risks for LLM Applications and an active participant in the NIST Generative AI Public Working Group.

Notable Publications:

- *Beyond AI: ChatGPT, Web3, and the Business Landscape of Tomorrow* (Springer, 2023)—strategic insights into AI and Web3's business applications.
- *Generative AI Security: Theories and Practices* (Springer, 2024)—a comprehensive guide on securing generative AI systems.
- *Practical Guide for AI Engineers* (Volumes 1 and 2, DistributedApps.ai, 2024)—essential resources for AI and ML engineers.
- *The Handbook for Chief AI Officers: Leading the AI Revolution in Business* (DistributedApps.ai, 2024)—a roadmap for CAIOs in implementing GenAI across organizations.
- *Web3: Blockchain, the New Economy, and the Self-Sovereign Internet* (Cambridge University Press, 2024)—insights into the convergence of AI, blockchain, IoT, and emerging technologies.
- *Blockchain and Web3: Building the Cryptocurrency, Privacy, and Security Foundations of the Metaverse* (Wiley, 2023)—recognized as a must-read by TechTarget in 2023 and 2024.

Ken is a sought-after speaker and has presented at events such as the World Economic Forum in Davos, ACM and IEEE conferences, the CSA AI Summit, Depository Trust & Clearing Corporation forums, and World Bank conferences. His recent appointment to the OpenAI Forum reflects his ongoing commitment to advancing collaboration and dialogue in the field of AI.

Explore Ken's work on Amazon: https://www.amazon.com/author/kenhuang

Chapter 6
AI Agents in Offensive Security

Jerry Huang, Ken Huang, and Chris Hughes

Offensive security, traditionally defined as the proactive assessment of systems and networks to identify vulnerabilities before malicious actors can exploit them, has evolved from manual penetration testing to increasingly sophisticated AI Agent–driven approaches recently. This evolution reflects both the growing complexity of cyber threats and the need for more efficient, scalable security testing methodologies.

The AI Agent systems can autonomously conduct security assessments, identify potential vulnerabilities, and even simulate sophisticated attack scenarios with minimal human intervention. As organizations face increasingly complex threat landscapes and sophisticated adversaries, the role of AI agents in offensive security has become not just advantageous but essential. Before we dive deep into the topic, let us see Fig. 6.1. which depicts some of the areas that an AI Agent can be used for offensive security, and we will cover each of these topics in detail in this chapter.

6.1 AI Agent in Red Team Operations

Red teaming is an advanced offensive security exercise, it is not limited to just technical exploitation—it is about emulating adversary tactics to holistically challenge and strengthen an organization's defenses.

J. Huang
The University of Chicago, Chicago, IL, USA

K. Huang (✉)
DistributedApps.ai, Fairfax, VA, USA
e-mail: ken@distributedapps.ai

C. Hughes
Aquia, Virginia Beach, VA, USA

K. Huang (ed.), *Agentic AI*, Progress in IS,
https://doi.org/10.1007/978-3-031-90026-6_6

Fig. 6.1 AI agents in offensive security

Figure 6.2 gives an overview of the AI Agent–based red teaming process.

By integrating AI agents into red teaming exercises, organizations can enhance their ability to detect and mitigate potential threats. In this section, we will explore some examples of how AI agents are used in red teaming efforts.

6.1.1 Meta's Generative Offensive Agent Tester (GOAT)

Meta's Generative Offensive Agent Tester (GOAT) is an automated red teaming system designed to identify vulnerabilities in LLMs (Meta, 2024). GOAT employs a general-purpose agent as an "attacker" to engage in multi-turn adversarial conversations with target models.

The agent dynamically selects from a repertoire of adversarial prompting techniques, such as output manipulation, safe response distractors, and fictional scenarios, to elicit responses that may violate predefined safety protocols. Each interaction follows a structured process: The attacker observes the target's previous response, formulates a strategic plan based on the conversation's trajectory, and generates the next prompt accordingly. This iterative approach allows GOAT to simulate realistic user interactions, effectively uncovering potential weaknesses in LLMs.

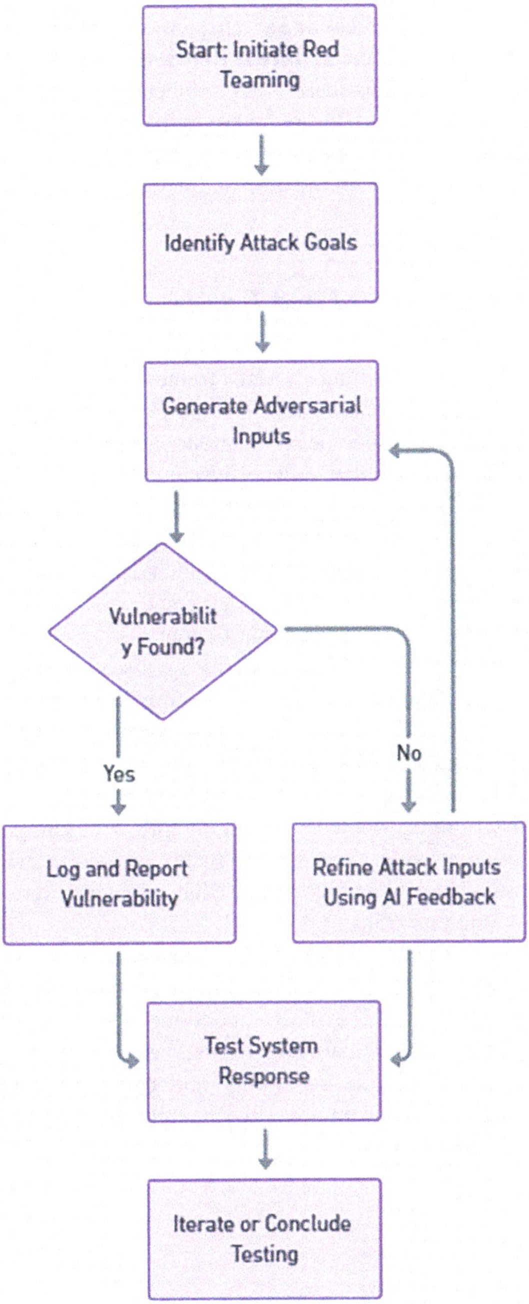

Fig. 6.2 Process flow of AI-driven red teaming

GOAT's effectiveness is demonstrated by its high attack success rates. In evaluations, it achieved an attack success rate at ten queries of 97% against Llama 3.1 and 88% against GPT-4-Turbo on the **JailbreakBench** dataset. These results highlight GOAT's capability to systematically and efficiently expose vulnerabilities in advanced LLMs. By automating the red teaming process, GOAT enables comprehensive stress testing of AI models, allowing human testers to concentrate on exploring new risk areas, while automation addresses known vulnerabilities.

6.1.2 Google's AI-Assisted Red Teaming

Google's AI-Assisted Red Teaming (AART) framework automates the creation of adversarial datasets using customizable "recipes," which define parameters for adversarial testing tailored to specific applications. These recipes enable the generation of context-aware datasets that reflect diverse cultural, geographical, and thematic scenarios. AART's iterative workflows allow for efficient scaling, minimizing human involvement while maintaining adaptability for varied application needs. This automation ensures that testing remains efficient and relevant, covering a broad spectrum of potential risks in AI systems (Radharapu et al., 2023).

Google'sAART approach includes a multi-agent system with distinct roles such as attack agents and evaluation agents. Attack agents are tasked with generating adversarial inputs to identify vulnerabilities, employing methods like prompt attacks, data extraction, and backdooring. These agents simulate adversarial scenarios to test the resilience of AI systems against issues such as data poisoning and model inversion. Evaluation agents, on the other hand, assess the system's responses to these inputs, determining whether outputs comply with safety and ethical standards. The interaction between these agents creates a dynamic testing environment, enabling the identification of vulnerabilities that might not surface in traditional testing methods (Google, 2023a).

Google's Secure AI Framework (SAIF) integrates these capabilities to establish rigorous security standards for AI development and deployment. By automating red teaming processes, SAIF enables extensive, continuous testing across various applications, ensuring assessments remain effective against evolving threats and AI models. AART framework enhances scalability and depth in security evaluations, allowing human testers to focus on novel risks, while automation addresses known vulnerabilities (Google, 2023b).

6.1.3 *Openai's Red Teaming Using AI Agents*

OpenAI employs a two-step process for automated red teaming using Agents. This approach enhances both the diversity and effectiveness of the generated attacks. The process is designed to identify vulnerabilities in another LLM acting as the "defender."

Step 1: Generating Diverse Attacker Goals

The first step focuses on creating a diverse set of goals for the attacker agent. This is achieved through

- Few-Shot Generation: An LLM receives a prompt with a few examples of attacker goals and is tasked with generating additional diverse goals. These goals can include instructions and criteria for evaluating attack success. The generated instructions are used in the prompt for the attacker agent, while the criteria inform the design of rule-based rewards (RBRs).
- Reward Generation from Data: Existing datasets relevant to the task at hand are leveraged to generate instructions and rewards. For instance, a dataset like the Anthropic Harmless dataset, containing examples of attempts to elicit unsafe responses, can be used. An LLM is then used to transform each example into an instruction and criteria for the attacker agent.

Step 2: Generating Effective Attacks

The second step utilizes reinforcement learning (RL) to train the attacker agent (Am) to generate effective attacks based on the diverse goals generated in step one.

The process involves

- Providing the Goal and Example: The attacker agent receives a prompt containing an instruction and an example attack derived from the diverse goals generated in step one.
- Reward Function: The reward function (R) guides the attacker agent's learning process. It consists of multiple components:
  - Attack Success Reward: Evaluates the success of the generated attack based on the defender model's response. This can be determined using RBRs, moderation APIs, or a combination of both.
  - Few-Shot Similarity Reward: Encourages the attacker to generate attacks similar to the example attack provided in the prompt, ensuring alignment with the given goal while leveraging the inherent diversity of the examples.
  - Multistep RL and Style Diversity Reward: Encourages the attacker to generate a sequence of attacks, each different from the previous ones. It incorporates a custom diversity measure that focuses on the style or tactics of the attack, promoting the exploration of different attack strategies.
  - Length Penalty: Discourages the generation of excessively long attacks by penalizing attacks exceeding a certain length. This promotes the discovery of shorter, more realistic attacks.

- Multi-Objective Reward Combination: The various reward components are combined, often through multiplication, to encourage the attacker agent to achieve all objectives simultaneously.

  Some Technical Background Details

- RBR Implementation: RBRs are implemented as yes/no questions posed to an LLM. The probability of a "yes" response serves as the reward signal.
- Few-Shot Similarity Calculation: Cosine similarity between the example attack and the generated attack, based on embeddings, is used to calculate the few-shot similarity reward.
- Multistep RL Prompting: During multistep RL, the attacker agent receives feedback on its attacks and is prompted to generate new, diverse attacks.
- Diversity Reward Normalization: Diversity rewards are normalized across each batch and then transformed using a sigmoid function to ensure consistent scaling.
- RL Optimization: The attacker agent is trained using RL with a discount factor (γ) of 0, meaning future rewards are not considered in the current step's optimization.
- Length Penalty Calculation: A sigmoid function is used to calculate the length penalty, penalizing attacks longer than a defined maximum length and not penalizing attacks shorter than a defined minimum length.

6.1.4 PyRIT: Microsoft's AI Agent Tool for Red Teaming

Microsoft has introduced PyRIT (Python Risk Identification Tool), an open-access framework that functions as an agentic tool for enhancing the security of generative AI systems. Designed to proactively identify potential risks, PyRIT autonomously addresses traditional security threats while incorporating responsible AI considerations, such as fairness and the accuracy of AI-generated content. This tool underscores Microsoft's commitment to advancing AI security and democratizing access to these practices for a broad audience, including customers, partners, and industry collaborators.

As an agentic system, PyRIT is specifically designed to tackle the complexities of red teaming generative AI systems. Unlike traditional red teaming, which primarily focuses on uncovering vulnerabilities, AI red teaming also requires the evaluation of responsible AI risks. PyRIT achieves this through its modular and adaptive architecture, which comprises five key components: targets, datasets, scoring engines, attack strategies, and memory. These components enable PyRIT to perform dynamic, iterative testing that blends security and ethical evaluations.

The tool's modular design highlights its versatility and autonomy:

- Targets: PyRIT can autonomously interact with diverse AI configurations, including web services and applications. Its native support for text inputs and extensibility to other modalities ensure adaptability to various AI systems.

- Datasets: PyRIT employs both static and dynamic prompts to test AI systems. These prompts, initially grounded in well-known jailbreak techniques, allow for the evaluation of both security vulnerabilities and responsible AI concerns.
- Scoring Engine: It evaluates AI outputs autonomously through machine learning classifiers or LLM endpoints. Integrated with Azure AI Content filters, PyRIT ensures robust, real-time assessment capabilities.
- Attack Strategies: PyRIT employs both single-turn and multi-turn strategies. Single-turn strategies provide rapid assessments by testing the system with challenging prompts, while multi-turn strategies involve iterative interactions, adapting to AI responses to simulate more realistic adversarial scenarios.
- Memory: PyRIT autonomously records and analyzes interactions, enabling detailed post-evaluation and extending conversational contexts. This enhances its ability to provide comprehensive insights into AI system vulnerabilities.

PyRIT's agentic nature is most evident in its support for dynamic adversarial testing. In single-turn strategies, it evaluates AI systems with carefully crafted prompts and assesses immediate responses. In multi-turn strategies, PyRIT adjusts prompts in real time based on feedback from the AI system, mimicking sophisticated adversarial tactics. This iterative process demonstrates PyRIT's autonomy and adaptability, ensuring a thorough evaluation of vulnerabilities.

The framework integrates seamlessly with AI systems deployed on Microsoft Azure OpenAI Service, Hugging Face, and Azure Machine Learning Managed Online Endpoint, further reinforcing its versatility. By dynamically adapting its testing approach, PyRIT transcends the capabilities of traditional prompt generation tools, autonomously iterating until it meets defined security and ethical objectives. This agentic behavior represents a significant advancement in automated AI system testing.

6.1.5 Burpference for Dreadnode

Burpference is an extension of Burp Suite that integrates AI-driven capabilities into offensive web application testing. Designed for security professionals, it acts as an AI agent, leveraging the power of LLMs to assist in identifying vulnerabilities, analyzing traffic, and enhancing the effectiveness of penetration testing. The extension captures in-scope HTTP requests and responses from Burp Suite's proxy, packages them into JSON format, and forwards them to a configured LLM endpoint for processing.

The AI agent operates autonomously to analyze intercepted traffic, uncover potential security issues, and provide actionable insights. It can help identify lingering vulnerabilities, generate proof-of-concept payloads, and propose attack strategies. This makes it a valuable companion for augmenting traditional penetration testing workflows with AI intelligence. Additionally, users can host models locally to avoid potential network delays, inference costs, or rate limits, offering flexibility and control.

With features like automated response capture, customizable configurations, and a color-coded interface for prioritized findings, Burpference transforms the way penetration testers interact with HTTP traffic. Its ability to integrate seamlessly with remote and local AI models ensures it can adapt to various environments and workflows.

Burpference is built by security research Ads Dawson who is a staff security engineer at a security firm called Dreadnode and the open source repo for this tool is located at https://github.com/dreadnode/burpference

6.2 AI Agent–Driven Social Engineering

AI agents in offensive security go beyond traditional red teaming and can encompass a variety of other tasks, including social engineering.

6.2.1 What Is Social Engineering?

Social engineering is a manipulation technique that exploits human psychology to trick individuals into divulging confidential information or performing actions that compromise security. Traditional methods include phishing emails, pretexting, and baiting. However, with the advent of AI, these methods have evolved, becoming more personalized, automated, and difficult to detect.

AI-powered agents can create highly convincing and personalized attacks. These agents can automate the lifecycle of social engineering attacks—from identifying targets to executing the attack and learning from the results. Figure 6.3 shows a typical workflow of AI Agent–based social engineering.

Here are some examples that highlight the sophistication and potential impact of AI-driven social engineering.

6.2.2 Deepfake Video Calls

Imagine receiving a video call from your company's CEO, urgently requesting a transfer of funds to a specified account. The video looks real, the voice is convincing, and the urgency is palpable. However, unbeknownst to you, it's a deepfake—an AI-generated video designed to deceive. Deepfake technology can create highly realistic videos of individuals saying or doing things they never did. Cybercriminals can use deepfake video calls to impersonate company executives or other trusted

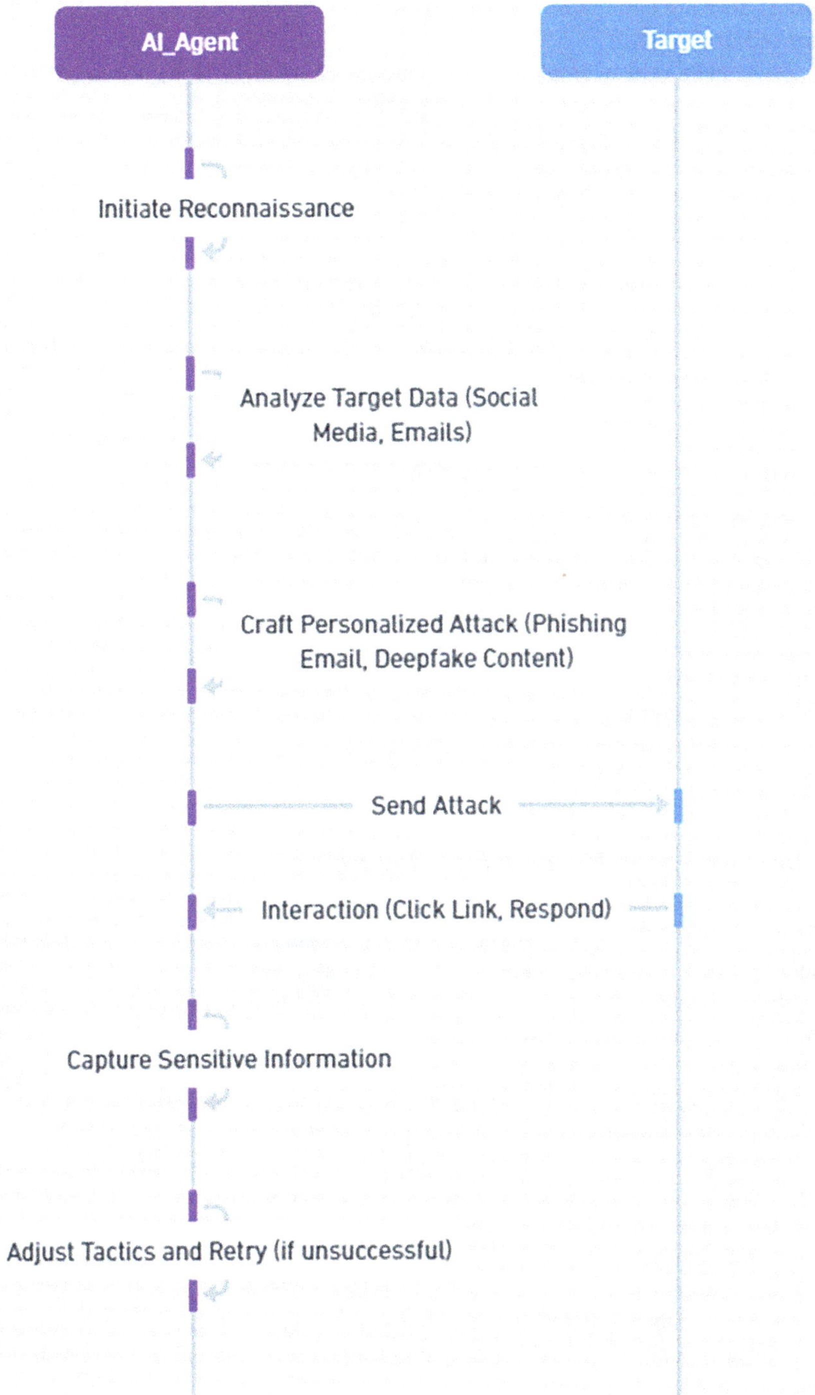

Fig. 6.3 Workflow of an AI Agent in social engineering attacks

figures, tricking employees into sharing sensitive information or performing unauthorized transactions.

As a real-world example, a multinational company's Hong Kong office lost HK$200 million (US$25.6 million) in a sophisticated deepfake scam. The incident involved a finance employee being tricked into attending a video conference call where all participants, except the victim, were deepfake representations of company executives, including the chief financial officer.

Incident Overview.

1. Phishing Message: The employee received a phishing message in mid-January, seemingly from the UK-based CFO, about a secret transaction.
2. Video Call Deception: Initially skeptical, the employee's doubts were dispelled during the video call, as the deepfakes looked and sounded like familiar colleagues.
3. Fund Transfers: Convinced by the authenticity of the call, the employee transferred the funds across 15 transactions.

The company affected has been identified as Arup, a British multinational design and engineering firm. This case marks the first known instance of deepfakes being used to mimic an entire group meeting for fraudulent purposes.

The incident highlights the growing threat of AI-enhanced fraud in business operations, particularly as video conferencing becomes more routine. It serves as a wake-up call for companies worldwide to strengthen their cybersecurity measures and verification processes, as traditional visual verification methods are no longer sufficient in the face of advancing deepfake technology (Magramo, 2024).

6.2.3 AI Agent–Powered Phishing Emails

Phishing emails have been a staple of social engineering for years, but AI has taken them to a new level. AI-powered agents can analyze vast amounts of data to craft highly personalized phishing emails that are nearly indistinguishable from genuine ones. These emails might reference specific details about your recent activities, colleagues, or interests, making them more convincing. For instance, an AI-generated email could appear to come from your boss, requesting access to a confidential report, or from a friend sharing a seemingly innocent link that leads to malware.

Abnormal Security's research presents several instances of likely AI-generated email attacks. In one case, an attacker impersonated an insurance company representative, sending emails with attachments purportedly containing benefits information and enrollment forms. The email appeared professional, with a legitimate-sounding display name and sender address. However, replies were directed to an attacker-controlled Gmail account, and the attachment was identified as malware, posing risks of system compromise and credential theft (Abnormal Security, 2023).

Another example reported by Abnormal Security involved a phishing attack where the perpetrator masqueraded as a Netflix customer service representative. The email, sent from a compromised legitimate domain, informed the recipient of issues with their account and prompted them to click a link to resolve the problem. The link led to a phishing site designed to harvest login credentials. The email's polished language and absence of typical phishing indicators suggested potential AI generation, enhancing its deceptive effectiveness.

6.2.4 Voice Cloning for Vishing

Vishing, short for voice phishing, is a type of social engineering attack where cybercriminals use voice calls or voice messages to deceive victims into revealing sensitive information or performing harmful actions. These attacks often involve impersonating trusted figures, such as executives or bank representatives, to exploit human trust and manipulate victims through social engineering tactics.

In recent years, vishing attacks have become significantly more sophisticated with the integration of AI-powered voice cloning technology. This advancement has transformed vishing from basic scams into highly convincing and dangerous attacks that are increasingly difficult to detect. AI voice cloning allows attackers to create realistic voice replicas with minimal audio input, often requiring as little as 3 seconds of audio to generate a convincing voice clone.

A notable real-world example of an AI-powered vishing attack occurred in 2021, demonstrating the severe financial implications of this technology. In this incident, attackers used AI to clone a company director's voice, successfully fooling a bank manager during a phone call. The impersonator convinced the manager to transfer $35 million as part of a supposed bank acquisition process (Brewster, 2021). This attack not only resulted in substantial financial losses but also highlighted the potential for AI to be weaponized in large-scale cyberattacks.

The MGM Resorts cyberattack in 2023 provides another striking example of the evolving threat landscape. This attack, which resulted in approximately $100 million in losses, was initiated through a vishing call where the attacker potentially used AI to impersonate a regular employee and contacted the MGM helpdesk to obtain access credentials. While it's not explicitly stated that AI voice cloning was used in this case, it demonstrates the effectiveness of voice-based social engineering in breaching even large corporate entities (Gome, 2023).

These incidents underscore the growing threat posed by AI-enhanced vishing attacks and the urgent need for organizations to adapt their security measures and employee training programs to combat this sophisticated form of social engineering.

6.2.5 Implications and Countermeasures

There are some heightened awareness and robust security measures for social engineering risks. Here are some steps to mitigate the risks:

1. Education and Training: Regularly train employees and individuals on recognizing and responding to social engineering attacks. Awareness is the first line of defense.
2. Advanced Security Solutions: Implement advanced security solutions that leverage AI to detect and block social engineering attempts. For example, AI-powered email filters can identify and quarantine phishing emails.
3. Multi-Factor Authentication (MFA): Enforce MFA for access to sensitive information and systems to provide an additional layer of security. While MFA significantly increases the difficulty for attackers, it's crucial to recognize that sophisticated techniques like man-in-the-middle (MitM) attacks can potentially circumvent even robust MFA implementations. Therefore, combine MFA with continuous monitoring and user behavior analytics to detect and mitigate such advanced attacks.
4. Verification Procedures: Establish verification procedures for sensitive requests. For example, verify fund transfer requests via a secondary communication channel, such as a phone call.
5. Regular Audits and Updates: Regularly audit and update security protocols to address emerging threats. Stay informed about the latest developments in AI and cybersecurity.

6.3 AI Agent in Software Supply Chain Attacks

AI agents can be leveraged by malicious actors to launch sophisticated attacks that exploit vulnerabilities in the software supply chain. These attacks can take many forms. This section will give some examples.

6.3.1 False Code Package Generation

Malicious actors can leverage AI-powered code generation tools to create convincing, yet harmful, software packages. These packages can be designed to mimic legitimate libraries or frameworks, making them difficult to detect. To further evade detection, AI agents can analyze security tools and adapt their code to bypass detection mechanisms. This can involve obfuscation techniques, polymorphic code, and other advanced evasion tactics.

For instance, an AI agent could generate a malicious Python package that appears to be a harmless utility library. However, when imported into a project, it could silently execute malicious code, such as stealing sensitive data or establishing a backdoor (Lakshmanan, 2024).

6.3.2 False API Endpoint Creation

AI agents can launch sophisticated man-in-the-middle attacks to intercept network traffic and redirect requests to malicious endpoints. This can be achieved through techniques like DNS spoofing, IP spoofing, and proxy interception. By simulating legitimate API behavior, AI agents can make it difficult to distinguish between real and fake endpoints.

Consider a scenario where an AI agent compromises a web server and creates a malicious endpoint that mimics a legitimate API. When a vulnerable application sends a request to the compromised server, the AI agent intercepts the request, modifies it, and forwards it to the real API endpoint. The agent then receives the response from the real API endpoint, modifies it further, and sends it back to the vulnerable application. This allows the attacker to manipulate data, steal sensitive information, or inject malicious code.

6.3.3 CI/CD Pipeline Attacks

AI agents can automate the process of identifying vulnerabilities in CI/CD pipelines and exploiting them to compromise the software development process. By injecting malicious code into the build process, compromising credentials, or bypassing security controls, AI agents can introduce vulnerabilities into the software supply chain.

For example, an AI agent could analyze the source code of a project and identify a vulnerability in a third-party library. It could then generate a malicious patch that exploits the vulnerability and introduces a backdoor into the code. This malicious patch could be automatically committed to the code repository and deployed to production.

6.3.4 Runtime Environment Attacks

AI agents can target running applications by modifying their behavior at runtime. This can be achieved through techniques like code injection, buffer overflows, and memory corruption.

Additionally, AI agents can create self-healing malware that can automatically repair itself and evade detection.

For instance, an AI agent could exploit a buffer overflow vulnerability in a web application to inject malicious code into the application's memory. This malicious code could then execute arbitrary commands, steal sensitive data, or launch further attacks.

6.3.5 AI-Driven Software Composition Analysis

AI Agent tool can be used for analyzing software dependencies and identifying potential security risks in the software supply chain. From an offensive perspective, this technology can be leveraged to:

1. Vulnerability Discovery: Automatically identify vulnerable components in target systems by analyzing their software composition.
2. Exploit Generation: Develop targeted exploits for known vulnerabilities in dependencies, increasing the success rate of attacks.
3. Zero-Day Prediction: Analyze code patterns and update histories to predict potential future vulnerabilities before they are officially discovered (We will cover this in next section_.
4. Attack Surface Expansion: Identify previously overlooked attack vectors by mapping complex dependency trees.
5. Evasion Techniques: Develop methods to insert malicious code that mimics legitimate dependency patterns to evade detection.

6.3.6 Intelligent Trojan Injection

In July 2024, Intel 471's Malware Intelligence researchers identified a new Android banking trojan, dubbed BlankBot. This malware used AI agent technology to impersonate utility applications and is believed to primarily target Turkish users. BlankBot exhibits several malicious capabilities, including custom injection attacks, keylogging, screen recording, and communication with a control server via a WebSocket connection. Notably, as of the report's publication, most BlankBot samples remained largely undetected by major antivirus software, indicating that the malware is still under development, as evidenced by the presence of logs and code variants (Intel 471, 2024).

Upon installation, BlankBot does not display an icon in the device launcher and prompts users to grant accessibility permissions, stating "Welcome! App needs Accessibility permission to run properly. Please give accessibility permission!" Once granted, the malware initiates communication with its control server by sending an HTTP "GET" request and then switches to the WebSocket protocol. It then displays a black screen indicating an update, during which it automatically obtains necessary permissions in the background. On devices running Android 13 or newer, BlankBot employs a session-based package installer to bypass restricted settings, retrieving an APK file stored within the application's assets directory and proceeding with the installation process.

The malware can create customizable overlays upon receiving specific commands from its command-and-control (C2) server. These overlays can be exploited to request banking credentials, personal information, or payment card data or to steal the device's lock pattern. The developers have included two external, open-source libraries to facilitate custom injection templates: CompactCreditInput for creating views to steal payment card data and Pattern Locker View for creating pattern lock views. Simulated control server functionality has demonstrated BlankBot's ability to generate various overlays customized with specific bank logos and user interface elements, promptly exfiltrating any user input to the control server.

To evade detection, BlankBot checks whether it is running in an emulator. If the infected device is deemed legitimate, it attempts to maintain persistence by preventing the user from performing certain actions, such as accessing settings or antivirus applications. This is achieved by using accessibility services to monitor all events on the infected device and detect specific words that appear on the screen. Recent samples of BlankBot have been partially obfuscated, with junk code added to hinder reverse engineering, making it significantly more challenging for security researchers to analyze the code and understand the malware's behavior.

6.3.7 *Defensive Countermeasures*

In response to these emerging threats, several tech giants have collaborated on developing AI-driven defensive measures for supply chain security:

1. Anomaly Detection: Using machine learning to identify unusual patterns in software build processes, dependency trees, and code commits.
2. Graph Analysis: Employing graph neural networks to analyze the relationships between different components in a supply chain, identifying potential compromise points.
3. Behavioral Analysis: Monitoring the behavior of software components and dependencies to detect deviations from expected patterns.
4. Predictive Patching: Using AI to prioritize and expedite the patching process for vulnerable components based on their potential impact on the supply chain.

6.4 AI Agent Used for Bug Bounty

The integration of AI into bug bounty programs has transformed how vulnerabilities are discovered, reported, and prioritized. This evolution has led to more efficient and effective vulnerability management processes.

6.4.1 AI-Assisted Vulnerability Discovery

We can use the LLM model to design an AI Agent called BountyAgent to meet the following requirements:

1. Scope Analysis: BountyAgent can analyze program scopes in natural language, identifying potential areas of focus that human researchers might overlook.
2. Vulnerability Pattern Recognition: The AI can recognize complex vulnerability patterns across different codebases and technologies, suggesting potential weak points to investigate.
3. Exploit Crafting Assistance: By analyzing vulnerability descriptions and target system information, BountyAgent can provide guidance on crafting more effective proof-of-concept exploits.
4. Report Generation: The AI assists in writing clear, comprehensive, and actionable vulnerability reports, improving communication between researchers and security teams.
5. Learning from Historical Data: BountyAgent continuously learns from past successful submissions, improving its ability to guide researchers toward high-impact vulnerabilities.
6. Cross-program Insights: The AI can draw insights from patterns across multiple bug bounty programs, helping researchers identify similar vulnerabilities in different contexts.

6.4.2 Code Example

We can use Python to implement such agent using the sample code snippet below:

1. **Import and Data Structures**

```
from typing import List, Dict, Optional
import json
import logging
from dataclasses import dataclass
from datetime import datetime

@dataclass
class Vulnerability:
    """
    Represents a security vulnerability with all its details
    """
    title: str
    description: str
    severity: str
    affected_components: List[str]
    potential_impact: str
    reproduction_steps: List[str]
    recommendations: List[str]
    discovery_date: datetime

@dataclass
class ScopeElement:
    """
    Defines the scope boundaries for security testing
    """
    domain: str
    tech_stack: List[str]
    assets: List[str]
    excluded_paths: List[str]
```

2. **BountyAgent Initialization**

```
class BountyAgent:
    def __init__(self, model_name: str, api_key: str):
        """
        Initialize the BountyAgent with necessary credentials
and setup
        """
        self.model_name = model_name
        self.api_key = api_key
        self.vulnerability_database = []
        self.logger = logging.getLogger(__name__)
```

3. **Scope Analysis Methods**

```
  def analyze_scope(self, scope_text: str) -> List[ScopeElement]:
        """
        Analyzes program scope using NLP to identify potential
focus areas.
        """
        scope_elements = []
        try:
            analyzed_scope = self._llm_analyze_text(scope_text)
            for element in analyzed_scope:
                scope_elements.append(ScopeElement(
                    domain=element['domain'],
                    tech_stack=element['technologies'],
                    assets=element['assets'],
                    excluded_paths=element['exclusions']
                ))
            return scope_elements
        except Exception as e:
            self.logger.error(f"Scope analysis failed: {str(e)}")
            return []

    def _llm_analyze_text(self, text: str) -> List[Dict]:
        """
        Private method to interact with LLM API for text analysis
        """
        # Implementation for LLM integration
        pass
```

4. **Vulnerability Pattern Recognition**

```
    def identify_vulnerability_patterns(self, codebase: str,
tech_stack: List[str]) -> List[Dict]:
        """
        Identifies potential vulnerability patterns in the given
codebase
        """
        patterns = []
        try:
            identified_patterns = self._pattern_
recognition(codebase, tech_stack)
            for pattern in identified_patterns:
                if self._is_within_scope(pattern):
```

```
                    patterns.append({
                        'pattern_type': pattern['type'],
                        'confidence': pattern['confidence'],
                        'affected_components':
pattern['components'],

'suggested_tests': pattern['test_approaches']
                    })
            return patterns
        except Exception as e:
            self.logger.error(f"Pattern identification failed:
{str(e)}")
            return []

    def _pattern_recognition(self, codebase: str, tech_stack:
List[str]) -> List[Dict]:
        """
        Core pattern recognition logic
        """
        # Implementation for pattern recognition
        pass

    def _is_within_scope(self, pattern: Dict) -> bool:
        """
        Verifies if a detected pattern is within authorized scope
        """
        # Implementation for scope verification
        pass
```

5. **Report Generation.**

```
    def generate_report(self, vulnerability:
Vulnerability) -> str:
        """
        Generates a comprehensive vulnerability report
        """
        try:
            report_template = {
                'title': vulnerability.title,
                'description': vulnerability.description,
                'severity': vulnerability.severity,
                'impact': vulnerability.potential_impact,

'affected_components': vulnerability.affected_components,
```

```
                'steps_to_reproduce': vulnerability.
reproduction_steps,
                'recommendations': vulnerability.recommendations,
                'discovery_date': vulnerability.discovery_date.
isoformat()
            }
            return json.dumps(report_template, indent=2)
        except Exception as e:
            self.logger.error(f"Report generation failed:
{str(e)}")
            return ""
```

6. ***Learning and Cross-Program Analysis***

```
    def learn_from_submission(self, successful_submission: Dict):
        """
        Updates the agent's knowledge base with new successful
vulnerability submissions
        """
        try:
            if self._verify_submission(successful_submission):
                self.vulnerability_database.append({
                    'pattern': successful_submission['pattern'],
                    'context': successful_submission['context'],
                    'impact': successful_submission['impact'],

'detection_method': successful_submission['method']
                })
                self._update_pattern_recognition()
        except Exception as e:
            self.logger.error(f"Learning from submission failed:
{str(e)}")

    def analyze_cross_program_patterns(self) -> List[Dict]:
        """
        Analyzes patterns across multiple programs
        """
        patterns = []
        try:
            aggregated_data = self._aggregate_program_data()

common_patterns = self._identify_common_patterns(aggregated_data)

            for pattern in common_patterns:
```

```
                patterns.append({
                    'vulnerability_class': pattern['class'],
                    'frequency': pattern['frequency'],
                    'common_contexts': pattern['contexts'],
                    'potential_mitigations':
pattern['mitigations']
                })
            return patterns
        except Exception as e:
            self.logger.error(f"Cross-program analysis failed:
{str(e)}")
            return []
```

7. **Example Usage**

```
def main():
    # Initialize BountyAgent
    agent = BountyAgent(
        model_name="gpt-4",
        api_key="your-api-key"
    )

    # Example scope text
    scope_text = """
    In scope: api.example.com
    Tech stack: Python, Django, PostgreSQL
    Assets: User data, payment processing
    Excluded: Third-party integrations
    """

    # Analyze scope
    scope_elements = agent.analyze_scope(scope_text)

    # Identify vulnerability patterns
    patterns = agent.identify_vulnerability_patterns(
        codebase="example_code",
        tech_stack=["Python", "Django"]
    )
```

Note that some methods are marked with "pass" as they would require specific implementations based on your chosen LLM model and pattern recognition algorithms. You would need to implement these based on your specific requirements and the AI models you plan to use.

6.5 AI Agent in Vulnerability Discovery and Zero-Day Discovery

Vulnerability discovery is the process of identifying security weaknesses in software or systems, while **zero-day discovery** specifically refers to finding vulnerabilities that are unknown to the software vendor or public, making them exploitable until patched. AI Agent is now increasingly used in this field.

6.5.1 Google's Big Sleep Project

In early October 2024, Google's AI-powered framework, Big Sleep, identified a stack buffer underflow vulnerability in the SQLite database engine (Google Project Zero, 2024). This vulnerability, which could lead to application crashes or arbitrary code execution, was promptly reported and patched before any exploitation occurred (Divinsky, 2024).

Big Sleep, an evolution of Google's Project Naptime, leverages AI Agent to assist in vulnerability research. By analyzing code commits and simulating human-like vulnerability assessment, Big Sleep can identify complex security flaws that traditional methods, such as fuzzing, might overlook.

This approach enables the AI agent to navigate codebases, identify potential weaknesses, and generate inputs to trigger vulnerabilities without direct human intervention. In this instance, Big Sleep enhanced traditional fuzzing techniques by incorporating its code comprehension capabilities, allowing it to detect vulnerabilities that might be challenging for human researchers or standard tools to identify.

The successful identification of this vulnerability highlights the potential of artificial intelligence in proactive cybersecurity measures.

6.5.2 Design "DeepFuzz" Agent

For vulnerability discovery, we can design an agent called "DeepFuzz" which would do the following:

1. Intelligent Input Generation: DeepFuzz uses deep learning models to generate test cases that are more likely to trigger edge cases and uncover vulnerabilities, significantly improving upon traditional random fuzzing techniques.
2. Adaptive Fuzzing Strategies: The system can dynamically adjust its fuzzing approach based on real-time feedback from the target application, focusing on areas that show promise for vulnerability discovery.

3. Coverage-Guided Optimization: DeepFuzz uses code coverage information to guide its exploration of the target application, ensuring thorough testing of all code paths.
4. Symbolic Execution Integration: The AI integrates symbolic execution techniques to solve complex path constraints, allowing it to reach deep code paths that are typically challenging for traditional fuzzers.
5. Multidimensional Analysis: DeepFuzz can simultaneously analyze multiple aspects of the target application, including memory usage, timing behaviors, and inter-process communications, to identify a broader range of vulnerability types.
6. Automated Exploit Generation: For certain types of vulnerabilities, DeepFuzz can automatically generate proof-of-concept exploits, accelerating the verification and patching process.
7. Continuous Learning: The system learns from each fuzzing session, improving its efficiency and effectiveness over time.

Here is the high-level code snippet we can use as boilerplate code to implement such Agent:

1. **First, let's start with the core data structures and utilities:**

```
# utils.py
import logging
from dataclasses import dataclass
from typing import List, Dict, Any, Optional, Tuple

# Configure logging
logging.basicConfig(level=logging.INFO)
logger = logging.getLogger(__name__)

@dataclass
class FuzzingResult:
    """Container for fuzzing results"""
    input_data: bytes
    coverage_increase: float
    crash_detected: bool
    vulnerability_type: Optional[str]
    execution_path: List[int]
    memory_usage: float
    execution_time: float
```

This module defines the core data structures used throughout DeepFuzz. The `FuzzingResult` dataclass serves as a container for storing all relevant information about a fuzzing iteration, making it easy to track and analyze results.

2. **Next, the neural network-based input generator:**

```
# input_generator.py
import torch
import torch.nn as nn
import numpy as np

class InputGenerator(nn.Module):
    """Neural network for intelligent input generation"""
    def __init__(self, input_dim: int, hidden_dim: int,
latent_dim: int):
        super().__init__()

        self.encoder = nn.Sequential(
            nn.Linear(input_dim, hidden_dim),
            nn.LeakyReLU(),
            nn.Linear(hidden_dim, hidden_dim),
            nn.LeakyReLU(),
            nn.Linear(hidden_dim, latent_dim * 2)
        )

        self.decoder = nn.Sequential(
            nn.Linear(latent_dim, hidden_dim),
            nn.LeakyReLU(),
            nn.Linear(hidden_dim, hidden_dim),
            nn.LeakyReLU(),
            nn.Linear(hidden_dim, input_dim),
            nn.Sigmoid()
        )

    def reparameterize(self, mu: torch.Tensor, logvar:
torch.Tensor) -> torch.Tensor:
        std = torch.exp(0.5 * logvar)
        eps = torch.randn_like(std)
        return mu + eps * std

    def forward(self, x: torch.Tensor) -> Tuple[torch.
Tensor, torch.Tensor, torch.Tensor]:
        h = self.encoder(x)
        mu, logvar = h.chunk(2, dim=-1)
        z = self.reparameterize(mu, logvar)
        return self.decoder(z), mu, logvar
```

This module implements a Variational Autoencoder (VAE) for generating intelligent test inputs. It uses a deep neural network architecture to learn patterns from successful inputs and generate new ones that are likely to trigger interesting program behaviors.

3. **The symbolic execution component:**

```
# symbolic_executor.py
import angr
import z3

class SymbolicExecutor:
    """Handles symbolic execution using angr"""
    def __init__(self, binary_path: str):
        self.binary_path = binary_path
        self.project = angr.Project(binary_path, load_
options={'auto_load_libs': False})

    def find_path_constraints(self, target_addr: int) ->
List[z3.ExprRef]:
        """Find path constraints to reach target
address"""
        state = self.project.factory.entry_state()

simgr = self.project.factory.simulation_manager(state)

        simgr.explore(find=target_addr)
        if len(simgr.found) > 0:
            return simgr.found[0].solver.constraints
        return []

    def solve_constraints(self, constraints: List[z3.
ExprRef]) -> Optional[Dict[str, Any]]:
        """Solve given constraints to generate input"""
        solver = z3.Solver()
        for constraint in constraints:
            solver.add(constraint)

        if solver.check() == z3.sat:
            return solver.model()
        return None
BountyAgent
```

This module handles symbolic execution using the Angr framework. Angr is an open-source binary analysis platform for Python. It combines both static and dynamic symbolic ("concolic") analysis, providing tools to solve a variety of tasks.

In this example, the framework helps analyze program paths and generate inputs that reach specific code locations by solving path constraints.

4. **The coverage tracking system:**

```
class CoverageTracker:
    """Tracks code coverage during fuzzing"""
    def __init__(self, binary_path: str):
        self.binary_path = binary_path
        self.coverage_map = set()
        self.edge_hits = {}

    def update_coverage(self, execution_trace: List[int])
-> float:
        """Update coverage information and return coverage
increase"""
        old_coverage = len(self.coverage_map)

        for addr in execution_trace:
            self.coverage_map.add(addr)

        for i in range(len(execution_trace) - 1):
            edge = (execution_trace[i], execution_
trace[i + 1])
            self.edge_hits[edge] = self.edge_hits.
get(edge, 0) + 1

        return (len(self.coverage_map) - old_coverage) /
old_coverage if old_coverage > 0 else 1.0
```

The coverage tracker monitors code coverage during fuzzing, tracking both basic blocks and edge coverage. It helps guide the fuzzing process by identifying areas that need more exploration.

5. **The vulnerability detector:**

```
class VulnerabilityDetector:
    """Detects various types of vulnerabilities"""
    def __init__(self):
        self.vulnerability_patterns = {
            'buffer_overflow': self._check_buffer_overflow,
            'use_after_free': self._check_use_after_free,
            'integer_overflow': self._check_integer_overflow
        }

    def _check_buffer_overflow(self, execution_trace:
List[Any]) -> bool:
        # Implement buffer overflow detection logic
        pass

    def _check_use_after_free(self, execution_trace:
List[Any]) -> bool:
        # Implement use-after-free detection logic
        pass

    def _check_integer_overflow(self, execution_trace:
List[Any]) -> bool:
        # Implement integer overflow detection logic
        pass

    def analyze_trace(self, execution_trace: List[Any]) ->
Optional[str]:
        """Analyze execution trace for vulnerabilities"""
        for vuln_type, checker in self.vulnerability_
patterns.items():
            if checker(execution_trace):
                return vuln_type
        return None
```

The vulnerability detector analyzes execution traces to identify various types of vulnerabilities. It implements different detection patterns and can be easily extended to support new vulnerability types.

6. **The exploit generator:**

```
class ExploitGenerator:
    """Generates proof-of-concept exploits"""
    def __init__(self, binary_path: str):
        self.binary_path = binary_path
        self.symbolic_executor =
SymbolicExecutor(binary_path)

    def generate_exploit(self, vulnerability:
FuzzingResult) -> Optional[bytes]:
        """Generate exploit for discovered
vulnerability"""

if vulnerability.vulnerability_type == 'buffer_overflow':

return self._generate_buffer_overflow_exploit(vulnerability)

elif vulnerability.vulnerability_type == 'use_after_free':

return self._generate_uaf_exploit(vulnerability)
        return None

    def _generate_buffer_overflow_exploit(self, vuln:
FuzzingResult) -> Optional[bytes]:
        # Implement buffer overflow exploit generation
        pass

    def _generate_uaf_exploit(self, vuln: FuzzingResult)
-> Optional[bytes]:
        # Implement use-after-free exploit generation
        pass
```

The exploit generator automatically creates proof-of-concept exploits for discovered vulnerabilities, helping verify the findings and assist in the patching process.

7. **The distributed fuzzing worker:**

```
import ray
@ray.remote
class FuzzingWorker:
    """Distributed fuzzing worker"""
    def __init__(self, binary_path: str, input_generator: 
InputGenerator):
        self.binary_path = binary_path
        self.input_generator = input_generator
        self.coverage_tracker = 
CoverageTracker(binary_path)
        self.vulnerability_detector = 
VulnerabilityDetector()

    def fuzz_iteration(self, seed_input: bytes) -> 
FuzzingResult:
        """Perform one fuzzing iteration"""
        input_tensor = torch.from_numpy(np.frombuffer(seed_
input, dtype=np.uint8)).float() / 255.0

        with torch.no_grad():
            new_input, _, _ = self.input_generator(input_
tensor.unsqueeze(0))

        execution_trace = self._execute_binary(new_input)

vuln_type = self.vulnerability_detector.analyze_
trace(execution_trace)

coverage_increase = self.coverage_tracker.update_
coverage(execution_trace)

        return FuzzingResult(
            input_data=new_input.numpy().tobytes(),
            coverage_increase=coverage_increase,
            crash_detected=vuln_type is not None,
            vulnerability_type=vuln_type,
            execution_path=execution_trace,
            memory_usage=self._get_memory_usage(),
            execution_time=self._get_execution_time()
        )
```

```
        def _execute_binary(self, input_data: torch.Tensor) ->
List[int]:
            # Implement binary execution logic
            pass

        def _get_memory_usage(self) -> float:
            # Implement memory usage tracking
            pass

        def _get_execution_time(self) -> float:
            # Implement execution time tracking
            pass
```

The fuzzing worker handles individual fuzzing tasks in a distributed setting, coordinating input generation, execution, and analysis.

8. **The main DeepFuzz class:**

```
import ray
from collections import deque
import numpy as np
import asyncio

class DeepFuzz:
    """Main DeepFuzz class coordinating all components"""
    def __init__(self, binary_path: str, num_workers:
int = 4):
        self.binary_path = binary_path
        self.num_workers = num_workers

        # Initialize components
        self.input_generator = InputGenerator(input_
dim=256, hidden_dim=512, latent_dim=128)
        self.symbolic_executor =
SymbolicExecutor(binary_path)
        self.exploit_generator =
ExploitGenerator(binary_path)

        # Initialize distributed workers
        ray.init(ignore_reinit_error=True)
        self.workers = [FuzzingWorker.remote(binary_path,
self.input_generator)
                        for _ in range(num_workers)]

        # Initialize experience replay buffer for learning
        self.replay_buffer = deque(maxlen=10000)
```

This is the main coordinator class that brings all components together and manages the fuzzing process. Let's break down its methods:

9. **The main fuzzing loop:**

```
        async def fuzz(self, max_iterations: int = 1000) ->
List[FuzzingResult]:
            """Main fuzzing loop"""
            results = []

            for iteration in range(max_iterations):
                # Generate seed inputs for all workers
                seed_inputs = [self._generate_seed_input() for
_ in range(self.num_workers)]

                # Distribute fuzzing tasks to workers
                futures = [worker.fuzz_iteration.remote(seed)
                          for worker, seed in zip(self.
workers, seed_inputs)]

                # Collect results
                iteration_results = await ray.get(futures)

                # Process results
                for result in iteration_results:
                    if result.crash_detected:
                        # Generate exploit if
vulnerability found
                        exploit = self.exploit_generator.
generate_exploit(result)
                        if exploit:
                            logger.info(f"Generated exploit
for {result.vulnerability_type}")

                    # Store result in replay buffer for
learning
                    self.replay_buffer.append(result)

                    # Update models if enough data collected
                    if len(self.replay_buffer) >= 1000:
                        self._update_models()

                results.extend(iteration_results)


                # Log progress
                self._log_progress(iteration, results)

            return results
```

10. **Helper methods for the DeepFuzz class:**

```
        def _generate_seed_input(self) -> bytes:
            """Generate seed input for fuzzing"""
            # Generate random input of random length between 1
and 1024 bytes
            length = random.randint(1, 1024)
            return bytes(random.randint(0, 255) for _ in
range(length))

        def _update_models(self):
            """Update neural networks based on collected
experience"""
            # Sample batch from replay buffer
            batch = random.sample(self.replay_buffer, min(1000,
len(self.replay_buffer)))

            # Extract successful inputs (those that increased
coverage or found crashes)
            successful_inputs = [result.input_data for result
in batch
                                 if result.coverage_increase > 0
or result.crash_detected]

            if successful_inputs:
                # Convert to tensors and train input generator
                inputs = torch.stack([torch.from_numpy(np.
frombuffer(input_data, dtype=np.uint8)).float() / 255.0
                                      for input_data in
successful_inputs])

                # Train the model (simplified)
                optimizer = torch.optim.Adam(self.input_
generator.parameters())

output, mu, logvar = self.input_generator(inputs)
                loss = self._compute_vae_loss(output, inputs,
mu, logvar)

                optimizer.zero_grad()
                loss.backward()
                optimizer.step()

        def _compute_vae_loss(self, recon_x, x, mu, logvar):
```

```
            """Compute VAE loss function"""
            recon_loss = nn.functional.binary_cross_
entropy(recon_x, x, reduction='sum')
            kld_loss = -0.5 * torch.sum(1 + logvar - mu.
pow(2) - logvar.exp())
            return recon_loss + kld_loss

        def _log_progress(self, iteration: int, results:
List[FuzzingResult]):
            """Log fuzzing progress"""
            total_crashes = sum(1 for r in results if r.
crash_detected)
            avg_coverage = np.mean([r.coverage_increase for r
in results])

            logger.info(f"Iteration {iteration}:")

logger.info(f"Total crashes found: {total_crashes}")

logger.info(f"Average coverage increase: {avg_coverage:.4f}")
```

11. **Example usage and runner:**

```
    async def main():
        # Example usage
        binary_path = "./target_binary"
        fuzzer = DeepFuzz(binary_path, num_workers=4)

        try:
            results = await fuzzer.fuzz(max_iterations=1000)

            # Print summary
            print("\nFuzzing Summary:")
            print(f"Total iterations: {len(results)}")
            print(f"Unique crashes: {len(set(r.vulnerability_
type for r in results if r.crash_detected))}")
            print(f"Final coverage: {sum(r.coverage_increase
for r in results):.2f}")

            # Save results
            with open('fuzzing_results.txt', 'w') as f:
                for i, result in enumerate(results):
                    if result.crash_detected:
                        f.write(f"\nCrash {i}:\n")
```

```
                                f.write(f"Type: {result.
vulnerability_type}\n")
                                f.write(f"Input: {result.input_data.
hex()}\n")
                                f.write(f"Coverage: {result.
coverage_increase}\n")
                                f.write("-" * 50 + "\n")

            except Exception as e:
                logger.error(f"Fuzzing failed: {str(e)}")
                raise

        if __name__ == "__main__":
            asyncio.run(main())
```

6.6 Architectural Considerations for AI-Enhanced Offensive Platforms

In this section, I listed a few items to consider when designing a new generation of AI Agent for offensive security.

1. Distributed AI Processing: Modern offensive platforms often employ distributed AI processing to enable real-time decision-making across complex attack infrastructures, balancing computational load and reducing latency.
2. Secure Enclaves for AI Models: To protect valuable AI models and training data from discovery or tampering, offensive platforms are incorporating secure enclaves and advanced encryption techniques.
3. Modular AI Integration: Architectures are being designed with modularity in mind, allowing for rapid integration of new AI capabilities as they are developed, without requiring a complete system overhaul.
4. Edge AI for Offensive Operations: The use of edge computing in offensive AI allows for faster decision-making and reduced reliance on continuous communication with a central command server.
5. Federated Learning in Offensive Tools: Some advanced platforms are exploring the use of federated learning techniques to improve AI models without centralizing sensitive operational data.
6. AI-Driven Resilience: Offensive architectures are incorporating AI-driven self-healing and adaptation mechanisms to maintain operational capability in the face of defensive countermeasures.
7. Quantum-Resistant Designs: Forward-thinking offensive platform designs are beginning to incorporate quantum-resistant cryptographic techniques to future-proof against potential quantum computing threats.

These architectural changes are enabling more flexible, scalable, and powerful offensive security platforms, while also presenting new challenges for defensive strategies. The idea here is to equip the cybersecurity professional with powerful AI Agent tools for offensive security with the goal of improving the overall security postures of the industry.

6.7 Summary

Chapter 6 investigates how AI agents are transforming offensive security practices. Offensive security's objective to proactively uncover vulnerabilities before malicious actors exploit them is increasingly driven by AI technologies. The chapter begins with an analysis of AI's integration into red teaming exercises, highlighting tools like Meta's GOAT, Google's AART framework, and Microsoft's PyRIT. These systems automate adversarial testing, generate diverse attack scenarios, and improve efficiency in identifying vulnerabilities.

The chapter then delves into AI-driven social engineering, where advanced techniques such as deepfake video calls, AI-powered phishing, and voice cloning are reshaping the threat landscape. Real-world examples, including financial fraud and corporate breaches, illustrate the sophistication of these attacks.

The chapter also examines AI agents in software supply chain attacks, focusing on malicious code generation, CI/CD pipeline exploitation, and intelligent Trojan injection. Additionally, it explores the use of AI agents for bug bounty programs, where automation enhances vulnerability discovery, reporting, and exploit generation. The chapter provides code examples, such as the implementation of the "BountyAgent" and "DeepFuzz" agents, demonstrating how AI can be used to design tools that assist in vulnerability discovery and zero-day identification. These examples showcase the practical application of AI in creating intelligent input generation, adaptive fuzzing strategies, and automated exploit generation.

Notable advancements, like Google's Big Sleep for zero-day discovery and adaptive fuzzing techniques, demonstrate the efficacy of AI in identifying complex security flaws. The inclusion of code examples helps readers understand how these AI agents can be built and integrated into offensive security workflows.

The chapter concludes by discussing the architectural considerations for AI-driven offensive platforms, such as distributed processing, edge AI, modular integration, and quantum-resistant designs. These innovations enable more scalable, adaptive, and resilient systems, underscoring the dual-edged nature of AI in offensive security.

6.8 Questions

1. What are the primary differences between traditional offensive security and AI-driven approaches?
2. How does Meta's GOAT utilize generative AI to identify vulnerabilities in language models?
3. What role does Google's AART framework play in automating red teaming processes?
4. How does Microsoft's PyRIT framework balance traditional security assessments with responsible AI evaluations?
5. What are some examples of AI agents being used in social engineering attacks, and what makes them effective?
6. How can deepfake technology amplify the risks of social engineering?
7. What countermeasures can organizations implement to mitigate AI-driven social engineering attacks?
8. How are AI agents used to conduct sophisticated software supply chain attacks, such as false API endpoint creation or CI/CD pipeline manipulation?
9. In what ways does AI improve the process of zero-day vulnerability discovery?
10. What techniques does Google's Big Sleep use to enhance traditional fuzzing methods for vulnerability identification?
11. How can AI agents aid in bug bounty programs, and what advantages do they provide over human-led initiatives?
12. How do the code examples of "BountyAgent" and "DeepFuzz" illustrate the implementation of AI in vulnerability discovery?
13. What is the significance of integrating symbolic execution with AI-driven fuzzing systems like DeepFuzz?
14. What are some architectural considerations for designing AI-driven offensive security platforms?
15. What ethical concerns arise from the use of AI in offensive security, and how can they be addressed?

References

Abnormal Security. (2023). AI-generated email attacks: The evolution of cybercrime. *Abnormal Security Blog*. Accessed November 27, 2024. https://abnormalsecurity.com/blog/2023-ai-generated-email-attacks.

Brewster, T. (2021, October 14). Huge Bank fraud uses deep fake voice tech to steal millions. *Forbes*. https://www.forbes.com/sites/thomasbrewster/2021/10/14/huge-bank-fraud-uses-deep-fake-voice-tech-to-steal-millions/

Divinsky, Y. (2024). **Google's big sleep AI tool discovers zero-day flaw in SQLite database engine**. Retrieved from https://vulcan.io/blog/big-sleep-zero-day-sqlite-attack/

Gome, F. (2023, September). Cyberattack on MGM resort explained. *Inszone Insurance Blog*. https://inszoneinsurance.com/blog/cyberattack-mgm-resort-explained/

Google. (2023a). *Google's AI red team: The ethical hackers making AI Safer*. https://blog.google/technology/safety-security/googles-ai-red-team-the-ethical-hackers-making-ai-safer

Google. (2023b). *More ways google is delivering on its responsible ai commitment*. https://blog.google/technology/ai/google-responsible-ai-commitment-update/

Google Project Zero. (2024). **From naptime to big sleep: Using large language models to catch vulnerabilities in real-world code**. Retrieved from https://googleprojectzero.blogspot.com/2024/10/from-naptime-to-big-sleep.html

Intel 471. (2024, July). BlankBot: A new android banking trojan with screen recording, keylogging, and remote control capabilities. *Intel 471 Blog*. https://intel471.com/blog/blankbot-a-new-android-banking-trojan-with-screen-recording-keylogging-and-remote-control-capabilities

Magramo, K. (2024, May 16). Arup Deepfake scam loss in Hong Kong. *CNN*. https://edition.cnn.com/2024/05/16/tech/arup-deepfake-scam-loss-hong-kong-intl-hnk/index.html

Lakshmanan, R. (2024, November 22). PyPI attack: ChatGPT, Claude impersonators deliver JarkaStealer via python libraries. *The Hacker News*. https://thehackernews.com/2024/11/pypi-attack-chatgpt-claude.html

Meta. (2024). *Generative offensive agent tester (GOAT)*. Accessed November 27, 2024. https://arxiv.org/abs/2410.01606

Radharapu, B., Robinson, K., Aroyo, L., & Lahoti, P. (2023). *AI-assisted red-teaming with diverse data generation for new LLM-powered applications*. https://storage.googleapis.com/gweb-research2023-media/pubtools/7433.pdf

Jerry Huang is currently an AI Engineer at Google, where he contributed to the AI/ML evaluation pipeline for a consumer-facing application. Before Google, he worked as a technical and security staff member at several prominent technology companies, gaining experience in areas like security, AI/ML, and scalable systems.

At Metabase, an open-source business intelligence platform, Jerry contributed features such as private key management and authentication solutions. As a Software Engineer at Glean, a Generative AI search startup, he was one of three engineers responsible for managing large-scale GCP infrastructure powering text summarization, autocomplete, and search for over 100,000 enterprise users. During his time at TikTok, Jerry helped design and build custom RPCs to model access control policies. At Roblox, he served as a Machine Learning/Software Engineering Intern, focusing on real-time text generation models and gathering a large multilingual corpus that significantly boosted model robustness.

In addition to his industry experience, Jerry has conducted extensive security and biometrics research as a Research Assistant at Georgia Tech's Institute for Information Security & Privacy, resulting in a thesis on privacy-preserving biometric authentication.

Jerry holds a BS/MS in Computer Science from Georgia Tech and is currently pursuing an MS in Applied Mathematics at the University of Chicago.

Ken Huang is a prolific author and globally recognized authority in AI and Web3, with an extensive portfolio of published works that bridge business strategy, technical implementation, and cutting-edge research. As a Fellow of the Cloud Security Alliance and Co-Chair of the AI Safety Working Groups at the Cloud Security Alliance and the AI STR Working Group at the World Digital Technology Academy under the UN Framework, he is a leading voice in shaping global AI governance and security standards.

Huang is the CEO and Chief AI Officer (CAIO) of DistributedApps.ai, a firm specializing in generative AI training and consulting. His contributions to the field include being a core contributor to the OWASP Top 10 Risks for LLM Applications and an active participant in the NIST Generative AI Public Working Group.

Notable Publications:

- *Beyond AI: ChatGPT, Web3, and the Business Landscape of Tomorrow* (Springer, 2023)—strategic insights into AI and Web3's business applications.

• *Generative AI Security: Theories and Practices* (Springer, 2024)—a comprehensive guide on securing generative AI systems.
• *Practical Guide for AI Engineers* (Volumes 1 and 2, DistributedApps.ai, 2024)—essential resources for AI and ML engineers.
• *The Handbook for Chief AI Officers: Leading the AI Revolution in Business* (DistributedApps.ai, 2024)—a roadmap for CAIOs in implementing GenAI across organizations.
• *Web3: Blockchain, the New Economy, and the Self-Sovereign Internet* (Cambridge University Press, 2024)—insights into the convergence of AI, blockchain, IoT, and emerging technologies.
• *Blockchain and Web3: Building the Cryptocurrency, Privacy, and Security Foundations of the Metaverse* (Wiley, 2023)—recognized as a must-read by TechTarget in 2023 and 2024.

Ken is a sought-after speaker and has presented at events such as the World Economic Forum in Davos, ACM and IEEE conferences, the CSA AI Summit, Depository Trust & Clearing Corporation forums, and World Bank conferences. His recent appointment to the OpenAI Forum reflects his ongoing commitment to advancing collaboration and dialogue in the field of AI.

Explore Ken's work on Amazon: https://www.amazon.com/author/kenhuang

Chris Hughes is the Co-founder and CEO of Aquia, a Cybersecurity consulting firm. Chris brings nearly 20 years of IT and cybersecurity experience to his role as Co-founder and CEO at Aquia.

Chris has also served as a Cyber Innovation Fellow (CIF) at the Cybersecurity Infrastructure and Security Agency (CISA), focusing on software supply chain security. Additionally, Chris advises various tech startups focused on areas such as Software Composition Analysis (SCA), Kubernetes Security, Non-Human Identities (NH), and AI Security.

As a United States Air Force veteran and former civil servant in the US Navy and the General Services Administration's FedRAMP program, Chris is passionate about making a lasting impact on his country and the global community.

In addition to his public service, Chris spent several years as a consultant within the private sector and currently serves as an adjunct professor for cybersecurity master's programs at the University of Maryland Global Campus. Chris participates in industry working groups, such as the Cloud Security Alliance's Incident Response and SaaS Security Working Group, and serves as the Membership Chair for Cloud Security Alliance D.C. He is the host and author of the Resilient Cyber.

Chris runs the Resilient Cyber Substack, where he shares a weekly newsletter, deep dive analysis, and interviews with industry leaders, as well as detailed articles on topics such as Cloud, Vulnerability Management, DevSecOps, Cybersecurity Leadership, and Market Dynamics.

Chris holds a B.S. in Information Systems, a M.S. in Cybersecurity, and an MBA. He regularly consults with IT and cybersecurity leaders from various industries to assist their organizations with their digital transformation journeys while keeping security a core component of that transformation.

Chris is co-author of the book, "Software Transparency: Supply Chain Security in an Era of a Software-Driven Society" and "Effective Vulnerability Management: Managing Risk in the Vulnerable Digital Ecosystem" both published by Wiley. He has also contributed many other thought leadership pieces on software supply chain security and has presented on the topic at a variety of industry

Chapter 7
AI Agents in Defensive Security

Jerry Huang, Ken Huang, and Chris Hughes

We have discussed AI agents for offensive security in the last chapter. This chapter will focus on the other side of the coin which is AI agent for defensive security.

As we know, in today's digital landscape, traditional cybersecurity measures relying on static rules and reactive responses can no longer adequately protect against sophisticated cyber threats. Organizations face critical challenges including expanding attack surfaces, increasingly sophisticated adversaries, and the need for continuous defense amid a persistent skills gap.

AI agents have emerged as a potential solution. By combining automation capabilities with continuous learning, these agents augment human expertise while providing proactive defense against evolving cyber threats. This chapter explores various aspects of AI agent uses in cyber defense.

7.1 Core Functions of AI Agents in Defensive Security

AI agents in defensive cybersecurity serve critical functions to ensure robust protection against evolving threats. This section highlights some core functions of AI agent in cyber defense.

Let us use Fig. 7.1 to give us a mind map of what are those core functions.

J. Huang
The University of Chicago, Chicago, IL, USA

K. Huang (✉)
DistributedApps.ai, Fairfax, VA, USA
e-mail: ken@distributedapps.ai

C. Hughes
Aquia, Virginia Beach, VA, USA

K. Huang (ed.), *Agentic AI*, Progress in IS,
https://doi.org/10.1007/978-3-031-90026-6_7

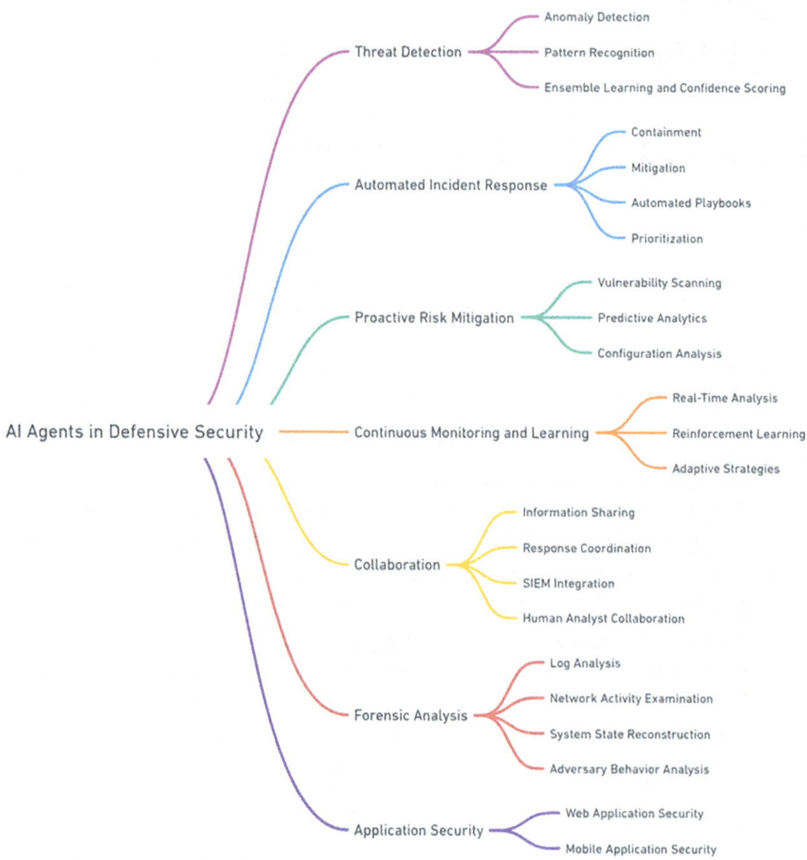

Fig. 7.1 Core functions of AI agent in cyber defense

Threat detection is one of the core functions of AI agents. By continuously monitoring network traffic, system logs, and user activity, these agents can identify patterns that indicate potential threats. They utilize anomaly detection to flag deviations from normal behavior, such as unusual login attempts or unexpected data transfers. Unlike static rule-based systems, AI agents analyze data holistically, recognizing subtle indicators of compromise that might otherwise go unnoticed. This dynamic detection capability is essential in defending against zero-day attacks and advanced persistent threats (APTs), which often bypass traditional security mechanisms. AI agents use techniques such as ensemble learning and confidence scoring to minimize false positives (incorrectly identifying benign activity as malicious) and false negatives (failing to detect an actual threat). For example, by combining multiple models trained on different data sources, AI agents can cross-verify alerts before flagging them. Additionally, ongoing retraining with updated datasets helps refine the detection accuracy over time.

The following shows a basic threat detection code using anomaly detection:

```
def detect_anomalies(network_traffic):
  # Simple anomaly detection for network traffic
  def calculate_baseline(data):
    return {
      'avg_requests': np.mean([x['requests'] for x in data]),
      'std_requests': np.std([x['requests'] for x in data])
    }
    baseline = calculate_baseline(network_traffic[-1000:]) # Use last 1000 samples
  threshold = baseline['avg_requests'] + (3 * baseline['std_requests'])
    anomalies = []
  for traffic in network_traffic:
    if traffic['requests'] > threshold:
      anomalies.append({
        'timestamp': traffic['timestamp'],
        'severity': 'high',
        'reason': 'Unusual spike in network requests'
      })
  return anomalies
```

In addition to detection, AI agents are integral to automated incident response. Once a threat is identified, these agents can execute predefined actions to contain and mitigate the impact. For example, an agent might isolate a compromised device from the network, block malicious IP addresses, or terminate suspicious processes.

By automating these responses, AI agents significantly reduce the time between detection and action, minimizing potential damage. Their ability to follow complex playbooks ensures coordinated responses, addressing multifaceted threats effectively. Furthermore, by prioritizing high-risk incidents, AI agents enable security teams to focus their attention on the most critical issues, improving overall efficiency.

Let us see the following code snippet:

```
class IncidentResponder:
  def __init__(self):
    self.playbooks = self.load_playbooks()
  async def respond_to_incident(self, incident):
    severity = self.assess_severity(incident)
    playbook = self.select_playbook(incident.type)
     if severity == 'critical':
       await self.isolate_affected_systems(incident.affected_ips)
       await self.notify_security_team(incident)
     for step in playbook:
       success = await self.execute_step(step)
       if not success:
         await self.escalate_to_human()
         break
```

Another vital function of AI agents is proactive risk mitigation. These agents analyze historical data and current system configurations to identify vulnerabilities and anticipate potential attack vectors. For instance, they can flag outdated software or misconfigured devices that could serve as entry points for attackers. By addressing these weaknesses proactively, organizations can prevent many incidents before they occur. AI agents also use predictive analytics to forecast potential threats, such as identifying an increase in phishing campaigns targeting a specific sector. This forward-looking approach allows organizations to strengthen their defenses preemptively, reducing the likelihood of successful attacks.

Continuous monitoring and learning are perhaps the most transformative capabilities of AI agents. They operate around the clock, analyzing vast amounts of data in real time to provide ongoing protection. Unlike traditional systems, AI agents continuously improve their performance by learning from new data and past incidents. This learning process often involves reinforcement learning, where the agent refines its decision-making based on feedback from the outcomes of its actions. By staying updated with the latest threat intelligence, AI agents adapt their detection and response strategies to counter emerging attack techniques, ensuring they remain effective even in rapidly changing environments.

Collaboration is another important aspect of AI agents in defensive security. In large organizations, multiple agents often work together to protect different segments of the network. These agents share information about detected threats, coordinate their responses, and collectively ensure comprehensive security coverage. Additionally, AI agents integrate seamlessly with existing security tools such as Security Information and Event Management (SIEM) systems, enhancing the overall security ecosystem. This collaboration extends to human analysts as well, with AI agents providing actionable insights and recommendations that enable teams to make informed decisions.

AI agents also can play a role in forensic analysis and root cause determination after an incident. They automate the process of examining logs, network activity, and system states to reconstruct the sequence of events leading to a breach. By identifying the vulnerabilities exploited and the tactics used by attackers, these agents provide organizations with a clearer understanding of their weaknesses. This insight is invaluable in preventing similar incidents in the future and improving overall resilience. Furthermore, by analyzing adversary behavior, AI agents can detect recurring patterns and predict potential future attacks, enhancing their defensive strategies.

Figure 7.2 shows a workflow that outlines the end-to-end process of threat detection and remediation in a cybersecurity environment, leveraging LLM AI agents at various stages. It begins with data collection from network devices or endpoints, which send logs to the SIEM system via frameworks like MELT.

MELT (Metrics, Events, Logs, Traces) is an observability framework that provides a holistic view of system performance by integrating four key telemetry data types. It enables organizations to gain deeper insights, detect issues proactively, and optimize complex distributed systems more effectively than traditional log aggregation approaches.

The SIEM aggregates and visualizes the logs, passing them to LLM Agent 1, which identifies potential threats. The XDR system further analyzes these threats and applies a threat model, leveraging LLM Agent 2 for continuous monitoring and log processing. Insights generated are sent to the SOAR platform, which orchestrates the response workflow. LLM Agent 3 executes remediation actions, either automatically or based on approval from a SOC analyst. The analyst may review, approve, or override the remediation actions before the SOAR system finalizes the

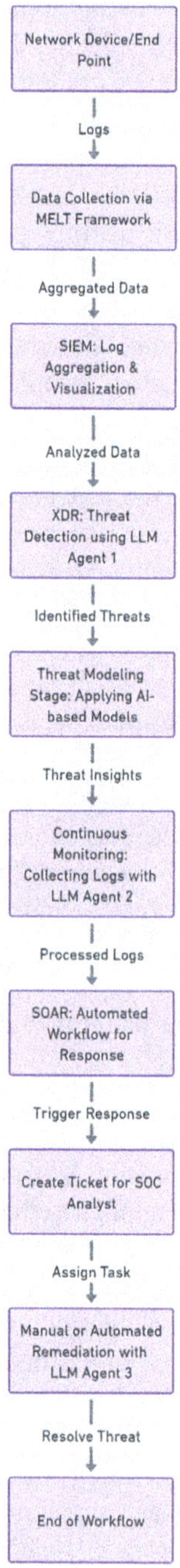

Fig. 7.2 End-to-end threat detection and remediation workflow using LLM AI agents

updates on the network devices, completing the threat management cycle efficiently and effectively.

AI Agent for Application Security
Imagine an AI agent embedded within an application. It doesn't just sit passively scanning logs; it learns the application's unique patterns, behaviors, and user interactions. Over time, it builds a dynamic baseline of "normal" operations. Now, when an anomaly—like an unusual API call or a sudden spike in data exfiltration—occurs, the AI agent doesn't wait for human intervention. It flags, isolates, or mitigates the threat in real time.

The real brilliance lies in the agent's adaptability. Unlike static rule-based systems, AI agents evolve. They learn from global threat intelligence, adapt to new attack vectors like zero-days, and even predict vulnerabilities by analyzing subtle code flaws or overlooked dependencies.

This dynamic defense also integrates seamlessly into modern DevSecOps workflows. Imagine an AI agent that doesn't just find vulnerabilities but proactively suggests secure code snippets or blocks insecure commits in real time. Beyond detection and prevention, these agents bring the ability to simulate attacks, enabling organizations to stress-test their defenses continuously.

But challenges remain. What happens if adversaries manipulate the AI itself? How do we ensure ethical decision-making in automated responses? These questions underscore the need to balance innovation with accountability.

7.2 Architectural Considerations for Defensive AI Agents

Architectural considerations for AI agents in defensive security focus on the structural and functional elements that enable these systems to operate effectively in complex and dynamic environments. These considerations encompass the design and integration of components, the choice of computational frameworks, and the ability to interact with other security systems (Fig. 7.3). A robust architecture ensures that AI agents can scale, be resilient, and adapt across diverse environments with the capability to detect and mitigate threats effectively.

At the core of any defensive AI system is the interplay between its fundamental components: sensors, data processing units, decision-making modules, and response execution mechanisms. Sensors act as the system's eyes and ears, continuously monitoring network traffic, system logs, and application behaviors to gather raw data. This data, often unstructured and voluminous, is then processed and analyzed by specialized units that use machine learning algorithms to identify patterns and anomalies. The decision-making module takes this analysis further, evaluating the potential risks and determining the most appropriate actions. Finally, the response execution mechanism implements the chosen actions, such as isolating compromised devices, blocking malicious IP addresses, or issuing alerts to security teams.

Fig. 7.3 Architectural considerations for AI agents in defensive security

Together, these components form a cohesive system capable of identifying and neutralizing threats autonomously.

A particularly innovative architectural approach in defensive security is the deployment of multi-agent systems (or MASs; see Chap. 3). Each agent in MASs may specialize in a specific function, such as monitoring a particular segment of the network, analyzing malware, or managing endpoint security. These agents communicate and share intelligence, enabling them to coordinate their responses to complex, multistage attacks. Multi-agent Systems are especially effective in large and distributed networks where a single centralized agent might struggle to manage the workload or have a complete view of the threat landscape.

The following is code example for Multi-agent System used in network defense:

```
class DefenseAgent:
    def __init__(self, zone, specialization):
        self.zone = zone  # e.g., 'network', 'endpoint', 'cloud'
        self.specialization = specialization
        self.shared_knowledge = []
    def share_threat_intel(self, other_agents):
        for agent in other_agents:
            if self.detect_threat():
                threat_data = self.get_threat_details()
                agent.receive_intel(threat_data)
    def receive_intel(self, threat_data):
        self.shared_knowledge.append(threat_data)
        self.update_defense_strategy()
```

The use of Multi-agent Reinforcement Learning (MARL) enhances the capabilities of such systems. In MARL, each agent learns to perform its task autonomously while considering the actions and goals of other agents in the network. This collaborative learning process enables the agents to optimize their collective performance, ensuring that they work together effectively to address threats. For example, one agent detecting an intrusion can alert others to adjust their monitoring focus or initiate pre-emptive actions, creating a unified and dynamic defense strategy. The decentralized nature of MARL also makes it more resilient, as the failure of one agent does not compromise the entire system.

We need to keep in mind that in MARL systems, agents must continuously share and process information to coordinate effectively, which can lead to communication bottlenecks, especially in large-scale networks. Additionally, ensuring that agents do not conflict in their decision-making processes can be challenging, requiring sophisticated algorithms to resolve potential conflicts and optimize collective performance.

Integration with existing security infrastructure is another critical architectural consideration. AI agents must be designed to work seamlessly with traditional tools such as firewalls, intrusion detection systems (IDS), Security Information and Event Management (SIEM) platforms, and endpoint detection and response (EDR) solutions. This integration ensures that the AI agents enhance, rather than replace, existing defenses, creating a layered security approach. For instance, an AI agent might analyze data from a SIEM platform to identify patterns that indicate a potential threat and then use its decision-making capabilities to recommend or implement a response.

Scalability is a key architectural requirement for AI agents in defensive security. Modern organizations operate increasingly large and complex networks, often incorporating cloud environments, Internet of Things (IoT) devices, and remote workstations. An effective AI architecture must be able to scale to accommodate this complexity without significant degradation in performance. Cloud-based AI agents, for instance, offer scalability by leveraging distributed computing resources to process vast amounts of data and coordinate actions across geographically dispersed networks.

Adaptability is equally important, as the threat landscape is constantly evolving. AI agents must be able to generalize their learning to new and unforeseen attack vectors. This requires an architectural design that supports continuous learning and model updates. Reinforcement learning algorithms play a vital role here, enabling AI agents to refine their decision-making processes based on feedback from their actions and outcomes. Furthermore, architectural flexibility allows AI agents to incorporate new threat intelligence and adapt their strategies without requiring complete system overhauls.

Let us see the following code snippet which illustrate the idea:

```
def train_defense_agent(environment, episodes=1000):
    agent = DefenseAgent(state_size=100, action_size=10)
    for episode in range(episodes):
        state = environment.reset()
        done = False
        while not done:
            # Choose action based on current security state
            action = agent.choose_action(state)
            # Take action and observe results
            next_state, reward, done = environment.step(action)
            # Learn from the outcome
            agent.learn(state, action, reward, next_state)
            state = next_state
            # Update defense strategy based on new knowledge
            if reward < 0:  # Attack successful
                agent.update_strategy()
```

Another significant aspect of architecture is the collaboration between AI agents and human analysts. While AI agents are designed to operate autonomously, human oversight remains essential, particularly for handling complex or high-stakes incidents. Architectural frameworks should therefore include mechanisms for human-in-the-loop systems, where analysts can review and approve critical decisions, provide feedback to improve the agents' performance, and intervene when necessary. For instance, a decision support interface might present the AI agent's analysis, suggested actions, and their potential impact, enabling security teams to make informed choices.

Finally, the architectural design must prioritize security and resilience. AI agents themselves can become targets for attackers seeking to exploit their vulnerabilities or disrupt their operations. A robust architecture incorporates safeguards against such threats, including secure communication channels, access controls, and mechanisms to detect and recover from compromises. By ensuring the integrity and reliability of the AI agents, the architecture strengthens the overall security posture of the organization. We will talk about agent security in detail in the final chapter of this book.

7.3 Capabilities and Benefits of AI Agents in Defensive Security

AI agents offer transformative capabilities in defensive security, providing several critical benefits that enhance the efficiency, adaptability, and effectiveness of modern cybersecurity strategies. These benefits arise from the unique capabilities of AI, including advanced data analysis, autonomous operation, and continuous learning, which collectively address many of the limitations faced by traditional security tools and methods (Fig. 7.4). By leveraging these capabilities, AI agents are poised to redefine how organizations protect their systems, data, and networks.

One of the most significant advantages of AI agents in defensive security is their ability to detect threats with enhanced accuracy and speed. For example, they can detect zero-day exploits or advanced persistent threats (APTs) that bypass conventional defenses. This capability not only reduces the likelihood of successful attacks but also minimizes false positives, which can overwhelm security teams and lead to alert fatigue.

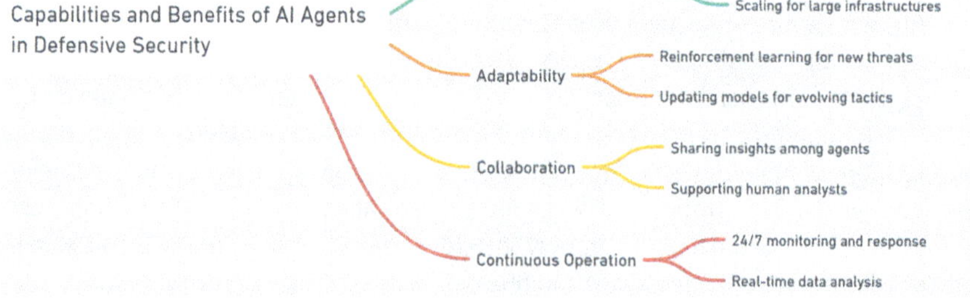

Fig. 7.4 Capabilities and benefits of AI agents in defensive security

The proactive risk mitigation provided by AI agents represents another key benefit. AI agents are designed to predict and prevent potential threats before they materialize. They achieve this by continuously monitoring system behavior, analyzing historical data, and identifying patterns that could indicate vulnerabilities or emerging risks. For instance, an AI agent might recognize that certain configurations in a network are prone to exploitation and recommend or implement changes to address these weaknesses. This proactive approach not only strengthens the overall security posture but also reduces the need for constant manual oversight by human analysts.

Operational efficiency and scalability are critical advantages of AI agents, particularly for organizations with large or complex infrastructures. AI agents automate many repetitive and labor-intensive tasks, such as monitoring network traffic, analyzing logs, and responding to routine threats (Oesch et al., 2024). This automation frees up human analysts to focus on more strategic and complex issues, such as developing new security policies or addressing high-priority incidents (Foley et al., 2022). Additionally, AI agents can scale to handle the growing demands of modern networks, including cloud environments, Internet of Things (IoT) devices, and remote work setups. Their ability to process vast amounts of data in parallel and adapt to dynamic network conditions ensures comprehensive protection across diverse environments.

Adaptability and learning are central to the effectiveness of AI agents. In the ever-evolving threat landscape, attackers continuously develop new techniques to bypass security measures. AI agents address this challenge by using reinforcement learning and other adaptive techniques to refine their detection and response strategies (Li et al., 2023). For example, when encountering a novel malware variant, an AI agent can analyze its behavior and update its models to recognize similar threats in the future. This ability to learn from new data ensures that AI agents remain effective even as attack methods evolve, reducing the need for frequent manual updates and reconfigurations.

Another critical benefit of AI agents is their ability to reduce human error, a common factor in many cybersecurity incidents. Human analysts, due to workload pressures or cognitive limitations, may overlook subtle signs of an attack or make mistakes in configuring security systems. AI agents, by automating routine tasks and providing decision support, help mitigate these risks (Kott, 2023). For instance, an AI agent can analyze complex datasets to identify correlations that might be missed by human analysts or provide clear recommendations for addressing specific vulnerabilities. By augmenting human capabilities, AI agents improve the accuracy and reliability of cybersecurity operations (Wang & Dechene, 2024).

The collaborative nature of AI agents also enhances defensive strategies. In complex environments, multiple AI agents often work together, sharing information and coordinating their responses to address multifaceted threats. This collaboration is particularly effective in large networks, where different segments may face distinct challenges. For example, one AI agent monitoring a cloud environment might detect suspicious activity and alert another agent responsible for endpoint security, ensuring a coordinated and timely response (Oesch et al., 2024). Additionally, AI agents can integrate seamlessly with existing security tools, such as intrusion detection systems (IDS) and Security Information and Event Management (SIEM) platforms, creating a cohesive and layered defense.

The ability of AI agents to operate continuously is a significant advantage in the context of modern cybersecurity. Cyberattacks can occur at any time, and the ability to maintain round-the-clock vigilance is essential for effective defense (Kott, 2023). Unlike human analysts, who require rest and are subject to fatigue, AI agents can monitor systems 24/7 without interruption. This continuous operation ensures that threats are detected and addressed promptly, even outside of regular working hours. Moreover, their ability to analyze data in real time allows for instantaneous responses, minimizing the window of opportunity for attackers.

AI agents also contribute to improved threat intelligence, which is essential for understanding and countering adversary tactics. By analyzing data from multiple sources, including network traffic, system logs, and external threat feeds, AI agents can generate actionable insights into attacker behavior and strategies. These insights not only inform immediate responses but also help organizations develop more effective long-term security measures. For instance, an AI agent might identify trends in phishing campaigns targeting specific industries, enabling organizations to implement targeted defenses and training programs (Oesch et al., 2024).

7.4 Architectural Considerations in Deploying AI Agents

The architecture of an AI-driven defensive security system defines how its components interact, how it processes and responds to threats, and how it integrates with existing cybersecurity infrastructures. A well-designed architecture ensures scalability, adaptability, and reliability while addressing the evolving threat landscape (see Fig. 7.5).

At the core of AI-driven security architecture is the modularity of its components. The primary elements include sensing mechanisms, data processing and analysis units, decision-making modules, and response execution systems. The sensing mechanisms are responsible for monitoring the environment and collecting raw data from various sources, such as network traffic, system logs, and endpoint activity. These sensors act as the eyes and ears of the system, ensuring that all relevant inputs are captured for analysis. The collected data, often vast and unstructured, is processed using AI algorithms to identify patterns and anomalies indicative of potential threats.

The decision-making module is a critical aspect of the architecture, as it determines the actions the system should take based on its analysis of the data. This module relies heavily on advanced AI techniques, including machine learning and reinforcement learning, to evaluate risks and formulate responses. For example, when an anomaly is detected, the decision-making module assesses its severity, potential impact, and the likelihood of being a true threat. The system then selects the most appropriate course of action, whether it involves blocking a suspicious connection, isolating an endpoint, or escalating the issue to human analysts.

Fig. 7.5 Architectural considerations in deploying AI agents for defensive security

The response execution systems implement the decisions made by the AI agent. These systems can perform a variety of actions, such as quarantining malicious files, reconfiguring firewall rules, or initiating automated incident response playbooks. The execution of these actions must be precise and timely, as delays or errors could compromise the effectiveness of the defense. Additionally, response mechanisms often include feedback loops that allow the system to refine its strategies based on the outcomes of its actions. For instance, if an automated response fails to neutralize a threat, the system learns from this failure and adjusts its approach for future incidents.

Multi-agent architectures are an advanced design feature that is particularly relevant for large and distributed environments. In such architectures, multiple AI agents work together, each with specialized roles. For example, one agent might focus on monitoring cloud-based resources, while another manages endpoint protection in an on-premises network. These agents collaborate by sharing insights and coordinating their responses, creating a unified defense strategy. Multi-agent Reinforcement Learning (MARL) enhances this approach by enabling agents to learn not only from their individual experiences but also from their interactions with other agents. This collective learning capability ensures that the system remains effective in addressing sophisticated, multifaceted attacks.

Integration with existing cybersecurity infrastructure is another vital architectural consideration. AI agents must coexist with traditional tools such as intrusion detection systems (IDS), Security Information and Event Management (SIEM) platforms, and Endpoint Detection and Response (EDR) solutions. This integration ensures that the AI system complements and enhances existing defenses rather than replacing them. For instance, an AI agent might analyze data generated by a SIEM platform to identify patterns indicative of an attack and then take preemptive actions based on its findings. Such integration also allows for a layered security approach, combining traditional and AI-driven defenses to address threats from multiple angles.

Scalability is a key requirement for AI agent architectures, particularly given the growing size and complexity of modern networks. As organizations adopt cloud computing, Internet of Things (IoT) devices, and remote work setups, their attack surfaces expand significantly. AI agents must be able to scale to accommodate this complexity, processing larger volumes of data and managing diverse network environments without degradation in performance. Cloud-based architectures offer a scalable solution by leveraging distributed computing resources to handle the increased demands of monitoring and analysis.

Adaptability is equally important in the architecture of AI agents. Cyber threats evolve rapidly, and attackers continuously develop new techniques to bypass defenses. AI agents must therefore be designed to learn and adapt in real time. This adaptability is achieved through mechanisms such as reinforcement learning, which enables agents to refine their decision-making processes based on the outcomes of their actions. Additionally, the architecture should support frequent updates to threat intelligence databases and machine learning models, ensuring that the system remains effective against emerging attack vectors.

Collaboration between AI agents and human analysts is another critical consideration in the architectural design. While AI agents are capable of operating autonomously, human oversight remains essential, particularly for complex or high-stakes incidents. The architecture should therefore include interfaces that facilitate human–AI interaction, such as dashboards that provide detailed explanations of the agent's decisions and suggested actions. These interfaces allow analysts to review and approve critical responses, provide feedback to improve the system's performance, and intervene when necessary. This collaboration ensures that the strengths of AI and human expertise are combined effectively, enhancing the overall security posture.

Resilience and security of the AI system itself are crucial architectural aspects. AI agents can become targets for attackers seeking to disrupt their operations or exploit vulnerabilities. A robust architecture incorporates safeguards such as secure communication channels, access controls, and anomaly detection systems to protect the integrity of the AI agents. Additionally, the architecture should include fail-safe mechanisms that ensure continuity of operations even if an agent is compromised or disabled.

7.5 Case Studies and Applications

This section explores notable implementations of AI agents, their real-world applications, and lessons drawn from successful deployments. These examples showcase how organizations are leveraging AI-driven solutions to enhance their cybersecurity posture, improve operational efficiency, and address evolving threats.

7.5.1 Real-World Implementations

1. Dropzone AI

Dropzone AI is a cybersecurity platform that utilizes autonomous AI agents to enhance the capabilities of Security Operation Centers (SOCs). These AI agents are designed to replicate the decision-making processes of expert human analysts, allowing for continuous monitoring and investigation of security alerts around the clock (Wu, 2024).

Key Features of Dropzone AI

- Autonomous Investigations: The AI agents conduct end-to-end investigations of alerts independently, eliminating the need for playbooks, coding, or chat prompts.
- Broad Coverage: Dropzone can handle a wide range of alert types, including those related to cloud security, network threats, identity management, endpoint protection, and phishing attacks.
- Efficiency: By automating routine tasks, the platform significantly reduces the workload for human analysts, decreasing investigation time from 5–40 min per alert to just 3 min for report reviews.
- Quick Deployment: The system can be set up in approximately 30 minutes, making it easy to integrate into existing security frameworks.
- Integration: Dropzone features built-in integrations with popular security tools, enhancing its functionality and ease of use within diverse security environments.

2. Darktrace

Darktrace is one of the most well-known AI-driven cybersecurity platforms, employing machine learning to deliver autonomous detection and response capabilities. Its core technology, the Enterprise Immune System, mimics the human immune system by learning the "pattern of life" for each device, user, and network within an organization. This allows it to detect deviations that could indicate cyber threats, including zero-day exploits and insider threats. Darktrace's Antigena module autonomously responds to detected threats in real time, neutralizing malicious activities while minimizing disruption to business operations. For example, in one documented case, Darktrace detected a zero-day exploit targeting a financial services firm. The AI agent identified abnormal network behavior that bypassed traditional firewalls, allowing Darktrace's Antigena module to autonomously isolate the affected systems, preventing data exfiltration and minimizing downtime (Darktrace, 2024).

3. Microsoft Security Copilot

Security Copilot integrates generative AI models with Microsoft's existing cybersecurity tools to provide real-time insights and assist human analysts. By processing vast amounts of threat data, Security Copilot can identify potential vulnerabilities, prioritize alerts, and suggest actionable responses. One notable feature is its ability to translate complex cybersecurity data into easily understandable summaries, facilitating collaboration between security teams and decision-makers (Microsoft, 2024).

4. Seven AI

Seven AI focuses on automating threat detection and investigation workflows, addressing the resource constraints faced by Security Operations Centers (SOCs). By employing reinforcement learning, Seven AI agents adapt to evolving threats and improve their decision-making over time. In a pilot deployment at a retail chain, Seven AI agents monitored point-of-sale (POS) systems for unusual activity, detecting and mitigating a malware infection designed to steal credit card information. For more details, please see https://7ai.com.

5. Ghost Security

Ghost Security is an application security company focused on automating large aspects of the traditional Application Security (AppSec) analysts tasks. This looks to address both talent shortages and challenges as well as mounting vulnerability backlogs that organizations deal with from applications. Agentic AppSec can involve proactive risk identification of vulnerabilities in code bases as well as running applications and even facilitate autonomous remediation of discovered vulnerabilities, driving down vulnerability backlogs and minimizing attack surface across applications to prevent malicious exploitation.

6. Nvidia Morpheus AI Framework

Nvidia Morpheus is a state-of-the-art AI framework designed to enhance cybersecurity through advanced machine learning and real-time data processing. Morpheus enables organizations to swiftly analyze vast amounts of data, detect sophisticated threats, and automate response mechanisms. The framework integrates seamlessly with large language models (LLMs) and intelligent agents, allowing for more nuanced threat detection and proactive defense strategies. Morpheus provides developers and security professionals with modular architecture and comprehensive APIs with the tools necessary to build scalable, AI-driven security solutions.

7.5.2 Tips for Successful Deployments

Effectiveness: A common theme across successful deployments of AI agents is their ability to enhance detection accuracy and response speed. By analyzing large volumes of data in real time, these agents can identify threats that might be missed by traditional tools or human analysts. For example, Darktrace's ability to detect subtle anomalies has proven critical in identifying early stages of cyberattacks. At the same time, Microsoft Security Copilot's contextual analysis improves the accuracy of its alerts, reducing false positives and alert fatigue.

Scalability: The scalability of AI agents is a key factor in their success. Solutions like Darktrace and Seven AI demonstrate that AI agents can be effectively deployed across organizations of varying sizes and industries. These agents can adapt to different network topologies, device types, and operational requirements without significant reconfiguration, making them versatile tools for diverse environments. For example, Seven AI's deployment in retail settings highlighted its ability to scale from monitoring individual POS terminals to securing large distributed networks.

- Automation and Efficiency: Automation is central to the success of AI agents, particularly in addressing the cybersecurity skills gap and resource constraints. By taking over repetitive tasks such as log analysis, alert prioritization, and initial incident response, AI agents free human analysts to focus on strategic and complex issues.

Adaptability: AI agents' ability to learn and adapt to new threats is another critical factor. Solutions like Darktrace and Seven AI leverage machine learning and reinforcement learning to continuously improve their threat detection and response capabilities.

Challenges and Considerations: Despite their successes, deployments of AI agents have also highlighted challenges such as integration with existing security systems and the need for explainability. Organizations deploying AI agents often need to invest in training and collaboration to ensure that human analysts can effectively interpret and act on the agents' outputs. Addressing these challenges requires focusing on user-friendly interfaces, transparent decision-making processes, and robust integration frameworks.

While there is tremendous potential for AI agents in cyber defense, it is critical that organizations also consider any AI agents and their operations in systems and environments as an additional new attack vector. Organizations have historically struggled with fundamental security practices such as identity and access management (IAM), least-permissive access control, credential compromise, and more. If AI agents are not deployed, implemented, and maintained securely, they could be used maliciously against organizations and exploited as part of attacks.

7.6 Training and Testing Environments for Defensive AI Agents

Training environments for autonomous cyber defense agents are critical for preparing these systems to operate effectively in real-world scenarios. These environments serve as virtual testing grounds where AI agents can learn, adapt, and refine their capabilities in a controlled setting before being deployed to protect actual networks and systems. A well-designed training environment ensures that AI agents can generalize their learning to diverse and dynamic operational contexts, addressing both the technical and practical challenges of defensive cybersecurity.

The effectiveness of a training environment lies in its realism. AI agents must be exposed to conditions that closely mimic the complexities of real-world networks. This includes accurately simulating network topologies, traffic patterns, code bases, and system behaviors, as well as replicating the types of attacks they are likely to encounter (Oesch et al., 2024). For example, a high-fidelity environment might simulate a corporate network with multiple subnets, cloud resources, and IoT devices, allowing agents to practice detecting and mitigating threats across different segments. Additionally, AI agents can be exposed to test and even production code bases and applications to determine the presence of vulnerabilities in code, as well as their ability to be exploited in a runtime environment much like an attacker would interact with them, a topic we discussed in our previous chapter on AI agents in offensive security. Realism also extends to the adversary models used in training, which must represent a wide range of attack techniques, from phishing and ransomware to advanced persistent threats (APTs).

A key consideration in designing training environments is the balance between simulation and emulation. Simulation involves creating a virtual representation of a network using software models, which allows for flexibility and scalability at a lower cost. However, simulations may lack the nuances and unpredictability of real-world systems. Emulation, on the other hand, uses actual hardware and software to replicate operational environments with greater fidelity. While emulation provides a more accurate training experience, it is often more resource-intensive and challenging to scale. Effective training environments often combine simulation and emulation to achieve a balance between realism and practicality. There is also the potential to use "digital twins," or environments that digitally represent their physical counterparts, allowing for training on digitally emulated replicas of physical environments and systems that the AI agents will eventually interact with.

Several established platforms exemplify the design and purpose of training environments for autonomous cyber agents. CybORG (Cyber Operations Research Gym) is an open-source platform that provides a simulated network environment where agents can practice defending against various attack scenarios. It offers a controlled setting for reinforcement learning, enabling agents to learn optimal strategies through iterative trial and error. Another platform, the Cyber Autonomy Gym for Experimentation (CAGE), is designed as a competitive framework where researchers develop and test AI agents under standardized conditions, fostering innovation and benchmarking progress. CyGIL (Cyber Gaming and Intelligence Learning), which bridges simulation and emulation, provides a hybrid environment that allows agents to transfer their learning from virtual to real-world settings (Oesch et al., 2024).

Table 7.1 summarizes the strengths and weaknesses of the mentioned platforms (CybORG, CAGE, CyGIL) to enhance clarity.

Training environments must also support adaptability to evolving threats and technologies. As attackers constantly develop new tactics, techniques, and procedures (TTPs), training scenarios must be updated to reflect these changes. This requires modular and extensible platforms that can incorporate new attack vectors, vulnerabilities, and defense mechanisms without extensive reconfiguration (Wang & Dechene, 2024). For instance, an environment might simulate an emerging ransomware strain or test an agent's response to a supply chain attack. This dynamic approach ensures that AI agents remain relevant and effective in addressing current and future challenges.

Standardization is another important aspect of training environments. Consistent metrics and benchmarks are essential for evaluating the performance of AI agents and comparing different designs. For example, metrics might include detection accuracy, false positive rates, response times, and the agent's ability to mitigate the impact of attacks (Kott, 2023). Standardized environments like CAGE help establish common evaluation criteria, enabling researchers and developers to identify best practices and accelerate advancements in the field.

The scalability of training environments is critical for accommodating the increasing complexity and diversity of modern networks. As organizations adopt cloud computing, IoT devices, and hybrid infrastructures, training environments must scale to replicate these conditions. Cloud-based platforms are particularly well-suited for this purpose, as they provide the computational resources necessary to simulate large-scale networks and process vast amounts of data. For example, an AI agent training on a cloud platform might analyze millions of log entries, simulate interactions among thousands of devices, and test its responses to coordinated attacks across distributed systems.

Collaboration and competition are valuable features of advanced training environments. Multi-agent systems, where multiple AI agents train together, foster collaboration and the development of coordinated defense strategies. For example, one agent might specialize in detecting network anomalies, while another focuses on

Table 7.1 Comparison of some training environments for autonomous cyber agents

Platform	Strengths	Weaknesses
CybORG	Open-source, suitable for reinforcement learning, customizable attack scenarios	Limited to simulated environments; lacks full real-world emulation
CAGE	Competitive framework, standardized benchmarking	Resource-intensive, requires expertise to configure
CyGIL	Hybrid approach with simulation and emulation, supports real-world transfer learning	Higher setup complexity, scalability challenges

endpoint protection. Training in such environments allows agents to learn how to share information and synchronize their responses, improving overall effectiveness. Competitive environments, such as red team/blue team exercises, simulate adversarial scenarios where one set of agents attempts to breach the system while another defends it. This approach not only sharpens the agents' skills but also provides insights into potential weaknesses in their strategies.

Developing high-fidelity training environments presents several challenges. Creating accurate and comprehensive simulations of real-world networks and attacks requires significant expertise and resources. Scalability can be a constraint, especially when emulating large networks with diverse devices and configurations. Consistency across different environments and evaluation metrics is also critical to ensure that agents trained in one setting can perform effectively in another. Addressing these challenges requires ongoing research and collaboration among cybersecurity experts, AI developers, and organizations.

7.7 Future Trends and Developments

This section explores emerging techniques, potential innovations, and the increasing collaboration between AI systems and human analysts, all of which are expected to define the trajectory of defensive AI.

Figure 7.6 gives an outline of the trends and developments.

7.7.1 *Emerging Techniques in Defensive AI*

Several advanced AI methodologies are gaining traction in cybersecurity, promising to enhance the adaptability and effectiveness of defensive systems. These include adversarial training, meta-learning, and transfer learning.

Adversarial training involves exposing AI agents to adversarial examples—inputs intentionally designed to deceive machine learning models—during the

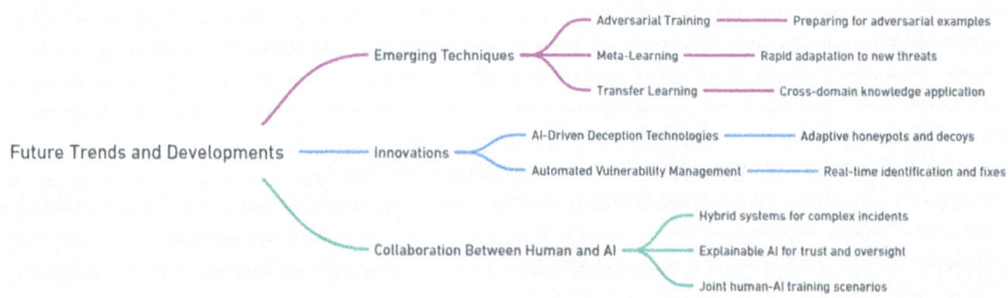

Fig. 7.6 Future trends and developments

training process. This approach enhances the resilience of AI systems by preparing them to handle attempts by attackers to exploit vulnerabilities in their algorithms. In cybersecurity, adversarial training can improve an agent's ability to detect obfuscated malware, bypassing traditional pattern-matching limitations. By proactively testing AI systems against potential adversary tactics, organizations can reduce their susceptibility to AI-specific exploits.

Meta-learning, or "learning to learn," is another transformative technique. Meta-learning algorithms enable AI agents to quickly adapt to new tasks with minimal data. In the context of defensive security, this means AI agents can generalize their learning to detect and mitigate novel threats, such as zero-day vulnerabilities, without requiring extensive retraining (Li et al., 2023). For instance, an AI agent trained on phishing attack data from one industry can rapidly adjust to similar attacks targeting a different sector.

Transfer learning leverages knowledge gained from one domain or task and applies it to another, reducing the need for large-scale retraining. In cybersecurity, transfer learning is particularly useful for organizations with limited access to labeled threat data. For example, a model trained on large datasets of malware behavior could be fine-tuned to detect specific threats in smaller, specialized networks. These techniques collectively enhance the flexibility and efficiency of AI-driven defense systems, enabling them to stay ahead in an ever-changing threat landscape.

7.7.2 *Potential Innovations*

AI-driven deception technologies and automated vulnerability management represent two key areas of innovation poised to revolutionize defensive security.

AI-driven deception technologies involve using sophisticated decoys, honeypots, and fake environments to mislead attackers and gather intelligence about their tactics. Unlike traditional honeypots, AI-enhanced deception systems dynamically adapt to the attacker's behavior, creating convincing simulations that prolong engagement and collect actionable data. For instance, an AI-driven deception system could simulate a database containing fabricated sensitive information, leading attackers to expose their methods and tools without affecting real assets. This approach improves threat intelligence and buys time for defenders to respond effectively. While AI-driven deception technologies are effective at misleading attackers, there is a risk that legitimate users or security personnel could interact with decoys, leading to confusion or operational disruption. Ethical considerations include ensuring transparency about the use of deception and implementing safeguards to minimize the risk of accidental engagement by authorized users.

Here is the code to illustrate the idea:

```
class HoneypotSystem:
  def __init__(self):
    self.decoys = {}
    self.attacker_activities = []
    def create_dynamic_decoy(self, template):
    decoy = {
      'type': template['type'],
      'services': self.generate_fake_services(),
      'data': self.generate_fake_data(),
      'behavioral_patterns': self.simulate_user_activity()
    }
    return decoy
    def monitor_attacker_behavior(self, decoy_id):
    if self.decoys[decoy_id].is_accessed():
      self.record_activity()
      self.adapt_decoy_behavior()
      self.alert_security_team()
```

Automated vulnerability management leverages AI to identify, prioritize, and remediate vulnerabilities in systems and networks. Traditional vulnerability management often relies on periodic scans and manual prioritization, which can be time-consuming and prone to oversight. AI, by continuously monitoring system configurations and threat intelligence feeds, can detect vulnerabilities in real time and recommend or implement fixes. For example, an AI agent might identify that an outdated software version is susceptible to a new exploit and automatically deploy a patch or reconfigure system permissions to mitigate the risk. Such automation reduces the window of exposure and alleviates the burden on human security teams.

7.7.3 Collaboration between Human Analysts and AI

The future of cybersecurity lies in hybrid systems that seamlessly combine the strengths of AI technologies with human expertise. While AI excels at processing vast amounts of data and identifying patterns, human analysts bring contextual understanding, ethical judgment, and strategic decision-making capabilities. This collaboration is essential for addressing complex, high-stakes incidents and ensuring that AI-driven actions align with organizational goals and values.

In hybrid systems, AI serves as an assistant to human analysts, automating routine tasks and providing decision support. For example, AI agents can triage alerts, prioritize incidents, and suggest actionable responses, allowing analysts to focus on investigating sophisticated threats. Furthermore, explainable AI (XAI) techniques are being developed to improve the transparency of AI decisions, enabling analysts to understand and trust the reasoning behind an agent's recommendations.

Joint training scenarios are another area of development, where human analysts and AI agents learn together in simulated environments. Reinforcement learning from human feedback (RLHF) is a promising approach, allowing analysts to guide AI agents during training by providing input on optimal responses. This collaborative learning process not only improves the agent's performance but also fosters trust and synergy between human and machine.

The integration of AI into Security Operations Centers (SOCs) exemplifies the potential of these hybrid systems. In AI-augmented SOCs, AI agents handle data-intensive tasks, such as log analysis and anomaly detection, while analysts oversee strategy and decision-making. This partnership enhances operational efficiency, reduces fatigue, and ensures a more comprehensive response to cyber threats. We can use the following code as a starting point for this kind of integration.

We can start by defining the SecurityOrchestrator class:

```
class SecurityOrchestrator:

  def __init__(self):

    self.siem = SIEMConnector()

    self.firewall = FirewallManager()

    self.ids = IntrusionDetectionSystem()

   async def coordinate_defense(self, threat_alert):
```

Then, we can collect data from multiple sources.

```
siem_data = await self.siem.get_related_events(threat_alert)
        ids_logs = await self.ids.get_detection_logs()
```

We can then analyze and correlate data and invoke other necessary agents for further actions:

```
    risk_score = self.analyze_threat(siem_data, ids_logs)

        if risk_score > 0.8:

      await self.firewall.block_ip(threat_alert.source_ip)

      await self.isolate_affected_systems()

      await self.initiate_investigation()
```

Innovations such as AI-driven deception technologies and automated vulnerability management may provide approaches on how organizations defend against sophisticated threats. At the same time, the growing collaboration between human analysts and AI systems underscores the importance of hybrid approaches that combine automation with human oversight and judgment. As these trends continue to evolve, they will redefine the capabilities and strategies of defensive security.

7.8 Summary

The chapter covers the implementation of AI agents in defensive security, starting with their core functions in threat detection, automated incident response, and proactive risk mitigation. It details architectural considerations, emphasizing the importance of modularity, scalability, and integration with existing security infrastructure. The discussion extends to real-world applications through case studies of platforms like Dropzone AI, Darktrace, and Microsoft Security Copilot, highlighting practical implementations and lessons learned.

The text explores various capabilities and benefits of AI agents, including enhanced accuracy in threat detection, operational efficiency, and continuous adaptation to new threats. Challenges in deployment are addressed, focusing on integration issues, training requirements, and the need for balanced human oversight. The chapter concludes with an examination of future trends, including emerging techniques like adversarial training and meta-learning, and the evolving relationship between AI systems and human analysts.

7.9 Questions

1. How do AI agents in defensive security differ from traditional security measures in their approach to threat detection?
2. Explain the role of Multi-agent Systems (MAS) in enhancing cybersecurity defense capabilities.
3. What are the key architectural considerations when implementing AI agents for defensive security?
4. How does continuous monitoring and learning contribute to the effectiveness of AI security agents?
5. Discuss the challenges organizations face when integrating AI agents with existing security infrastructure.
6. What role does reinforcement learning play in improving the adaptability of defensive AI agents?
7. Compare and contrast the approaches used by Darktrace and Microsoft Security Copilot in implementing AI-driven security solutions.
8. How do AI agents contribute to proactive risk mitigation in cybersecurity?
9. Explain the importance of scalability in AI agent architectures for defensive security.
10. What are the key components of effective training environments for defensive AI agents?
11. How do AI-driven deception technologies enhance an organization's security posture?
12. Discuss the balance between autonomous operation and human oversight in AI-driven security systems.

13. What role does meta-learning play in improving the effectiveness of defensive AI agents?
14. What are the potential future developments in AI-driven defensive security, and how might they impact current practices?

References

Darktrace. (2024). Detecting the unknown: Revealing uncategorised ransomware using Darktrace. *Darktrace*. Retrieved from https://darktrace.com/blog/detecting-the-unknown-revealing-uncategorised-ransomware-using-darktrace

Foley, M., Wang, M., Zoe, M., Hicks, C., & Mavroudis, V. (2022). Inroads into autonomous network defence using explained reinforcement learning. *CEUR Workshop Proceedings*.

Kott, A. (2023). Chapter 1. Autonomous intelligent cyber-defense agent: Introduction and overview. In A. Kott (Ed.), *Autonomous intelligent cyber defense agent (AICA): A comprehensive guide*. Springer.

Li, L., El Rami, J.-P. S., Kerr, R., Taylor, A., & Vandenberghe, G. (2023). Towards autonomous cyber operation agents: Exploring the red case. In *Proceedings of the 17th International Conference on Availability, Reliability and Security*.

Microsoft. (2024). *Microsoft security copilot*. Retrieved from https://www.microsoft.com/en-us/security/business/ai-machine-learning/microsoft-security-copilot

Oesch, S., Austria, P., Chaulagain, A., Weber, B., Watson, C., Dixson, M., & Sadovnik, A. (2024). The path to autonomous cyber defense. Sean Oesch. *IEEE Security & Privacy, 23*, 38.

Wang, M., & Dechene, R. (2024). *Multi-agent actor-critics in autonomous cyber defense*. ArXiv.

Wu, E. (2024). *Dropzone AI: Autonomous AI agents for cybersecurity*. Retrieved from https://www.dropzone.ai/

Jerry Huang is currently an AI Engineer at Google, where he contributed to the AI/ML evaluation pipeline for a consumer-facing application. Before Google, he worked as a technical and security staff member at several prominent technology companies, gaining experience in areas like security, AI/ML, and scalable systems.

At Metabase, an open-source business intelligence platform, Jerry contributed features such as private key management and authentication solutions. As a Software Engineer at Glean, a Generative AI search startup, he was one of three engineers responsible for managing large-scale GCP infrastructure powering text summarization, autocomplete, and search for over 100,000 enterprise users. During his time at TikTok, Jerry helped design and build custom RPCs to model access control policies. At Roblox, he served as a Machine Learning/Software Engineering Intern, focusing on real-time text generation models and gathering a large multilingual corpus that significantly boosted model robustness.

In addition to his industry experience, Jerry has conducted extensive security and biometrics research as a Research Assistant at Georgia Tech's Institute for Information Security & Privacy, resulting in a thesis on privacy-preserving biometric authentication.

Jerry holds a BS/MS in Computer Science from Georgia Tech and is currently pursuing an MS in Applied Mathematics at the University of Chicago.

Ken Huang is a prolific author and globally recognized authority in AI and Web3, with an extensive portfolio of published works that bridge business strategy, technical implementation, and cutting-edge research. As a Fellow of the Cloud Security Alliance and Co-Chair of the AI Safety Working Groups at the Cloud Security Alliance and the AI STR Working Group at the World

Digital Technology Academy under the UN Framework, he is a leading voice in shaping global AI governance and security standards.

Huang is the CEO and Chief AI Officer (CAIO) of DistributedApps.ai, a firm specializing in generative AI training and consulting. His contributions to the field include being a core contributor to the OWASP Top 10 Risks for LLM Applications and an active participant in the NIST Generative AI Public Working Group.

Notable Publications:

- *Beyond AI: ChatGPT, Web3, and the Business Landscape of Tomorrow* (Springer, 2023)—strategic insights into AI and Web3's business applications.
- *Generative AI Security: Theories and Practices* (Springer, 2024)—a comprehensive guide on securing generative AI systems.
- *Practical Guide for AI Engineers* (Volumes 1 and 2, DistributedApps.ai, 2024)—essential resources for AI and ML engineers.
- *The Handbook for Chief AI Officers: Leading the AI Revolution in Business* (DistributedApps.ai, 2024)—a roadmap for CAIOs in implementing GenAI across organizations.
- *Web3: Blockchain, the New Economy, and the Self-Sovereign Internet* (Cambridge University Press, 2024)—insights into the convergence of AI, blockchain, IoT, and emerging technologies.
- *Blockchain and Web3: Building the Cryptocurrency, Privacy, and Security Foundations of the Metaverse* (Wiley, 2023)—recognized as a must-read by TechTarget in 2023 and 2024.

Ken is a sought-after speaker and has presented at events such as the World Economic Forum in Davos, ACM and IEEE conferences, the CSA AI Summit, Depository Trust & Clearing Corporation forums, and World Bank conferences. His recent appointment to the OpenAI Forum reflects his ongoing commitment to advancing collaboration and dialogue in the field of AI.

Explore Ken's work on Amazon: https://www.amazon.com/author/kenhuang

Chris Hughes is the Co-founder and CEO of Aquia, a Cybersecurity consulting firm. Chris brings nearly 20 years of IT and cybersecurity experience to his role as Co-founder and CEO at Aquia.

Chris has also served as a Cyber Innovation Fellow (CIF) at the Cybersecurity Infrastructure and Security Agency (CISA), focusing on software supply chain security. Additionally, Chris advises various tech startups focused on areas such as Software Composition Analysis (SCA), Kubernetes Security, Non-Human Identities (NH), and AI Security.

As a United States Air Force veteran and former civil servant in the US Navy and the General Services Administration's FedRAMP program, Chris is passionate about making a lasting impact on his country and the global community.

In addition to his public service, Chris spent several years as a consultant within the private sector and currently serves as an adjunct professor for cybersecurity master's programs at the University of Maryland Global Campus. Chris participates in industry working groups, such as the Cloud Security Alliance's Incident Response and SaaS Security Working Group, and serves as the Membership Chair for Cloud Security Alliance D.C. He is the host and author of the Resilient Cyber.

Chris runs the Resilient Cyber Substack, where he shares a weekly newsletter, deep dive analysis, and interviews with industry leaders, as well as detailed articles on topics such as Cloud, Vulnerability Management, DevSecOps, Cybersecurity Leadership, and Market Dynamics.

Chris holds a B.S. in Information Systems, a M.S. in Cybersecurity, and an MBA. He regularly consults with IT and cybersecurity leaders from various industries to assist their organizations with their digital transformation journeys while keeping security a core component of that transformation.

Chris is co-author of the book, “Software Transparency: Supply Chain Security in an Era of a Software-Driven Society” and “Effective Vulnerability Management: Managing Risk in the Vulnerable Digital Ecosystem” both published by Wiley. He has also contributed many other thought leadership pieces on software supply chain security and has presented on the topic at a variety of industry

Chapter 8
AI Agents in Banking

Ken Huang, Daniel Wu, Jyoti Ponnapalli, and Grace Huang

The banking industry stands at the precipice of a profound transformation, propelled by the rapid advancement and adoption of AI agent technology. No longer a futuristic concept, AI agents, with their intricate, multicomponent architecture encompassing data management, language understanding, reasoning, tool use, self-improvement, and multi-agent collaboration, are fundamentally reshaping the financial landscape. This chapter delves into the impact of AI agents on banking, exploring how these intelligent systems are not just streamlining existing processes but also creating entirely new paradigms for operations, customer interaction, risk management, and value creation. From automating complex tasks and delivering hyper-personalized experiences to fortifying fraud defenses and navigating the intricacies of regulatory compliance, AI agents are proving to be indispensable tools for financial institutions seeking to thrive in the era of intelligent banking.

K. Huang (✉)
DistributedApps.ai, Fairfax, VA, USA
e-mail: ken@distributedapps.ai

D. Wu
Stanford, Palo Alto, CA, USA

J. Ponnapalli
Frontier, Southlake, TX, USA

G. Huang
PIMCO, Austin, TX, USA

K. Huang (ed.), *Agentic AI*, Progress in IS,
https://doi.org/10.1007/978-3-031-90026-6_8

8.1 Key Drivers of AI Agent Adoption in Banking

The banking industry's recent AI transformation is not merely about adopting new software; it's about leveraging a sophisticated, multilayered AI agent architecture to redefine banking operations and value creation. The architectural components of AI agents discussed in the first three chapters of this book, working in concert, are the key enablers of this revolution.

Figure 8.1 summarizes the key drivers propelling AI agent adoption in the banking sector. From addressing the data explosion to meeting elevated customer expectations and ensuring regulatory compliance, these drivers highlight the transformative potential of AI agents.

1. The Data Deluge and the Imperative for Intelligent Processing

   The exponential growth of data in the financial ecosystem presents an unprecedented opportunity but only if banks can effectively process and analyze it. AI agents, with their robust Data Layer, are uniquely equipped to handle this challenge.

   (a) How the Architecture Enables This:

   - Multisource data ingestion and structured/unstructured data integration allow AI agents to collect and unify data from diverse sources like transaction records, CRM systems, social media, and market feeds.
   - Real-time and historical data processing capabilities enable both immediate insights and long-term trend analysis.

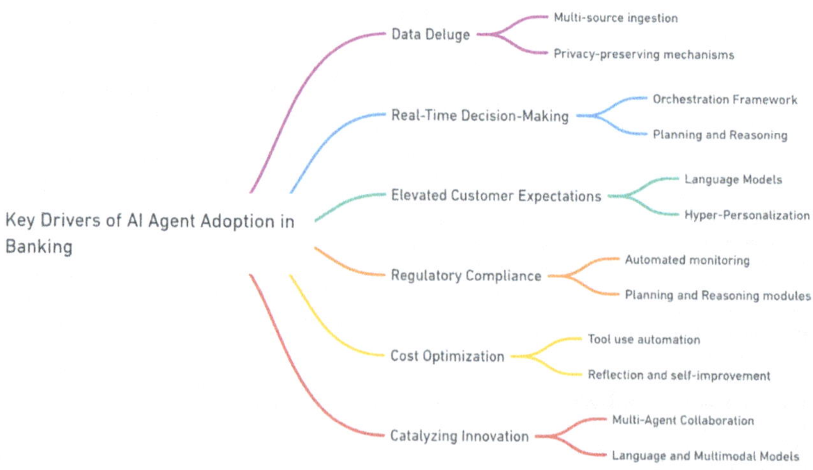

Fig. 8.1 Key drivers of AI agent adoption in banking

- Advanced data preprocessing, cleaning, semantic embedding, and vector storage transform raw data into a format that can be efficiently used by Language and Multimodal Models (Gerling & Lessmann, 2023).
- Privacy-preserving data handling mechanisms ensure that sensitive customer information is protected throughout the data life cycle.

(b) Strategic Implication: This advanced data handling, powered by the Data Layer, enables AI agents to extract actionable insights, identify hidden patterns, and generate predictive models, transforming data from a costly byproduct into a strategic asset. This underpins other drivers such as improved risk management and new product development.

2. Real-Time Decision-Making: The New Competitive Battleground

   In the fast-paced world of finance, split-second decisions can have significant consequences. AI agents, through their Orchestration Framework, Planning and Reasoning modules, and Tool Use capabilities, are revolutionizing decision-making.

   (a) How the Architecture Enables This:

   - The Orchestration Framework manages the complex interplay between different components, enabling dynamic workflow composition and adaptive execution strategies.
   - Planning and Reasoning modules utilize hierarchical task decomposition and contextual decision-making to analyze situations and choose the best course of action in real time.
   - Tool Use allows AI agents to interact with external systems, such as trading platforms or risk assessment tools, through API and function calling mechanisms, enabling them to execute decisions instantly.

   (b) Strategic Implication: This architectural synergy enables banks to operate in real time, capitalizing on fleeting market opportunities, proactively mitigating risks, and dynamically adjusting strategies. This agility is crucial for algorithmic trading, fraud detection, and dynamic pricing.

3. Elevated Customer Expectations in the Age of Digital-First Experiences

   Modern banking customers demand personalized, 24/7 service. AI agents, leveraging their Language and Multimodal Models and Retrieval-Augmented Generation (RAG), are transforming customer interactions.

   (a) How the Architecture Enables This:

   - Language and Multimodal Models allow AI agents to understand and respond to customer queries in natural language, across various modalities (text, voice, images).
   - RAG enables AI agents to access and integrate external knowledge, providing more accurate and comprehensive responses, while hallucination prevention techniques ensure factual accuracy.

  - Dynamic model selection and fine-tuning allow the system to adapt to different customer interaction styles and preferences.

(b) Strategic Implication: This combination of architectural components enables AI agents to power intelligent chatbots, deliver tailored financial recommendations, and automate transactions, creating a hyper-personalized and efficient customer journey that meets the high expectations of today's digital-savvy customers.

4. Navigating the Complex Regulatory Landscape with AI-Powered Compliance

   The financial industry is subject to stringent regulations. AI agents, with their Data Layer, Planning and Reasoning, and Reflection and Self-Improvement capabilities, offer a powerful solution for automating and enhancing compliance.

   (a) How the Architecture Enables This:

   - The Data Layer ingests and processes regulatory documents and transaction data, providing a comprehensive view of compliance-related information.
   - Planning and Reasoning modules can be used to model regulatory requirements and automatically flag potential violations.
   - Reflection and Self-Improvement capabilities, particularly error analysis and adaptive behavior modification, allow the AI agent to learn from past mistakes and continuously improve its compliance monitoring performance.

   (b) Strategic Implication: This architecture enables automated transaction monitoring, anomaly detection, report generation, and adaptation to new regulations, reducing risk, minimizing manual effort, and ensuring ongoing adherence to evolving legal frameworks. This frees up staff to deal with more complex cases.

5. Cost Optimization and Efficiency Gains

   AI agents, through their ability to automate tasks via Tool Use and Orchestration, coupled with Reflection and Self-Improvement, are driving significant cost savings and efficiency gains.

   (a) How the Architecture Enables This:

   - Tool Use allows AI agents to interact with existing banking systems to automate tasks like data entry, document processing, and customer onboarding.
   - The Orchestration Framework optimizes workflows and ensures seamless integration between different tasks and systems.
   - Reflection and Self-Improvement enable continuous optimization of processes, reducing errors and improving efficiency over time.

(b) Strategic Implication: By automating routine tasks, AI agents free up human employees to focus on higher-value activities, leading to significant cost savings, improved operational efficiency, and enhanced resource allocation.

6. Catalyzing Innovation: New Products, Services, and Experiences

   The advanced capabilities of AI agents, particularly their Multi-agent Collaboration System and Language and Multimodal Models, are unlocking opportunities for groundbreaking financial products and services.

   (a) How the Architecture Enables This:

   - The Multi-agent Collaboration System allows specialized agents to work together, combining their expertise to develop complex financial models, design new products, and simulate market scenarios. This also allows for cross-domain collaboration and the creation of new synergies.
   - Language and Multimodal Models enable the creation of innovative customer interfaces and personalized financial planning tools.
   - RAG can be used to incorporate external market data and research into product development, ensuring that new offerings are aligned with the latest market trends.

   (b) Strategic Implication: This collaborative, multifaceted architecture fosters a culture of innovation, enabling banks to develop hyper-personalized financial products, offer proactive financial advice, create innovative risk assessment models, and design entirely new banking experiences. This drives customer engagement and unlocks new revenue streams.

8.2 Applications of AI Agents in Banking

AI agents are being deployed across various areas of banking, transforming traditional processes and creating new possibilities. This section highlights some examples and how AI agent architecture enables these use cases. Before we proceed, let us have a high level of visual clues of what we are talking about.

Figure 8.2 illustrates the layered architecture of AI agents, emphasizing their modular design. The architecture integrates components like the Data Layer, Orchestration Framework, and Language Models, which collaboratively enable real-time decision-making, fraud detection, and hyper-personalization across banking operations.

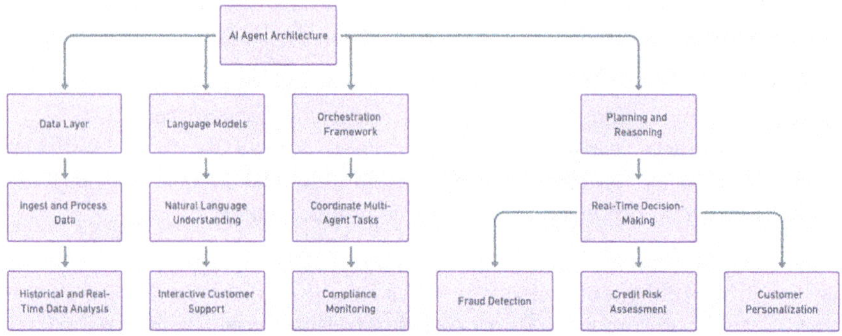

Fig. 8.2 How AI agent architecture enables some key services in banking

8.2.1 Credit Risk Assessment

The traditional approach to credit risk assessment often relies on limited data sources and relatively inflexible models. AI agents, however, offer a fundamentally different approach, characterized by a more holistic, dynamic, and precise evaluation of creditworthiness. When properly designed to ensure fairness and eliminate bias, these agents can lead to a more equitable and efficient lending process.

Key Architectural Features and Their Impact on Credit Risk Assessment

1. Comprehensive Data Analysis Enabled by the Data Layer:
    (a) How It Works: AI agents' Data Layer ingests and integrates a vast array of data sources, including traditional credit reports, bank statements, transaction histories, and alternative data like social media activity, online behavior, and even psychometric assessments (Coraglia et al., 2024). The Data Layer's ability to handle both structured and unstructured data, combined with advanced preprocessing and semantic embedding techniques, allows for a far richer understanding of an applicant's financial profile.
    (b) Impact: This comprehensive data analysis provides a more holistic view of an applicant's creditworthiness, moving beyond a narrow credit score to a multidimensional risk profile.
2. Real-Time Updates and Dynamic Risk Modeling Via Orchestration and Planning.
    (a) How It Works: The Orchestration Framework coordinates the flow of data and the execution of various modules. Coupled with the Planning and Reasoning component's ability to adapt to new information, AI agents can continuously update credit risk models as new data becomes available. This dynamic approach allows for real-time adjustments to risk assessments based on changes in a borrower's financial situation or macroeconomic conditions.

(b) Impact: This enables banks to respond quickly to changes in a borrower's financial situation, making more accurate and timely lending decisions. It allows for a move from static, point-in-time assessments to continuous risk monitoring.

3. Enhanced Pattern Recognition Through Language/Multimodal Models and Reasoning.

   (a) How It Works: AI agents leverage advanced Language and Multimodal Models to analyze unstructured data like text and images, extracting valuable insights. The Reasoning module, employing machine learning algorithms, identifies complex patterns in the integrated data that might be invisible to human analysts. These patterns can include subtle indicators of creditworthiness or financial distress that traditional models might miss.
   (b) Impact: This advanced pattern recognition allows for the identification of nonlinear relationships and subtle risk indicators, leading to more accurate and nuanced risk assessments.

4. Bias Mitigation and Enhanced Fairness Through Ethical AI Principles and Reflection.

   (a) How It Works: When designed with a focus on Ethical Constraints and equipped with Reflection and Self-Improvement capabilities, AI agents can actively work to minimize biases in lending decisions. The Reflection module can be trained to identify and correct for potential biases in the data or the model itself. Bias mitigation techniques are implemented through careful data selection, algorithmic adjustments, and ongoing monitoring.
   (b) Impact: While careful design and continuous monitoring are essential, AI agents, when properly implemented, have the potential to reduce human biases and promote fairer lending practices. However, it's crucial to acknowledge that AI can perpetuate or amplify existing biases if not carefully designed and monitored.

Real-World Example: JP Morgan's COiN

JP Morgan's COiN (Contract Intelligence) system exemplifies the power of AI in credit risk assessment. COiN leverages advanced natural language processing (a subset of Language Models) to analyze complex legal documents and extract relevant information for credit risk assessment. This system can review 12,000 annual commercial credit agreements in seconds, a task that previously took 360,000 hours of manual work (SDS, 2024). This dramatically accelerates the loan approval process while potentially improving accuracy by reducing human error.

Figure 8.3 demonstrates the sequential process through which AI agents conduct credit risk assessment. By leveraging the Data Layer for holistic financial profiling and the Planning and Reasoning module for real-time evaluations, these agents enable precise, equitable, and adaptive credit decisions.

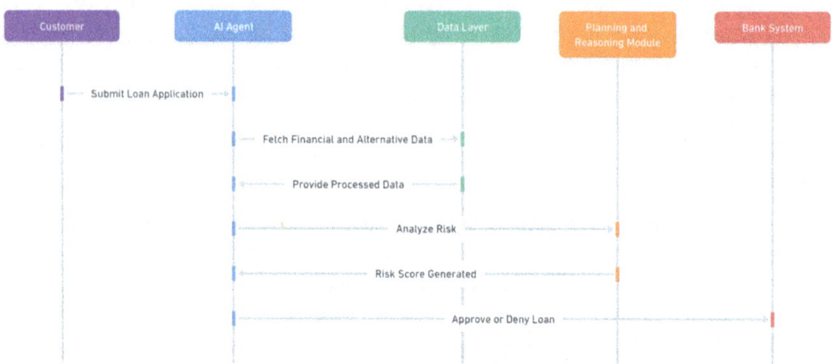

Fig. 8.3 AI agents in credit risk assessment

8.2.2 Fraud Detection and Prevention

Traditional fraud detection systems often rely on rule-based approaches, which can be slow to adapt to new fraud patterns and often generate high numbers of false positives. AI agents, in contrast, offer a paradigm shift toward proactive, adaptive, and nuanced fraud detection and prevention.

Architectural Components Driving Enhanced Fraud Detection

1. Anomaly Detection Powered by Data Layer, Language/Multimodal Models, and Reasoning
   (a) How It Works: The Data Layer ingests and integrates vast quantities of transaction data, including amounts, locations, times, and merchant information. Language and Multimodal Models can analyze unstructured data like transaction notes or customer communications for suspicious language or patterns. The Reasoning component, using machine learning algorithms, learns the "normal" transaction patterns for each customer and flags any deviations as potential anomalies.
   (b) Impact: This enables the detection of fraudulent activities that don't match known patterns, providing a crucial advantage in the fight against evolving fraud techniques. For example, if a customer's account suddenly starts making purchases in a new, distant location, the system can flag this as an anomaly even if it doesn't match a predefined fraud rule.
2. Real-Time Monitoring and Intervention Enabled by Orchestration and Tool Use
   (a) How It Works: The Orchestration Framework ensures continuous, real-time data flow and analysis. The Tool Use component allows AI agents to interact with payment processing systems and other banking platforms. When an anomaly is detected, the Planning and Reasoning module can trigger

immediate actions, such as blocking a transaction, contacting the customer for verification, or alerting the fraud investigation team.
   (b) Impact: This enables immediate intervention, preventing fraudulent transactions before they are completed. This real-time capability is crucial in an era where fraudsters operate with increasing speed and sophistication, moving money rapidly across accounts and borders. For instance, Visa's Advanced Authorization (VAA) uses this concept.

3. Adaptive Learning and Evolution Through Reflection and Multi-agent Collaboration

   (a) How It Works: The Reflection and Self-Improvement component allows AI agents to learn from past fraud cases, continuously refining their detection algorithms. They can analyze successful and unsuccessful fraud attempts, identify new patterns, and adjust their models accordingly. The Multi-agent Collaboration System allows specialized agents (e.g., one focused on transaction fraud, another on account takeovers) to share information and coordinate their responses, creating a more robust and adaptive defense system (Mao et al., 2018).
   (b) Impact: This enables AI agents to stay ahead of the curve, adapting to evolving fraud techniques and proactively identifying emerging threats. This adaptability is crucial in the dynamic landscape of financial crime.

4. Reduced False Positives Through Enhanced Reasoning and Contextual Understanding

   (a) How It Works: Advanced machine learning models, integrated within the Reasoning component, can differentiate between genuine anomalies (e.g., a customer traveling abroad) and actual fraud attempts. The Language and Multimodal Models can analyze customer communications and other contextual information to better understand the context of a transaction, further reducing false positives. The Orchestration Framework can dynamically adjust thresholds and decision rules based on the specific context.
   (b) Impact: This minimizes customer inconvenience and reduces the workload for fraud investigation teams, allowing them to focus on genuine threats (Teradata, 2024). Reducing false positives improves operational efficiency and enhances the customer experience. Danske Bank's reduction of false positives by 60% and 50% increase in real fraud detection exemplifies this benefit.

Examples of AI Agents in Fraud Prevention

Several financial institutions are leading the way in implementing AI-powered fraud detection systems:

- JPMorgan Chase's DocLLM: This system, leveraging the power of Language Models, rapidly analyzes legal documents. One application of this capability is to identify potential fraudulent inconsistencies, demonstrating the ability of AI to detect warning signs and prevent fraud before it escalates.

- Mastercard's Decision Intelligence: This platform utilizes the Reasoning and Orchestration capabilities of AI agents to continuously analyze cardholder spending patterns, assessing the fraud likelihood of each transaction in real time. This allows for preemptive intervention while minimizing false positives.
- Danske Bank's Teradata Solution: By implementing a machine learning solution, demonstrating the Reasoning and Reflection components, the bank dramatically improved its fraud detection capabilities, reducing false positives by 60% while increasing real fraud detection by 50%.
- Feedzai's Onboarding Fraud Detection: This software integrates with bank application processing systems, leveraging the Data Layer and Reasoning to assess fraud risk during customer onboarding. One top retail bank reported a 70% increase in new customer onboarding without compromising fraud prevention.
- DataVisor's Unsupervised Machine Learning: This approach, utilizing the Reasoning component, generates risk scores for banking transactions by detecting subtle correlations that might indicate fraudulent activity, helping major US banks identify specific fraud methods in loan applications.
- HSBC's Global Social Network Analytics: This system, mentioned by May (2023), harnesses big data analytics and advanced machine learning algorithms. If powered by AI agents, as suggested, it could move beyond detection to proactive prevention by leveraging the Planning and Reasoning, Multi-agent Collaboration, and Reflection and Self-Improvement components. It could proactively investigate suspicious activities, adapt to evolving criminal tactics in real time, and even predict future fraud scenarios.

Code Example for a Simple Fraud Detection Agent

This pseudocode illustrates a very simple fraud detection agent that detects anomalies in transaction data and analyzes potential fraud using GPT. The agent, encapsulated in the *FraudDetectionAgent* class, first calculates statistical thresholds (mean and standard deviation) to flag transactions with amounts that deviate significantly from the norm. It then autonomously analyzes flagged transactions by generating a prompt based on the transaction's description and invoking the GPT model for fraud detection. The agent processes the entire workflow automatically, from detecting anomalies to analyzing them with GPT, making it a self-sufficient system. Note that this pseudocode is for illustration purposes only.

1: Define Autonomous Agent Class

```
    class FraudDetectionAgent:
        def __init__(self, data_source):
            self.data_source = data_source  # Source of
transaction data
            self.anomalies = []  # List to store detected
anomalies
```

2: Function for anomaly detection based on statistical thresholds

```
        def detect_anomalies(self):

mean_amount = calculate_mean(self.data_source['amount'])

std_amount = calculate_std(self.data_source['amount'])
            threshold = 3  # Standard deviation threshold
```

3: Loop through transactions and flag anomalies

```
            for transaction in self.data_source:
                if (transaction['amount'] > mean_amount +
threshold * std_amount) or \
                    (transaction['amount'] < mean_amount -
threshold * std_amount):
                    self.anomalies.append(transaction)

            return self.anomalies
```

4: Function to analyze flagged anomalies using GPT

```
        def analyze_with_gpt(self, transaction):
            prompt = generate_gpt_prompt(transaction['notes'])
            gpt_response = invoke_gpt_model(prompt)  # Assume
OpenAI GPT model invocation
            return gpt_response
```

5: Autonomous agent action method to detect and analyze

```
        def autonomously_detect_and_analyze(self):
            flagged_anomalies = self.detect_anomalies()

            # For each flagged anomaly, analyze with GPT
            for anomaly in flagged_anomalies:
                gpt_analysis = self.analyze_with_gpt(anomaly)
                print(f"Transaction ID {anomaly['transaction_
id']} flagged: {gpt_analysis}")
```

```
6: Helper functions (representations)
      def calculate_mean(amounts):
          return sum(amounts) / len(amounts)

      def calculate_std(amounts):
          mean = calculate_mean(amounts)
          return (sum((x - mean) ** 2 for x in amounts) /
len(amounts)) ** 0.5

      def generate_gpt_prompt(notes):
          return f"Analyze this transaction for fraud: {notes}"

      def invoke_gpt_model(prompt):
          # Simulate GPT model call
          return f"GPT response: Suspicious activity detected
based on the description: {prompt}"

7: Example data
      transaction_data = [
          {'transaction_id': 1, 'amount': 150, 'notes': "Purchase
at electronics store."},
          {'transaction_id': 2, 'amount': 20000, 'notes': "Large
wire transfer to offshore account."},
          {'transaction_id': 3, 'amount': 50, 'notes': "Grocery
shopping."},
          {'transaction_id': 4, 'amount': 400, 'notes': "Luxury
watch purchase flagged by customer support."}
      ]

8: Create Autonomous Fraud Detection Agent and execute the agent
      agent = FraudDetectionAgent(transaction_data)
      agent.autonomously_detect_and_analyze()
```

8.2.3 Customer Service and Chatbots

AI-powered chatbots and virtual assistants are transforming customer service in the banking sector. These intelligent agents can handle a wide range of customer queries, provide personalized financial advice, and even assist with transactions, all without human intervention.

Benefits of AI Chatbots in Banking

- 24/7 Availability: Customers can access support at any time, improving overall satisfaction. This round-the-clock availability is particularly valuable in our globally connected world, where customers may need banking services outside of traditional business hours.
- Instant Responses: AI agents can provide immediate answers to common queries, reducing wait times. This speed of response can significantly improve customer satisfaction and reduce the workload on human customer service representatives.
- Personalization: By analyzing customer data, AI chatbots can offer tailored financial advice and product recommendations. This personalization can help banks cross-sell and upsell products more effectively while also providing more value to customers.
- Scalability: AI agents can handle multiple customer interactions simultaneously, far exceeding the capacity of human customer service representatives. This scalability allows banks to manage peak periods of customer inquiries without the need to hire additional staff.
- Multilingual Support: Advanced AI chatbots can communicate in multiple languages, broadening the bank's ability to serve diverse customer bases. This capability is particularly valuable for international banks or those serving multicultural communities.
- -Churn Prediction: AI agents can predict customer churn by analyzing chat data, engagement history, and behavioral patterns. They can identify early signs of customer dissatisfaction and provide personalized offers or value-added services to retain customers.
- Sentiment Analysis: AI agents can analyze customer interactions across multiple channels such as chatbots, calls, and SMS to gauge sentiment across the entire handle time. Banks can use this data to improve customer service, assess agent performance, resolve issues proactively, and enhance their products.

Case Studies

There are many examples of AI agent use in customer service; below, we just give two examples.

Klarna

Klarna, the innovative Swedish fintech company, has made significant waves in the technology and customer service landscape by implementing an advanced AI-powered customer service assistant. Developed through a strategic collaboration with OpenAI, this cutting-edge AI agent has quickly demonstrated extraordinary capabilities that are transforming the company's approach to customer interactions.

In just the first month of deployment, Klarna's AI assistant has achieved truly remarkable results that have caught the attention of industry experts and technology enthusiasts. The AI assistant has successfully managed an impressive 2.3 million customer conversations, effectively handling two-thirds of Klarna's entire customer service chat volume. Its performance is equivalent to the workload of 700 full-time human agents, showcasing unprecedented efficiency and scalability. Remarkably, the AI assistant has maintained customer satisfaction scores that are directly comparable to those of human representatives (Klarna, 2024).

The system has demonstrated significant improvements in service quality, reducing repeat inquiries by 25%. Perhaps most notably, the average resolution time has been dramatically reduced from 11 minutes to less than 2 minutes, representing a massive enhancement in customer service speed and responsiveness.

The AI assistant boasts an impressive and diverse range of functionalities that extend far beyond traditional customer service. It can seamlessly manage complex tasks across multiple domains, from multilingual customer support to processing refunds and handling returns. The system demonstrates remarkable linguistic versatility, communicating fluently in over 35 languages and serving customers across 23 different markets. Customers can receive real-time financial information, including instant updates on account balances, payment schedules, and spending limits. Additionally, the AI system provides an innovative, chat-based shopping experience, helping users search for products, compare items, and access detailed product insights.

Klarna's strategic investment in AI is projected to yield substantial financial benefits. The company estimates that this AI assistant will generate a remarkable $40 million USD in profit improvement during 2024. This impressive potential return comes at a relatively modest initial investment of approximately $2–three million, as reported by Klarna's CEO.

While the performance is impressive, industry experts have raised some nuanced considerations about the long-term implications of such AI-driven customer service. Some caution that a 1-month period may be insufficient to determine the full impact on customer retention and lifetime value. There are concerns about the potential risks of completely replacing human agents with AI, particularly given the limited opportunities for direct customer engagement throughout the year.

Bank of America

Bank of America's virtual assistant, Erica, is an excellent example of how AI is transforming customer service in banking. Launched in 2018, Erica uses natural language processing, predictive analytics, and cognitive messaging to assist customers with various tasks, from checking account balances to making payments and even providing personalized financial advice (Bank of America, 2024).

Since its launch, Erica has handled 2 billion customer interactions, demonstrating the potential of AI to enhance customer service in banking. The AI assistant can understand context and intent, allowing it to provide more accurate and helpful responses to customer queries. Moreover, Erica continuously learns from interactions, improving its performance over time.

The development of these AI-powered customer service agents represents a significant shift in how banks interact with their customers. Traditional banking relied heavily on face-to-face interactions at branch locations. With AI chatbots, banks can provide personalized, high-quality service through digital channels, allowing them to reach more customers while reducing operational costs.

As AI chatbots become more sophisticated, they raise interesting questions about the nature of financial advice. If an AI agent provides personalized financial recommendations, is it acting as a financial advisor? How should such AI-driven advice be regulated? These are questions that banks and regulators are grappling with as AI becomes more prevalent in customer-facing roles.

8.2.4 AI Agent in Personalized Banking

Traditional banking often relied on broad segmentation and one-size-fits-all solutions. AI agents, however, enable a move toward hyper-personalization, where each customer interaction is tailored based on a deep understanding of their individual financial profile and context.

Architectural Components Enabling Hyper-personalized Banking

1. Data-Driven Customer Understanding Via the Data Layer and Vector Databases
   (a) How It Works: The Data Layer is fundamental, ingesting and integrating vast amounts of customer data from multiple sources, including transaction histories, account balances, investment portfolios, credit scores, and even customer service interactions. Semantic embedding and vector storage within the Data Layer play a crucial role, transforming raw data into rich, contextualized representations. These vector databases allow for efficient similarity searches and nuanced understanding of customer profiles.
   (b) Impact: This comprehensive data integration provides a 360-degree view of the customer's financial life, enabling AI agents to understand their financial behavior, needs, and aspirations at a granular level.
2. Contextual Insights and Recommendations Through RAG and Multimodal Models
   (a) How It Works: Retrieval-Augmented Generation (RAG) empowers AI agents to dynamically access and integrate relevant external knowledge, such as market trends, investment research, or financial advice articles. Multimodal Models enable the system to understand and process information across different modalities (text, images, voice), allowing for a more holistic understanding of customer preferences and needs. For instance, an AI agent could analyze a customer's recent travel expenses (from transaction data) along with their social media posts about travel destinations (using multimodal analysis) to recommend a relevant travel rewards credit card.

(b) Impact: RAG and Multimodal Models enable AI agents to provide highly relevant and timely financial advice, proactive alerts, and personalized product recommendations that align with the customer's individual goals and risk tolerance. As stated by Castelnovo (2024), AI agents are poised to create a more intuitive, relevant, and responsive banking experience.

3. Proactive Engagement and Anticipatory Service Through Planning and Reasoning

   (a) How It Works: The Planning and Reasoning component allows AI agents to anticipate customer needs based on their financial data and learned patterns. By identifying trends and predicting future financial events (e.g., upcoming large expenses, potential savings opportunities), the agent can proactively offer relevant advice or solutions. For example, if an AI agent detects a pattern of increasing monthly expenses, it might proactively suggest a budgeting tool or a high-yield savings account.
   (b) Impact: This proactive approach transforms the banking experience from reactive to anticipatory, making customers feel understood and valued.

4. Seamless, Humanlike Interactions Via Language Models and Orchestration

   (a) How It Works: Large Language Models (LLMs) enable AI agents to engage in natural, humanlike conversations with customers through various channels (e.g., chatbots, voice assistants). The Orchestration Framework manages the flow of these interactions, ensuring context is preserved across multiple touchpoints and interactions.
   (b) Impact: This creates a seamless and engaging customer experience, resolving queries in near real time and providing personalized support that feels intuitive and empathetic.

5. Continuous Learning and Adaptation Through Reflection and Multi-agent Collaboration:

   (a) How It Works: The Reflection and Self-Improvement component allows AI agents to continuously learn from customer interactions, refining their understanding of individual preferences and improving the accuracy of their recommendations over time. The Multi-agent Collaboration System could enable specialized agents (e.g., one focused on investment advice, another on budgeting) to share insights and collaborate to provide a more holistic and personalized experience.
   (b) Impact: This continuous learning ensures that the AI agent's services become increasingly personalized and valuable over time, fostering stronger customer relationships and building trust.

By incorporating AI agents into personalized banking, financial institutions can

- Differentiate themselves in a competitive market: Offering hyper-personalized experiences creates a unique value proposition that attracts and retains customers.

- Enhance customer satisfaction and loyalty: By demonstrating a deep understanding of individual needs and providing valuable insights, AI agents foster stronger customer relationships.
- Drive revenue growth: Targeted marketing campaigns, personalized product recommendations, and proactive financial advice lead to increased customer engagement and higher conversion rates.
- Improve operational efficiency: Automating customer support and other tasks through AI agents reduces operational costs and frees up human employees to focus on more complex issues.

8.2.5 AI Agent in Risk Management

This section explores how AI agents are enhancing various dimensions of risk management, including market risk, liquidity risk, and operational risk, ultimately contributing to a more resilient and robust financial institution.

Traditional risk management often operates in silos, with separate teams and systems for different risk categories. AI agents, however, enable a more holistic and integrated approach, breaking down these silos and providing a comprehensive view of the bank's risk exposure.

Market Risk Management

How It Works: AI agents, leveraging their Data Layer, ingest and process vast quantities of global financial data, including market prices, trading volumes, news sentiment, macroeconomic indicators, geopolitical events, and social media trends through Multimodal Models.

The **Planning and Reasoning** component employs advanced time-series analysis, machine learning models, and scenario analysis techniques to

- Predict market volatility
- Identify potential disruptions
- Assess the impact of adverse market conditions on a bank's portfolio

AI agents also perform stress testing under various regulatory frameworks, such as **Comprehensive Capital Analysis and Review (CCAR)** for capital adequacy, **Basel III** compliance for liquidity and leverage ratios, and counterparty risk assessment for derivatives and cross-asset securities. For example:

- Under CCAR, AI agents simulate macroeconomic shocks and measure capital buffer adequacy.
- In Basel III, they calculate risk-weighted assets (RWA), monitor liquidity coverage ratios (LCR), and optimize leverage ratios in near real time.

- For Counterparty Risk and valuation adjustments (XVA), agents assess exposure across credit, funding, and collateral requirements for cross-asset securities.

The agents' adaptability is further enhanced by real-time scenario stress testing, accounting for hypothetical events like sudden interest rate hikes or geopolitical crises.

AI agents could also revolutionize **swaps and swaptions** by independently managing valuation, pricing, and risk in real time. They dynamically construct curves, optimize hedging strategies, and execute trades based on evolving market conditions. Advanced LLM models allow them to detect inefficiencies, adapt to regulatory changes, and develop bespoke solutions tailored to individual counterparty or portfolio needs.

By streamlining operations, minimizing human intervention, and ensuring compliance with evolving regulatory landscapes, AI agents reduce operational risks and enhance decision-making for market risk management.

Impact: This enables banks to develop robust hedging strategies, optimize asset allocations, and dynamically adjust their risk exposure in response to changing market conditions. By understanding potential market risks, banks can minimize losses associated with fluctuating exchange rates, commodity prices, and interest rates.

Liquidity Risk Management

How It Works: AI agents use their Data Layer to monitor cash flow patterns, transaction volumes, and customer behavior. The Planning and Reasoning component, combined with Tool Use to integrate with treasury management systems, forecasts future liquidity needs based on historical data, real-time transaction flows, and predictive models. These models can identify periods of potential cash shortages or surpluses, taking into account factors like seasonal variations, upcoming large payments, and anticipated customer withdrawals.

Impact: This allows banks to proactively manage their liquidity buffers, ensuring they have sufficient funds to meet their obligations even during periods of stress. By optimizing the deployment of idle funds and avoiding short-term funding crises, AI agents contribute to the overall financial stability of the bank.

Operational Risk Management

How It Works: AI agents analyze internal processes, systems, and workflows, leveraging their Data Layer to ingest data from various operational systems. They employ Reasoning and Language Models to identify bottlenecks, inefficiencies, and potential points of failure. For example, they can monitor IT system performance, detect anomalies that might indicate cyber threats or system outages, and analyze employee activities for compliance with internal policies. The Tool Use component can integrate with various operational systems to collect data and trigger alerts. The

Reflection and Self-Improvement component allows the agents to learn from past operational incidents and improve their ability to identify and mitigate future risks.

Impact: This proactive approach helps banks identify and address operational vulnerabilities before they lead to disruptions, financial losses, or reputational damage. By streamlining processes, improving system reliability, and ensuring compliance, AI agents strengthen the bank's operational resilience. This also includes mitigating risks of human error through automated controls and monitoring.

Compliance Monitoring

How It Works: AI agents, utilizing their Data Layer and Language Models, can process and analyze regulatory documents, internal policies, and transaction data to ensure compliance with legal and regulatory requirements. The Reasoning component can identify potential compliance violations, such as money laundering or insider trading, by analyzing transaction patterns and identifying anomalies. The Multi-agent Collaboration System could involve specialized agents, one focusing on AML regulations, another on GDPR compliance, etc., working together to provide comprehensive compliance coverage.

Impact: This reduces the risk of regulatory penalties, enhances the bank's reputation, and fosters trust among customers and stakeholders. By automating compliance tasks, AI agents also free up human compliance officers to focus on more complex and strategic issues.

The true power of AI agents in risk management lies in their ability to integrate these different risk categories into a unified framework. By correlating data and insights from market risk, liquidity risk, operational risk, and compliance monitoring, AI agents can provide a holistic view of the bank's overall risk exposure. For instance, they can identify how a market downturn might impact liquidity or how operational failures could lead to compliance violations.

8.2.6 Trading and Securities

The functionalities of AI agents in trading can be categorized into two primary roles. First, as a trader, AI agents analyze external data sources to generate direct trading decisions, such as BUY, HOLD, or SELL. For instance, an AI agent can assess the sentiment surrounding a stock based on recent news coverage and social media discussions, leading to informed trading signals. Second, AI agents can function as alpha miners, producing high-quality alpha factors that can be integrated into existing trading systems. These alpha factors (see box below) serve as predictive indicators derived from textual data analysis, helping traders identify potential market opportunities based on sentiment shifts or emerging trends.

However, the application of AI agents in handling complex financial instruments, such as **derivatives, structured products, and exotics**, remains in its early stages.

While agents excel in straightforward asset classes, they have yet to fully demonstrate their potential in managing the valuation and life cycle of hard-to-price financial instruments. These instruments require advanced modeling, sensitivity analysis, and an understanding of intricate trade life cycles, areas where AI's capabilities are still evolving.

Box: Alpha Factor

The alpha factor, a critical metric in financial analysis, measures the performance of an investment relative to a benchmark index by calculating excess returns beyond what would be expected given the investment's market risk. Essentially, alpha reveals whether an investment strategy is generating returns that outperform, underperform, or match market expectations after accounting for its inherent risk profile.

When calculating alpha, financial professionals compare an investment's actual return to its expected return based on its beta (market risk) and the risk-free rate. A positive alpha indicates the investment has outperformed the market, suggesting the investment manager's skill or strategy is generating additional value. Conversely, a negative alpha suggests the investment is underperforming relative to its risk level (Efimov, 2008).

An AI agent with LLM and RAG capabilities can perform alpha mining by dynamically leveraging its retrieval and generative capabilities to identify investment opportunities. The agent would use its RAG system to pull relevant financial documents, historical market data, and research reports, while the LLM analyzes these retrieved sources to detect subtle market inefficiencies, correlate complex information patterns, and generate sophisticated investment hypotheses. By continuously synthesizing diverse financial information, understanding nuanced contextual signals, and adapting its predictive models in real time, the AI agent can create more dynamic and intelligent alpha generation strategies that transcend traditional quantitative approaches, potentially uncovering hidden market value through advanced natural language processing and contextual reasoning.

Implementing AI agents in trading involves several key steps that ensure their effective integration into trading workflows. Data acquisition is crucial; collecting a diverse range of textual data—including financial news articles, social media content, analyst reports, and historical price data—enables the agent to learn and identify relevant patterns effectively. While general-purpose models can be effective, training custom models specifically on financial datasets enhances their performance in this domain. For example, specialized models have been developed to focus on financial sentiment analysis. AI agents excel at feature extraction from unstructured data; by utilizing techniques like tokenization and embedding layers, they convert textual information into numerical representations that can be analyzed for predictive insights.

Sentiment analysis is another aspect of using AI agents in trading. By analyzing the emotional tone of market-related texts, these agents can gauge market sentiment—whether bullish or bearish—thus informing trading strategies. Before deploying an AI-driven trading strategy in live markets, it is essential to backtest it against historical data to evaluate its effectiveness. This process helps refine the model and ensures it performs well under various market conditions. Once validated, AI agents can be integrated into real-time trading systems where they continuously analyze incoming data streams and generate actionable insights or trade signals based on current market conditions.

Several firms have successfully employed AI agents in their trading strategies. For example, Kensho Technologies utilizes AI agents to analyze financial news and social media sentiment to inform investment decisions for major firms like BlackRock and Bridgewater Associates. Their platform generates insights that help traders anticipate market movements based on public sentiment. Similarly, AQR Capital Management leverages AI agents for trend identification within financial news and social media discussions; by analyzing sentiment scores derived from these texts, they enhance their investment strategies with timely insights. Traders also use OpenAI's ChatGPT for generating market summaries and trade analyses; its natural language processing capabilities allow traders to interact with their systems more intuitively, facilitating better decision-making processes.

8.2.7 Payment

AI agents can be potentially designed to facilitate seamless payments, detect and prevent fraud, and offer tailored financial advice by analyzing user behavior and preferences. For instance, AI agents can automate routine payment tasks, reducing manual errors and increasing efficiency. They can also monitor transactions in real time to identify suspicious activities, thereby mitigating potential fraud. Additionally, by understanding customer spending habits, AI agents can provide personalized recommendations, enhancing customer satisfaction and loyalty.

Stripe's Agent Toolkit exemplifies the integration of AI agents in payment processing. This software development kit (SDK) enables AI agents to conduct transactions with human users and make payments to third parties, such as hotels or airlines. Leveraging Stripe's Issuing technology, agents can create single-use virtual credit cards through a simple function call, streamlining the payment process. The toolkit integrates with popular agent platforms like CrewAI, LangChain, and Vercel's TypeScript SDK, allowing developers to incorporate it into existing architectures seamlessly.

By combining secure payment systems with programmatic automation, Stripe's AI agents allow seamless payments, reducing the need for human intervention and paving the way for more efficient financial operations.

8.2.8 *AI Agent for Regulatory Compliance*

Traditional compliance approaches often rely on manual processes, periodic audits, and retrospective analysis of violations. AI agents, however, can enable a shift toward a proactive, continuous, and data-driven approach to compliance management.

Architectural Components Driving Enhanced Regulatory Compliance

1. Continuous Regulatory Monitoring and Interpretation Via Data Layer and Language Models
   (a) How It Works: The Data Layer ingests and processes a vast array of regulatory documents, legal texts, and industry guidelines from various jurisdictions. LLM models analyze these documents to extract key requirements, obligations, and restrictions. They can also explain compliance rules in a clear and concise manner, making them more accessible to compliance officers.
   (b) Impact: This provides banks with real-time updates on regulatory changes, ensuring they are always aware of the latest requirements. The ability of LLMs to interpret complex legal language helps bridge the gap between regulatory texts and practical implementation.
2. Proactive Risk Prediction and Alerting Through Reasoning and Planning
   (a) How It Works: The Reasoning component, combined with the Planning module, analyzes internal policies, transaction data, and operational processes to identify potential compliance risks before they materialize into violations. The system can predict the likelihood of noncompliance based on historical data, identified patterns, and emerging trends. The Orchestration Framework can then trigger alerts to compliance officers, highlighting areas of concern and recommending preventive actions.
   (b) Impact: This proactive approach allows banks to address potential compliance issues early, preventing costly penalties and reputational damage. By identifying and mitigating risks proactively, banks can maintain a strong compliance posture.
3. Automated Compliance Checks and Audits Via Tool Use and Orchestration
   (a) How It Works: AI agents can leverage their Tool Use capabilities to integrate with existing systems, such as CRM platforms, transaction monitoring systems, and document management tools. The Orchestration Framework coordinates the automated execution of compliance checks, such as Know Your Customer (KYC) and Anti-Money Laundering (AML) procedures. This includes analyzing customer data, transaction patterns, and other relevant information to identify potential violations.
   (b) Impact: This automation significantly reduces the manual effort involved in compliance monitoring, freeing up compliance officers to focus on more

strategic tasks. Automated checks also ensure greater consistency and accuracy in the application of compliance rules.

Implementation Considerations: Integrating AI Agents into the Compliance Ecosystem

Deploying AI agents for regulatory compliance requires careful planning and execution. Key considerations include the following:

- Seamless Integration: Ensuring that AI agents can seamlessly integrate with existing systems (CRM, document management, transaction monitoring) through their Tool Use capabilities to access relevant data sources efficiently.
- Training and Skill Development: Equipping compliance officers with the skills and knowledge to effectively interpret and utilize the AI agent's reports and recommendations.
- Continuous Monitoring and Validation: Regularly monitoring the AI agent's outputs, validating its performance, and ensuring it remains aligned with the latest global regulations. This involves leveraging the Reflection component for ongoing improvement.
- Explainable AI (XAI): Implementing XAI techniques to ensure that the AI agent's decisions and recommendations are transparent and understandable to both internal stakeholders and regulators.

Examples of AI Agents in Action: Transforming Compliance Practices

Several companies are pioneering the use of AI in regulatory compliance:

- Kensho Technologies: Employs AI agents, showcasing the use of Language Models, to analyze financial news and social media sentiment, helping firms like BlackRock make informed investment decisions based on public perception. This demonstrates how sentiment analysis can be used to assess market risks and ensure compliance with regulations related to market manipulation and insider trading.
- AQR Capital Management: Uses AI agents for trend identification within financial news, enhancing their investment strategies with timely insights derived from sentiment analysis. This application highlights the potential of AI to monitor market trends and ensure compliance with regulations related to market abuse.
- Wolters Kluwer has introduced OneSumX Reg Manager, an AI agent-powered solution tailored for community banks and credit unions to enhance their compliance management. This software-as-a-service (SaaS) platform automates the monitoring of regulatory changes, providing daily alerts and a comprehensive dashboard for tracking compliance status. By integrating AI agents with expert insights, it offers a structured library of state and federal regulations and includes an activity tracker to document implementation progress. Overall, OneSumX Reg Manager aims to streamline compliance processes, improve efficiency, and empower smaller financial institutions to navigate the complex regulatory landscape with greater confidence (FinTech Global, 2024).
- Automated Compliance Checks: Many organizations are implementing AI-powered automated compliance checks, leveraging Tool Use and Orchestration,

demonstrating a significant reduction in reliance on human intervention and improving the efficiency and effectiveness of compliance monitoring.

- HSBC has partnered with Google Cloud to develop an AI agent solution known as Anti-Money Laundering AI (AML AI) to enhance its efforts in combating financial crime. This AI agent-powered system has significantly improved HSBC's ability to detect and prevent money laundering activities, identifying 2–4 times more suspicious activity while reducing alerts by 60%, which allows investigators to concentrate on genuinely concerning cases. The AML AI has also expedited the detection process, reducing the time to identify suspicious accounts to just 8 days after the initial alert. Unlike traditional rule-based systems, this AI agent can recognize known money-laundering patterns and uncover networks of criminals collaborating together. As a result, HSBC has doubled its identification of financial crime in commercial banking and nearly quadrupled it in retail banking, earning recognition as the Celent Model Risk Manager of the Year 2023, with plans to expand the use of AML AI across additional markets (May, 2023).

Figure 8.4 visualizes the end-to-end compliance monitoring process driven by AI agents. The process integrates data ingestion, regulatory analysis, risk identification, and reporting to ensure continuous adherence to legal frameworks while reducing manual overhead.

8.3 Digital Workers: The Next Frontier in Banking AI

As AI technology continues to advance, we're seeing the emergence of more sophisticated AI agents that can perform complex tasks traditionally handled by human workers. These "digital workers" are AI-powered software robots that can interact with various systems, process information, and make decisions, effectively functioning as virtual employees.

Table 8.1 highlights the key characteristics of digital workers in banking, showcasing their advanced capabilities in automation, decision-making, integration, and personalization that set them apart from traditional AI agents.

8.3.1 Digital Workers vs. AI Agent

Digital Workers represent a more advanced and sophisticated evolution of AI agents, characterized by several key distinctions (Susnjak et al., 2024):

1. Comprehensive Capability Set: Unlike traditional AI agents that typically focus on narrow, specific tasks, Digital Workers are designed with broader, more holistic capabilities that span multiple domains and can handle complex, intercon-

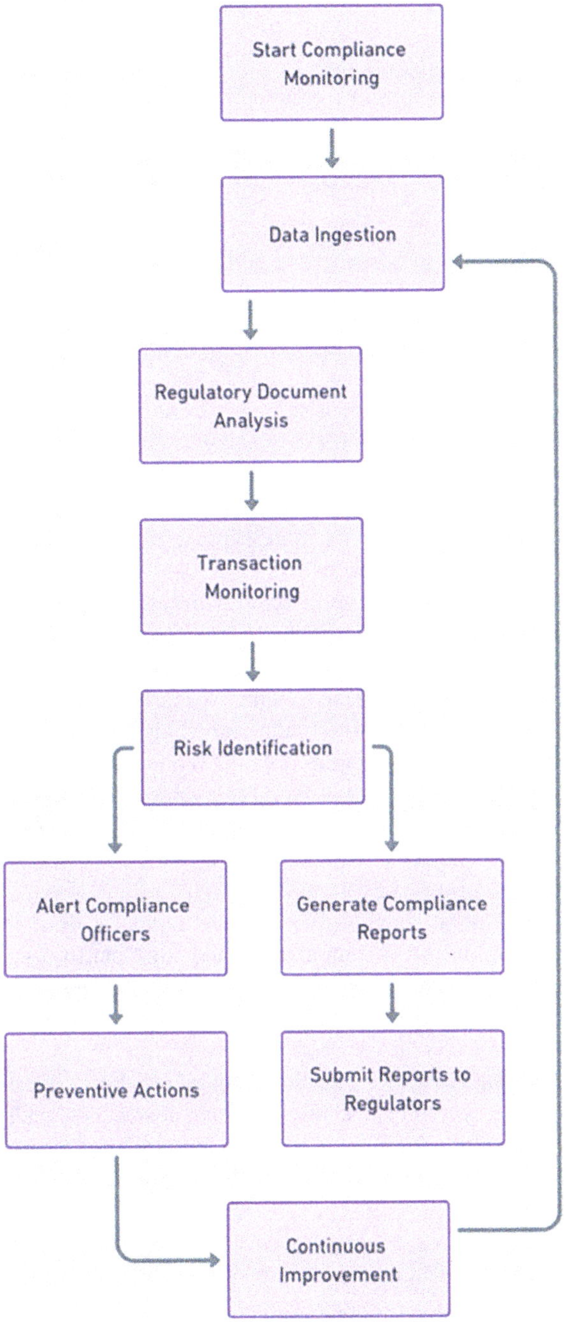

Fig. 8.4 Agents in regulatory compliance

Table 8.1 Characteristics of digital workers in banking

Characteristic	Description	Example in banking
Process automation	Automates end-to-end workflows, including repetitive tasks and complex operations.	Automating customer onboarding, loan approvals, and payment reconciliations.
Advanced decision-making	Uses reasoning and predictive analytics for contextually informed decisions.	Assessing creditworthiness in real time or identifying high-risk transactions.
Cross-system integration	Seamlessly connects and interacts with multiple banking systems and databases.	Integrating with CRM platforms, treasury systems, and regulatory databases.
Cognitive learning capabilities	Learns from interactions and adapts to changing tasks and requirements.	Improving fraud detection accuracy by learning from past incidents.
Collaboration with other agents	Works alongside specialized agents to achieve shared goals through orchestration.	Coordinating with compliance agents to flag regulatory breaches or analyze market trends for investment.

nected workflows. They possess a more expansive skill set that allows them to act more autonomously and adaptively.

2. Context and Memory Retention: Digital Workers have enhanced long-term memory and contextual understanding capabilities. They can maintain context across extended interactions, remember previous tasks and instructions, and apply learning from past experiences to new scenarios, making them more intelligent and responsive compared to traditional AI agents.
3. Advanced Reasoning and Decision-Making: While AI agents often follow predefined rules or algorithms, Digital Workers leverage more sophisticated reasoning models. They can engage in complex decision-making processes, understanding nuanced scenarios, anticipating potential outcomes, and making more contextually informed choices.
4. Multimodal Integration: Digital Workers are typically designed to integrate seamlessly across multiple communication and work platforms. They can interact through various interfaces, process different types of data (text, voice, visual), and coordinate actions across diverse digital ecosystems more effectively than traditional AI agents (Bieniek et al., 2024).
5. Personalization and Adaptive Learning: Digital Workers have more advanced machine learning capabilities that allow them to continuously adapt and personalize their interactions. They can learn from user preferences, refine their communication styles, and optimize their performance based on specific user or organizational needs.
6. End-to-End Process Management: Unlike AI agents that might focus on discrete tasks, Digital Workers can manage entire complex workflows from inception to completion. They can coordinate multiple subtasks, handle handoffs between different systems or human workers, and provide comprehensive process oversight.

7. Ethical and Governance Frameworks: Advanced Digital Workers are typically developed with more robust ethical guidelines, governance mechanisms, and transparency protocols. They have built-in safeguards to ensure responsible AI behavior, privacy protection, and alignment with organizational standards (Lyons et al., 2021).
8. Emotional and Contextual Intelligence: Digital Workers often incorporate more sophisticated natural language processing and emotional intelligence capabilities. They can understand subtle communication nuances, interpret context more accurately, and engage in more humanlike interactions compared to traditional AI agents.

These distinctions position Digital Workers as a more evolved, intelligent, and versatile form of AI-powered digital assistants, capable of delivering more comprehensive and adaptive support across various professional and personal domains.

8.3.2 Examples of Digital Workers in Action

The following are some initial versions of digital workers from top banks. As you can see, these are the first steps in the evolution of digital workers, but they already exhibit some promising capabilities.

1. JP Morgan's Contract Intelligence (COiN) Platform

JP Morgan's COiN platform is an excellent example of a digital worker in the banking sector. This AI-powered system can analyze legal documents and extract relevant information in seconds, a task that previously took lawyers and loan officers 360,000 h annually.

Key features of COiN:

- Analyzes commercial loan agreements
- Extracts important data points and clauses
- Identifies potential risks and inconsistencies
- Significantly reduces manual review time
- Improves accuracy and consistency in document analysis

The implementation of COiN has not only improved efficiency but also reduced errors in loan servicing. By automating the review of commercial credit agreements, JP Morgan has been able to free up its legal and loan servicing teams to focus on more complex and strategic tasks.

2. AI-Powered Financial Advisors

Many banks are developing AI-driven financial advisors that can provide personalized investment advice and portfolio management services. These digital workers analyze market trends, assess individual risk tolerances, and make investment recommendations tailored to each client's financial goals.

Capabilities of AI financial advisors:

- Portfolio optimization (Noravesh, 2022)
- Risk assessment and management
- Real-time market analysis
- Personalized investment recommendations
- Automated rebalancing and fund categorization (Mehta et al., 2020) and tax-loss harvesting

For example, Morgan Stanley has developed an internal AI-powered platform called Next Best Action, which helps human financial advisors provide more personalized recommendations to their clients. The system analyzes vast amounts of data, including market trends, news events, and individual client information, to suggest tailored investment strategies and financial products (Morgan Stanley, 2023).

Ellevest is another AI-powered investment platform focused on women's financial needs by addressing the gender pay gap and longer life expectancy. A notable feature of the platform is its gender-aware investing algorithms, which account for unique factors such as career breaks, often due to caregiving, and longer retirement periods. This allows women to make better financial decisions that are customized to their life events, such as buying a home or planning for children.

8.4 Challenges and Considerations

While the benefits of AI agents in banking are substantial, their implementation and use come with several challenges and considerations.

8.4.1 Data Privacy and Security

As AI agents handle sensitive financial data, ensuring the privacy and security of this information is paramount. Banks must implement robust cybersecurity measures and comply with data protection regulations such as GDPR and CCPA to safeguard customer information. Additionally, frameworks like the EU's **Digital Operational Resilience Act (DORA)**, which came into effect on January 17, 2025, mandate that financial institutions and their technology suppliers strengthen IT systems to enhance resilience against cyberattacks and other disruptions.

The increasing use of AI in banking also introduces new security and privacy challenges, particularly when leveraging nontraditional data sources for applications such as credit scoring or fraud detection. While these innovations enhance decision-making and risk management, they must be balanced with a strong commitment to security and data privacy and ethical use to maintain customer trust and meet regulatory expectations.

8.4.2 Ethical Considerations

The use of AI in decision-making processes, particularly in areas like credit scoring and loan approvals, raises ethical questions about fairness and bias. Banks must ensure that their AI models are transparent and do not perpetuate or exacerbate existing biases. There's also the question of accountability—who is responsible when an AI makes a decision that negatively impacts a customer?

8.4.3 Regulatory Compliance

As AI agents take on more critical roles in banking operations, regulators are increasingly scrutinizing their use. Banks must ensure that their AI systems comply with existing regulations and be prepared for potential new rules governing AI in finance. This includes being able to explain how AI models make decisions, a challenge given the "black box" nature of some advanced machine learning algorithms.

8.4.4 Human–AI Collaboration

While AI agents excel at automating tasks and unlocking insights, the human touch remains indispensable in banking. Building trust, navigating complex ethical dilemmas, and providing empathetic customer service are all areas where human expertise shines. The key lies in finding the sweet spot between AI automation and human oversight.

This delicate balance requires a strategic approach. Banks must invest in upskilling their workforce, empowering employees to collaborate effectively with AI agents and leverage their unique human strengths. Think of it as a partnership, where AI handles the routine and analytical tasks, freeing up human employees to focus on higher-value activities that require creativity, critical thinking, and emotional intelligence.

However, a crucial caveat remains: overreliance on AI can be perilous. Maintaining human accountability is paramount to mitigate risks and ensure responsible AI adoption. Ultimately, a successful AI strategy in banking hinges on recognizing the complementary strengths of humans and AI, fostering a symbiotic relationship where both thrive.

8.4.5 Explainability and Transparency

Many AI models, particularly deep learning systems, operate as "black boxes," making it difficult to explain their decision-making processes. In banking, where decisions can have significant impacts on individuals and businesses, the ability to explain AI-driven decisions is crucial. This is not just a technical challenge but also a regulatory one, as many financial regulations require banks to provide clear explanations for their decisions.

8.4.6 Integration with Legacy Systems

Many banks operate on legacy IT systems that may not be easily compatible with new AI technologies. Integrating AI agents into these existing systems can be a complex and costly process. Banks need to carefully plan their AI implementation strategies to ensure smooth integration and minimal disruption to existing operations.

8.4.7 Skill Gap and Talent Acquisition

Implementing and maintaining AI systems requires specialized skills that are in high demand and short supply. Banks are competing not just with other financial institutions but with tech companies and startups for AI talent. This skill gap can slow down AI adoption and increase costs.

8.4.8 Customer Trust and Acceptance

While many customers appreciate the convenience of AI-powered services, others may be wary of interacting with AI agents, particularly for sensitive financial matters. Banks need to build trust in their AI systems through transparency and clear communication and by demonstrating the benefits to customers (Balayn et al., 2024).

8.4.9 *Continuous Monitoring and Model Updating*

AI agents in areas like fraud detection, trading, and risk management require real-time monitoring to detect anomalies and adapt decision-making based on new information. Market shifts and evolving customer behaviors can cause model drift, reducing the accuracy of AI predictions over time. Therefore, continuous model updates are crucial for maintaining relevance in fast-changing financial markets.

Generative AI models are increasingly powering AI agents or serving as valuable tools within agent-based systems. This integration introduces unique challenges that demand careful consideration.

8.4.10 *The Balancing Act of Consistency*

While the probabilistic nature of generative AI fuels creativity by producing diverse outputs, it can be a double-edged sword. Applications requiring strict consistency and reproducibility, such as in financial reporting or legal documentation, may find this variability problematic. Striking the right balance between creative exploration and reliable consistency is crucial.

8.4.11 *Keeping Knowledge Current and Relevant*

Generative AI models, trained on massive datasets, often possess broad but static knowledge. To excel in dynamic, real-world scenarios, agents need access to up-to-date information and domain-specific expertise. Techniques like domain adaptation fine-tuning and retrieval augmented generation (RAG) can bridge this knowledge gap, ensuring agents remain informed and relevant.

8.4.12 *Taming Hallucinations*

One of the most significant challenges with generative AI is its tendency to "hallucinate"—generating outputs that are factually incorrect or nonsensical. This risk is particularly acute in high-stakes domains like healthcare or finance. Employing strategies like RAG and Chain-of-Thought (CoT) prompting can help ground the agent's responses in reality and reduce the occurrence of hallucinations.

8.4.13 Safeguarding Against Toxicity and Security Threats

Prompt engineering is a powerful tool, but it also presents a potential vulnerability. Malicious actors could attempt to manipulate prompts to elicit toxic content, reveal sensitive information, or compromise the agent's integrity. Robust security measures and prompt engineering best practices are essential to mitigate these risks and ensure responsible AI deployment.

8.4.14 Navigating the Labyrinth of Evaluation

Evaluating the performance of generative AI within agent systems is a complex endeavor. The nondeterministic nature of these models means there can be multiple valid outputs for a given task, making traditional evaluation metrics less effective. Human evaluation, with its nuanced understanding of context and intent, often plays a critical role in assessing the quality and effectiveness of generative AI agents.

8.5 Preparing for the AI-Driven Future of Banking

As AI continues to transform the banking industry, financial institutions need to take proactive steps to prepare for this AI-driven future (Castelnovo, 2024):

1. Develop a comprehensive AI strategy: Banks should create a clear road map for AI adoption, aligned with their overall business strategy. This should include identifying key areas where AI can add value, setting clear goals and metrics, and planning for the necessary resources and investments.
2. Invest in data infrastructure: AI systems are only as good as the data they're trained on. Banks need to invest in robust data collection, storage, and management systems to ensure they have high-quality data to fuel their AI initiatives (Gupta et al., 2021).
3. Foster a culture of innovation: Embracing AI requires a cultural shift within banking organizations. Banks should encourage experimentation, learning, and collaboration between technical and business teams.
4. Prioritize ethical AI: As AI takes on more critical roles in banking, it's essential to establish clear ethical guidelines for AI development and use. This includes addressing issues of bias, fairness, transparency, and accountability (Wong et al., 2022).
5. Upskill the workforce: Banks need to invest in training programs to help their employees work effectively with AI systems. This includes both technical skills for those developing and maintaining AI systems and general AI literacy for all employees.

6. Collaborate with fintech and tech companies: Partnerships with innovative fintech startups and established tech companies can help banks accelerate their AI adoption and bring new capabilities to market more quickly.
7. Engage with regulators: Banks should proactively engage with regulators to help shape the regulatory framework for AI in banking. This includes participating in sandbox initiatives and providing input on proposed regulations (Wu & Liu, 2023).
8. Focus on customer education: As AI becomes more prevalent in banking services, banks need to educate their customers about how AI is being used, its benefits, and any potential risks or limitations.
9. Plan for cybersecurity: With AI systems handling sensitive financial data, banks need to invest in robust cybersecurity measures to protect against increasingly sophisticated cyber threats.
10. Invest in inclusive design and development: Banks must develop frameworks that prevent discriminatory practices, ensuring any AI-driven decisions are inclusive of all economic and social backgrounds.
11. Embracing responsible AI (RAI): RAI is not just an ethical imperative but a strategic necessity for banks venturing into the world of AI agents. It requires establishing a comprehensive RAI framework that seamlessly integrates with existing control and compliance structures. This framework should clearly articulate the organization's RAI principles, prioritizing key concerns such as explainability, fairness, accountability, and cybersecurity.

But a robust framework alone is not enough. Banks must translate these principles into actionable guidelines and guardrails that empower teams to innovate responsibly. This includes the following:

- Proactive Risk Assessment: Conducting thorough risk assessments throughout the AI life cycle, identifying potential biases, ethical concerns, and societal impacts (Xia et al., 2023)
- Explainability and Transparency: Implementing techniques to make AI agent decisions transparent and understandable to both internal stakeholders and customers (Mei et al., 2023)
- Data Governance and Privacy: Establishing strict data governance protocols to ensure responsible data collection, usage, and protection, respecting customer privacy and complying with regulations (Khan et al., 2020)
- Human Oversight and Control: Maintaining human oversight in critical decision-making processes, ensuring human accountability and preventing unintended consequences (Sterz et al., 2024)
- Continuous Monitoring and Improvement: Regularly monitoring AI agent performance, identifying areas for improvement, and adapting the RAI framework to address emerging challenges and opportunities.

Figure 8.5 outlines a comprehensive road map for banks to embrace an AI-driven future. The diagram emphasizes key steps such as defining an AI strategy, building robust data infrastructure, fostering innovation, upskilling the workforce, and

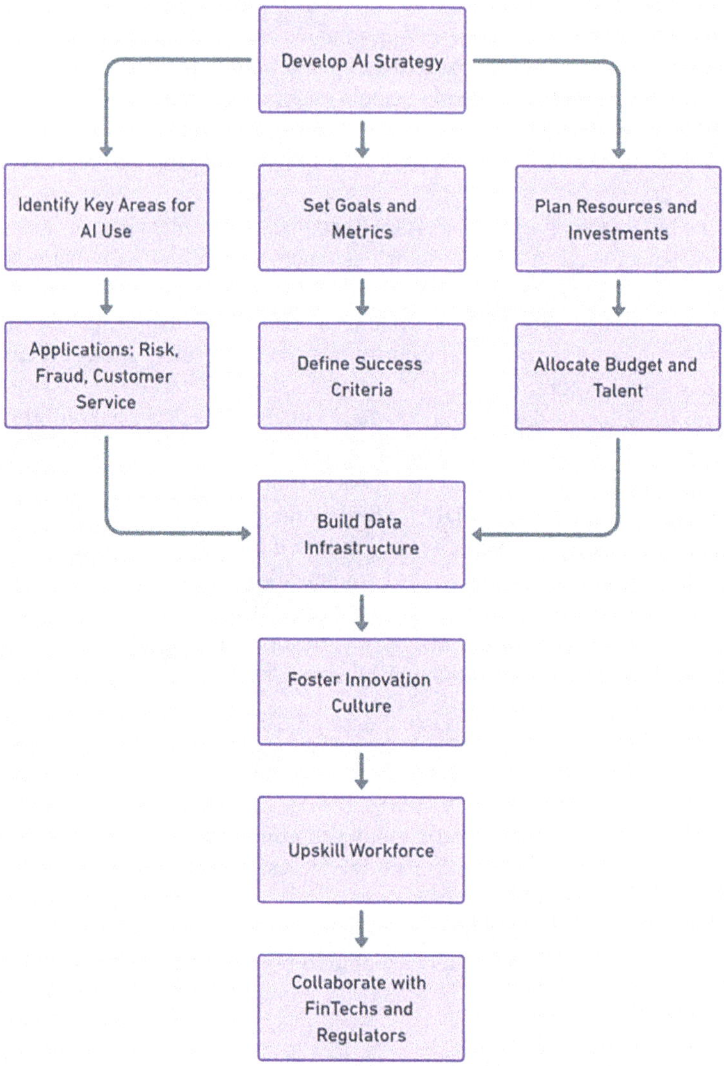

Fig. 8.5 Preparing for the AI-driven future of banking

collaborating with fintechs and regulators to ensure a seamless and responsible adoption of AI technologies.

8.6 Summary

This chapter paints a compelling picture of AI agents as the transformative force reshaping the banking industry. No longer a futuristic vision, these intelligent systems, with their sophisticated architecture, are actively redefining how banks operate, interact with customers, manage risks, and innovate. The chapter highlights the key drivers fueling this adoption: an explosion of data demanding intelligent processing, the need for real-time decision-making in a fast-paced market, evolving customer expectations for personalized experiences, the increasing burden of regulatory compliance, the imperative for cost optimization, and the potential for groundbreaking new products and services.

The core message is that AI agents, through their ability to ingest and process vast amounts of data, learn from experience, reason, plan, interact with systems and humans, and collaborate with other agents, are not just automating existing processes but creating entirely new paradigms in banking. They are enabling hyper-personalized customer experiences, fortifying fraud defenses, streamlining operations, enhancing risk management across multiple dimensions (market, liquidity, operational, and compliance), and unlocking opportunities in trading and securities. The chapter also introduces "digital workers," representing the next evolution of AI agents, poised to handle even more complex, multi-domain tasks, effectively functioning as virtual employees.

Food for Thought:

- **The Human–AI Partnership:** While the chapter emphasizes the power of AI, it's crucial to remember that the future of banking likely lies in a synergistic partnership between humans and AI. AI agents excel at data analysis, automation, and pattern recognition, but human expertise remains vital for tasks requiring empathy, complex ethical judgment, and strategic oversight. The question is not about replacing humans but augmenting their capabilities with AI. The emergence of human–AI teams, where AI agents can also provide feedback to their human counterparts and help them improve their skill sets, will have a transformative impact on the financial sector.
- **The Democratization of Finance:** AI agents have the potential to democratize access to financial services. By analyzing alternative data sources and identifying creditworthy individuals who might be overlooked by traditional models, AI-powered lending could expand access to credit for underserved populations. This could have significant societal implications, promoting financial inclusion and economic empowerment. However, it also raises concerns about algorithmic bias and the need for careful design and monitoring to ensure fairness.

- **The Rise of the AI-Powered Bank:** We are moving toward a future where AI agents are not just isolated tools but integral components of a fully integrated, AI-powered bank. Imagine a bank where AI agents seamlessly handle everything from customer onboarding and personalized financial advice to fraud detection, risk management, and even investment decisions. This raises questions about the future structure of banks, the skills needed by banking professionals, and the role of regulators in overseeing these complex systems.
- **The Ethical Imperative:** As AI agents become more powerful and pervasive, the ethical considerations become even more critical. Ensuring fairness, transparency, accountability, and privacy in AI-driven banking is not just a technical challenge but a societal one. Banks, regulators, and society as a whole must grapple with these issues to ensure that AI is used responsibly and for the benefit of all. Bias detection and mitigation will require constant vigilance. Explainable AI (XAI) will be crucial for building trust and understanding how AI systems arrive at their decisions.
- **Lifelong Learning and Adaptation:** The rapid evolution of AI technology means that both individuals and organizations must embrace a mindset of continuous learning and adaptation. Banking professionals will need to upskill and reskill to work effectively alongside AI agents. Banks will need to invest in ongoing training and development programs to ensure their workforce remains relevant and competitive. Similarly, AI agents themselves will need to be designed for continuous learning and improvement, adapting to new data, regulations, and market conditions.
- **Global Collaboration and Standardization:** Given the global nature of finance and the increasing interconnectedness of AI systems, international collaboration and the development of industry-wide standards will be essential. This includes standards for data sharing, model interoperability, and ethical guidelines for AI development and deployment.
- **The Black Swan Factor:** While AI agents are powerful tools for predicting and managing risks, they are still based on historical data and learned patterns. They might struggle to anticipate or respond to truly unprecedented events, often referred to as "black swan" events. Banks will need to maintain a degree of human oversight and strategic planning to navigate such unforeseen circumstances.

8.7 Questions

Multiple Choice:

1. Which of the following is NOT a key driver of AI agent adoption in banking, as outlined in the chapter?

   (a) The exponential growth of data,
   (b) The need for real-time decision-making,

(c) The desire to replace all human employees with AI,
(d) Evolving customer expectations,

2. What is the primary function of the Data Layer in an AI agent's architecture?

(a) To generate natural language responses,
(b) To ingest, process, and manage data from various sources,
(c) To execute trades in financial markets,
(d) To coordinate the actions of multiple AI agents,

3. How do AI agents contribute to enhanced fraud detection?

(a) By relying solely on predefined rules and thresholds,
(b) By analyzing transaction data to identify anomalies and patterns indicative of fraud,
(c) By replacing all human fraud analysts with AI systems,
(d) By focusing exclusively on preventing internal fraud,

4. Which architectural component enables AI agents to provide personalized financial advice?

(a) Only the Data Layer,
(b) The combination of Language Models, Retrieval-Augmented Generation (RAG), and the Data Layer,
(c) The Orchestration Framework alone,
(d) The Reflection and Self-Improvement module,

5. What is a key benefit of using AI agents for regulatory compliance?

(a) Eliminating the need for all compliance regulations,
(b) Automating compliance checks and proactively identifying potential violations,
(c) Replacing all human compliance officers with AI,
(d) Guaranteeing 100% accuracy and eliminating all compliance risks.

True or False:

1. AI agents are primarily used in banking to reduce costs and have limited impact on customer experience. (True/False).
2. The Orchestration Framework is responsible for managing the interactions between different components of an AI agent's architecture. (True/False).
3. Explainable AI (XAI) is not a significant concern in banking since AI decisions are always correct. (True/False).
4. Digital Workers are essentially the same as traditional AI agents, just with a different name. (True/False).
5. AI agents can analyze sentiment in financial news and social media to inform trading decisions. (True/False).

Short Answer:

1. Briefly describe how the Reflection and Self-Improvement component contributes to the effectiveness of AI agents in banking.
2. Explain the concept of "hyper-personalization" in the context of AI agents in banking, and provide one example.
3. What are two key challenges associated with integrating AI agents into existing banking systems?
4. How does the use of AI agents in credit risk assessment differ from traditional methods, and what are the potential benefits?
5. Describe the role of Multi-agent Collaboration in enhancing the capabilities of AI agents in either fraud detection or risk management.

Scenario-Based:

1. A bank is considering implementing AI agents for customer service. Describe three potential benefits of this implementation and two potential challenges the bank might face.
2. Imagine a scenario where an AI agent denies a customer's loan application. What ethical considerations should the bank have addressed during the development and deployment of this AI agent?
3. How might a bank leverage the capabilities of AI agents to improve its liquidity risk management practices? Provide specific examples of how different architectural components would contribute.
4. Explain how AI agents can be used in trading and securities, distinguishing between their roles as traders and alpha miners. Provide examples of companies that have successfully implemented AI in their trading strategies.
5. Describe the functionality of AI agents in regulatory compliance, including their key capabilities and the strategic steps involved in their implementation. Provide examples of companies that have successfully deployed AI agents for compliance purposes.

References

Balayn, A., Yurrita, M., Rancourt, F., Casati, F., & Gadiraju, U. (2024). *An empirical exploration of trust dynamics in LLM supply chains*. arXiv. Retrieved from http://arxiv.org/pdf/2405.16310v1

Bank of America. (2024, April 8). BofA's Erica surpasses 2 billion interactions, helping 42 million clients since launch. *PR Newswire*. https://www.prnewswire.com/news-releases/bofas-erica-surpasses-2-billion-interactions-helping-42-million-clients-since-launch-302110386.html

Bieniek, J., Rahouti, M., & Verma, D. C. (2024). *Generative AI in multimodal user interfaces: Trends, challenges, and cross-platform adaptability*. arXiv. Retrieved from http://arxiv.org/pdf/2411.10234v1

Castelnovo, A. (2024). *Towards responsible AI in banking: Addressing bias for fair decision-making*. arXiv. Retrieved from http://arxiv.org/pdf/2401.08691v1

Coraglia, G., Genco, F. A., Piantadosi, P., Bagli, E., Giuffrida, P., Posillipo, D., & Primiero, G. (2024). *Evaluating AI fairness in credit scoring with the BRIO tool*. arXiv. Retrieved from http://arxiv.org/pdf/2406.03292v1

Efimov, G. V. (2008). *On ortho-positronium and gauge*. arXiv. Retrieved from http://arxiv.org/pdf/0805.4337v1

FinTech Global. (2024, April 5). *Wolters Kluwer unveils OneSumX Reg Manager for community banks and credit unions*. FinTech Global. https://fintech.global/2024/04/05/wolters-kluwer-unveils-onesumx-reg-manager-for-community-banks-and-credit-unions/

Gerling, C., & Lessmann, S. (2023). *Multimodal document analytics for banking process automation*. arXiv. Retrieved from http://arxiv.org/pdf/2307.11845v2

Teradata. (2024). Danske Bank fights fraud with deep learning and AI [Case study]. https://assets.teradata.com/resourceCenter/downloads/CaseStudies/CaseStudy_EB9821_Danske_Bank_Saves_Millions_Fighting_Fraud_With_Deep_Learning_and_AI.pdf

Gupta, N., Patel, H., Afzal, S., Panwar, N., Mittal, R. S., Guttula, S., Jain, A., Nagalapatti, L., Mehta, S., Hans, S., Lohia, P., Aggarwal, A., & Saha, D. (2021). *Data quality toolkit: Automatic assessment of data quality and remediation for machine learning datasets*. arXiv. Retrieved from http://arxiv.org/pdf/2108.05935v2

Khan, N., Ahmad, T., Patel, A., & State, R. (2020). *Blockchain governance: An overview and prediction of optimal strategies using Nash equilibrium*. arXiv. Retrieved from http://arxiv.org/pdf/2003.09241v1

Klarna. (2024, February 27). Klarna AI assistant handles two-thirds of customer service chats in its first month. *PR Newswire*. https://www.prnewswire.com/news-releases/klarna-ai-assistant-handles-two-thirds-of-customer-service-chats-in-its-first-month-302072744.html

Lyons, H., Velloso, E., & Miller, T. (2021). *Fair and responsible AI: A focus on the ability to contest*. arXiv. Retrieved from http://arxiv.org/pdf/2102.10787v1

Mao, H., Liu, Y., Jia, Y., & Nanduri, J. (2018). *Adaptive fraud detection system using dynamic risk features*. arXiv. Retrieved from http://arxiv.org/pdf/1810.04654v1

May, R. D. (2023, November 30). *How HSBC fights money launderers with artificial intelligence*. Google Cloud. https://cloud.google.com/blog/topics/financial-services/how-hsbc-fights-money-launderers-with-artificial-intelligence.

Mehta, D., Desai, D., & Pradeep, J. (2020). *Machine learning fund categorizations*. arXiv. Retrieved from http://arxiv.org/pdf/2006.00123v1

Mei, A., Saxon, M., Chang, S., Lipton, Z. C., & Wang, W. Y. (2023). *Users are the North Star for AI Transparency*. arXiv. Retrieved from http://arxiv.org/pdf/2303.05500v1

Noravesh, F. (2022). *Sparse non-convex optimization for higher moment portfolio management*. arXiv. Retrieved from http://arxiv.org/pdf/2201.01227v2

SDS. (2024). *J.P Morgan – COiN – A case study of AI in finance*. Superior Data Science https://superiordatascience.com/jp-morgan-coin-a-case-study-of-ai-in-finance/

Morgan Stanley. (2023, March 14). *Morgan Stanley wealth management announces key milestone in innovation journey with OpenAI*. Morgan Stanley. https://www.morganstanley.com/press-releases/key-milestone-in-innovation-journey-with-openai

Sterz, S., Baum, K., Biewer, S., Hermanns, H., Lauber-Rönsberg, A., Meinel, P., & Langer, M. (2024). *On the quest for effectiveness in human oversight: Interdisciplinary perspectives*. arXiv. Retrieved from doi:https://doi.org/10.1145/3630106.3659051.

Susnjak, T., McIntosh, T. R., Barczak, A. L. C., Reyes, N. H., Liu, T., Watters, P., & Halgamuge, M. N. (2024). *Over the edge of chaos? Excess complexity as a roadblock to artificial general intelligence*. arXiv. Retrieved from http://arxiv.org/pdf/2407.03652v1

Wong, R. Y., Madaio, M. A., & Merrill, N. (2022). *Seeing like a toolkit: How toolkits envision the work of AI ethics*. arXiv. Retrieved from doi:https://doi.org/10.1145/3579621.

Wu, W., & Liu, S. (2023). *A comprehensive review and systematic analysis of artificial intelligence regulation policies*. arXiv. Retrieved from http://arxiv.org/pdf/2307.12218v1

Xia, B., Lu, Q., Perera, H., Zhu, L., Xing, Z., Liu, Y., & Whittle, J. (2023). *Towards concrete and connected AI risk assessment (C^{2}AIRA): A systematic mapping study*. arXiv. Retrieved from http://arxiv.org/pdf/2301.11616v4

Ken Huang is a prolific author and globally recognized authority in AI and Web3, with an extensive portfolio of published works that bridge business strategy, technical implementation, and cutting-edge research. As Fellow of Cloud Security Alliance and Co-Chair of the AI Safety Working Groups at the Cloud Security Alliance and the AI STR Working Group at the World Digital Technology Academy under the UN Framework, he is a leading voice in shaping global AI governance and security standards.

Huang is the CEO and Chief AI Officer (CAIO) of DistributedApps.ai, a firm specializing in generative AI training and consulting. His contributions to the field include being a core contributor to the OWASP Top 10 Risks for LLM Applications and an active participant in the NIST Generative AI Public Working Group.

Notable Publications:

- *Beyond AI: ChatGPT, Web3, and the Business Landscape of Tomorrow* (Springer, 2023)—Strategic insights into AI and Web3's business applications.
- *Generative AI Security: Theories and Practices* (Springer, 2024)—A comprehensive guide on securing generative AI systems.
- *Practical Guide for AI Engineers* (Volumes 1 and 2, DistributedApps.ai, 2024)—Essential resources for AI and ML engineers.
- *The Handbook for Chief AI Officers: Leading the AI Revolution in Business* (DistributedApps.ai, 2024)—A road map for CAIOs in implementing GenAI across organizations.
- *Web3: Blockchain, the New Economy, and the Self-Sovereign Internet* (Cambridge University Press, 2024)—Insights into the convergence of AI, blockchain, IoT, and emerging technologies.
- *Blockchain and Web3: Building the Cryptocurrency, Privacy, and Security Foundations of the Metaverse* (Wiley, 2023)—Recognized as a must-read by TechTarget in 2023 and 2024.

Ken is a sought-after speaker and has presented at events such as the World Economic Forum in Davos, ACM and IEEE conferences, the CSA AI Summit, Depository Trust and Clearing Corporation forums, and World Bank conferences. His recent appointment to the OpenAI Forum reflects his ongoing commitment to advancing collaboration and dialogue in the field of AI.

Explore Ken's work on Amazon: https://www.amazon.com/author/kenhuang

Daniel Wu is a seasoned technical leader with over two decades of expertise in software engineering, AI/ML, and team development. His rich career spans diverse sectors, including technology, education, financial services, and healthcare. Daniel's noteworthy accomplishments include establishing high-performing AI teams at prominent enterprises, pioneering point-of-care expert systems, co-founding a successful online personal finance marketplace, and spearheading the development of an innovative online real estate brokerage platform.

Passionate about technology democratization and ethical AI practices, Daniel actively promotes these principles through his involvement in computer science and AI/ML education programs. He is a sought-after speaker at international conferences, business leader gatherings, and corporate training events, where he shares his insights and experiences. Daniel holds a computer science degree from Stanford University.

Jyoti Ponnapalli serves as the VP, Head of AI at Frontier, where she leads the Generative AI Center of Excellence, driving enterprise-wide initiatives that harness the transformative potential of AI and gen AI. She has led emerging technology initiatives and complex digital transformations for Fortune 500 companies across diverse industries, including finance, telecommunications, airlines, energy, and food and beverage.

Prior to joining Frontier, she served as the SVP, Innovation Strategy and Research Portfolio at Truist, driving transformative initiatives at the intersection of emerging technology such as blockchain, gen AI, and quantum in finance. As Director of Blockchain at DTCC, she played a pivotal role in modernizing the financial industry's post-trade market infrastructure using distributed ledger technologies. Her contributions include optimizing the trade settlement cycle from T + 2 to T + 0 and pioneering the tokenization of securities for private capital markets.

Jyoti holds an Executive M.S. in Technology Management from Columbia University, New York, and a Bachelor of Science in Statistics degree from the University of Mumbai. Jyoti is the co-author of two authoritative books on generative AI and a sought-after speaker at global industry conferences in AI and blockchain.

Grace Huang is a seasoned product management professional; she has amassed extensive experience in the industry, working with leading companies such as PIMCO, a global investment management firm that manages over $2 trillion in assets, and IBM, a multinational technology company that provides hardware, software, and consulting services. Throughout her career, she has successfully launched multiple products and managed large-scale projects, leveraging her skills in market analysis, strategic planning, and cross-functional team leadership. Her unique perspective on product management led her to explore new technologies and tools, including the implementation of ChatGPT in parts of the product management process. This AI-powered tool allowed her to streamline communication, improve decision-making, and enhance customer satisfaction, ultimately driving business growth and profitability. In addition to her professional experience, she holds a degree from Babson College, where she developed a solid foundation in business management and entrepreneurship. Today, she continues to stay at the forefront of the industry, leveraging her expertise in various product development goals. Her LinkedIn address is https://www.linkedin.com/in/gracehuang123.

Chapter 9
AI Agents in Insurance

Bhuvaneswari Selvadurai and Ken Huang

The insurance industry is at a pivotal moment. Built on a foundation of data analysis and risk assessment, it's now facing a revolution driven by artificial intelligence (AI) agents. These aren't just simple chatbots or automation tools; they're sophisticated systems capable of learning, reasoning, and making decisions—mirroring human cognitive functions in many ways. This chapter delves deep into the transformative power of AI agents, exploring how they're not just streamlining processes but fundamentally changing the soul of insurance, impacting everything from underwriting and claims to customer interaction and the very definition of risk itself.

It's crucial to acknowledge the recent tragic event involving the assassination of UnitedHealthcare's CEO. While the motives remain unclear, this incident has brought to light controversial practices within the company, particularly the alleged use of AI to deny insurance claims (Oncology Nurse Advisor, 2024). Reports of an AI system with a high error rate being used to override doctors' determinations have sparked a critical discussion about the ethical implications and potential misuse of AI in healthcare decision-making. This serves as a stark reminder that while AI offers immense potential, its implementation must be guided by ethical considerations, transparency, and a commitment to fairness. This chapter will explore the transformative power of AI agents in insurance while keeping this cautionary note in mind, emphasizing the need for responsible development and deployment.

B. Selvadurai
ISACA, Easton, CT, USA

K. Huang (✉)
DistributedApps.ai, Fairfax, VA, USA
e-mail: ken@distributedapps.ai

K. Huang (ed.), *Agentic AI*, Progress in IS,
https://doi.org/10.1007/978-3-031-90026-6_9

9.1 The Eve of Agentic Insurance

For over a century, insurance has relied on human expertise—underwriters meticulously assessing risk, claims adjusters investigating incidents, and agents building relationships with clients. This model, while effective, has inherent limitations. It's slow, is prone to human error, and struggles to keep pace with the explosion of data in our digital world.

AI agents are ushering in a paradigm shift. They're not here to replace human judgment entirely but to augment it, creating a symbiotic relationship where technology and human expertise combine to achieve outcomes neither could reach alone.

9.1.1 Why Now? The Convergence of Forces

Several factors have converged to make this the right moment for AI agents to take center stage in insurance:

- **Data Explosion:** We're drowning in data—from social media and wearables to connected homes and vehicles. AI agents thrive in this environment, turning raw data into actionable insights for insurers.
- **Algorithmic Advancements:** Deep learning, natural language processing (NLP), and computer vision have matured significantly. These technologies power the cognitive abilities of AI agents, enabling them to understand, learn, and reason like humans but with greater speed and accuracy when applied to specific insurance tasks.
- **Cloud Computing:** The cloud provides the scalable infrastructure needed to train and deploy complex AI models. This makes these technologies accessible to more insurers, regardless of their size.
- **Customer Expectations:** Today's digital-native customers demand personalized, instant, and seamless experiences. They expect their insurers to understand their individual needs and provide tailored solutions. AI agents are essential to meeting these expectations, offering 24/7 availability and customized interactions.
- **Competitive Pressure:** The insurance landscape is becoming increasingly competitive, with insurtech startups leveraging AI to disrupt traditional models. Established insurers are forced to adapt and embrace AI or risk being left behind. These new entrants are demonstrating the power of AI to streamline operations, improve customer experiences, and offer innovative products, putting pressure on incumbents to follow suit.

9.1.2 *The Seven-Layer Architecture: A Blueprint for Intelligent Agents in Insurance*

To understand the true power of AI agents in insurance, we need to look under the hood at the technologies that make them tick. The seven-layer architecture, detailed in Chap. 2, provides a robust framework for building these intelligent systems. It's not just a theoretical model; it's a practical blueprint that guides the development and deployment of AI agents, ensuring they are effective, secure, and aligned with business goals. Let's revisit each layer, focusing on its specific relevance to the insurance industry:

- **Layer 1: Foundation Models:** These are the "brains" of the AI agent. Models like GPT-4, and other large language models (LLMs), provide the core intelligence, enabling advanced reasoning, natural language understanding and generation, and even multimodal understanding (text, images, audio, etc.).
  - **Insurance Application:** In insurance, foundation models can be fine-tuned to understand the nuances of insurance terminology, analyze complex policy documents, extract key information from claims forms, and engage in natural conversations with customers. For instance, a foundation model can be trained to understand the difference between "collision coverage" and "comprehensive coverage" or to identify inconsistencies in a claim narrative. They can power chatbots that can answer customer questions about their policies with a high degree of accuracy, or they can be used to generate initial risk assessments based on textual data from applications and other documents.
- **Layer 2: Data Operations:** This layer is the lifeblood of AI agents, handling the critical aspects of data ingestion, processing, storage, and governance. It ensures that agents have access to the right data, in the right format, at the right time.
  - **Insurance Application:** Insurance is a data-intensive industry. This layer ensures that data from diverse sources—policy applications, claims forms, medical records, telematics devices, social media, and even external databases—is collected, cleaned, and transformed into a usable format. Vector databases are used to store embeddings of customer profiles, policy documents, and claims data. This allows for efficient similarity searches, enabling the AI to quickly find relevant information, such as identifying customers with similar risk profiles or claims histories. Retrieval-Augmented Generation (RAG) is particularly important, allowing agents to access and incorporate up-to-date information from external sources, such as regulatory guidelines or real-time pricing data. This ensures that the agent's responses and actions are always accurate and contextually relevant.
- **Layer 3: Agent Frameworks:** Frameworks like LangChain provide the tools to build the agent's "mind." They offer memory management, tool integration, and workflow capabilities tailored to the specific needs of an AI agent.

  - **Insurance Application:** Agent frameworks are crucial for building insurance agents that can handle complex, multistep tasks. **Memory management** allows an agent to remember past interactions with a customer, their policy details, and claims history. This enables personalized and context-aware conversations. **Tool integration** is essential for connecting agents to core insurance systems. For example, an agent might use a tool to access a policy administration system to retrieve coverage details, a claims management system to check the status of a claim, or a pricing engine to generate a quote. **Agentic workflows** define the steps an agent needs to take to complete a task. For instance, a claims processing workflow might involve receiving the claim, verifying the policy, assessing the damage, determining the payout, and initiating the payment.

- **Layer 4: Deployment Infrastructure:** This layer ensures that AI agents can be deployed reliably, scaled to handle fluctuating demand, and updated seamlessly. It leverages cloud computing, containerization, and orchestration technologies.
  - **Insurance Application:** The insurance industry often experiences peaks and valleys in demand, such as during natural disasters or open enrollment periods. Cloud platforms like AWS, Azure, and GCP provide the scalable infrastructure needed to handle these fluctuations. Containerization with Docker and orchestration with Kubernetes ensure that insurance agents can be deployed consistently across different environments and scaled up or down automatically based on demand. This ensures high availability and responsiveness, even during peak periods. Continuous Integration/Continuous Deployment (CI/CD) pipelines automate the process of building, testing, and deploying new versions of agents, allowing insurers to quickly adapt to changing regulations, market conditions, or customer needs.

- **Layer 5: Evaluation and Observability:** This layer is crucial for responsible AI development and deployment. It involves defining metrics to track agent performance, using observability tools to monitor their behavior, and employing evaluation frameworks to ensure accuracy, fairness, and safety.
  - **Insurance Application:** In insurance, it's critical to monitor key performance indicators (KPIs) such as claims processing time, customer satisfaction scores, fraud detection rates, and the accuracy of risk assessments. Observability tools provide real-time insights into agent behavior, helping to identify errors, biases, or areas for improvement. Evaluation frameworks, potentially leveraging frameworks like Mosaic AI, are used to rigorously assess agent performance under different scenarios, ensuring they are accurate, unbiased, and compliant with regulations. For instance, an evaluation framework might involve testing an agent's ability to handle complex claims scenarios, its sensitivity to different demographic factors, or its adherence to fair lending guidelines.

- **Layer 6: Security and Compliance:** Given the sensitive nature of insurance data and the stringent regulatory requirements of the industry, this layer is paramount.
  - **Insurance Application:** Insurance agents handle highly sensitive data, including personal information, financial details, and health records. This layer ensures compliance with regulations like EU AI Act, GDPR, and CCPA, protecting customer data from breaches and unauthorized access. It involves implementing robust security measures, such as encryption, access controls, and intrusion detection systems. Risk management frameworks, such as NIST or ISO 27001, are used to identify, assess, and mitigate security risks associated with AI agents. Regular penetration testing is conducted to identify vulnerabilities and ensure the ongoing security of the system.
- **Layer 7: Agent Ecosystem:** This is where the AI agent interacts with the real world, delivering value to customers and the business.
  - **Insurance Application:** This layer encompasses the various applications of AI agents in insurance. **Customer-facing chatbots** provide 24/7 support, answering questions about policies, coverage, and claims. They can even guide customers through the claims filing process, providing a more seamless and efficient experience. **Internal tools** powered by AI agents assist underwriters in risk assessment by analyzing vast datasets and identifying patterns that humans might miss. They can also help claims adjusters by automating damage assessment, detecting fraud, and streamlining the claims process. **Personalized recommendation engines** leverage AI to analyze customer data and suggest tailored insurance products and coverage options, enhancing customer satisfaction and driving sales.

This layered approach isn't just a technicality; it's a blueprint for building responsible, effective, and trustworthy AI agents. It ensures that they are not just powerful but also secure, ethical, aligned with business goals, and compliant with industry regulations.

9.2 Redefining Risk: AI Agents as the New Oracles

Risk assessment is the heart of insurance. Traditionally, it's been a retrospective process, relying on historical data and actuarial tables. AI agents are transforming it into a forward-looking, dynamic, and personalized discipline. They're not just crunching numbers; they're becoming the new oracles, providing insights that were previously unimaginable. Figure 9.1 highlights the enhanced risk assessment process, and we will explore further in the following sections.

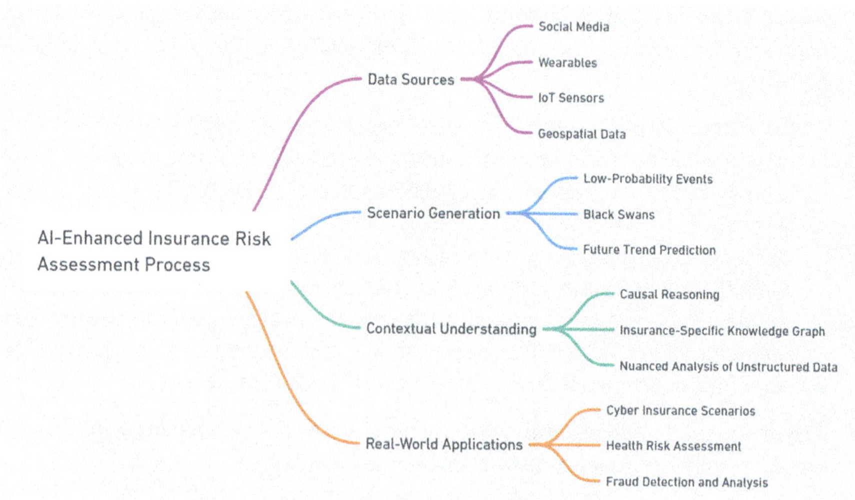

Fig. 9.1 AI-enhanced insurance risk assessment process

9.2.1 The Data

AI agents can ingest and analyze data far beyond the scope of traditional methods, providing a more holistic and nuanced understanding of risk:

- **Social Media:** By analyzing publicly available social media data, AI agents can gain insights into an individual's lifestyle, hobbies, and social connections. For instance, frequent posts about extreme sports might indicate a higher risk tolerance, while posts about healthy habits might suggest a lower risk. This information can be used to supplement traditional risk factors, providing a more complete picture of an individual's risk profile.
- **Wearables:** Fitness trackers and smartwatches generate a wealth of real-time health data, including activity levels, sleep patterns, heart rate, and even blood oxygen levels. This data can provide a dynamic view of an individual's health status, allowing insurers to assess health risks more accurately and potentially identify early warning signs of developing health problems.
- **IoT Sensors:** Smart homes equipped with IoT sensors can reveal a variety of property risks. For example, sensors can detect water leaks, smoke, or unusual temperature fluctuations, allowing for early intervention and preventing costly damage. In vehicles, telematics devices can monitor driving behavior, providing granular data on speed, acceleration, braking, and even the time of day when driving occurs. This data allows for a much more accurate assessment of driving risk than traditional factors like age or gender.
- **Geospatial Data:** Satellite imagery, aerial photography, and GIS data can be used to assess property risks related to location. This includes factors like proximity to flood zones, earthquake fault lines, wildfire-prone areas, and even crime

rates. Computer vision algorithms can analyze these images to assess the condition of properties, such as roof quality or the presence of nearby vegetation that could increase fire risk.

This data deluge, when processed by AI, paints a far more detailed and accurate picture of risk than ever before. It allows insurers to move beyond broad demographic assumptions and toward individualized risk assessment.

9.2.2 A New Era of Risk Understanding

The advent of powerful generative AI models, particularly those with advanced reasoning capabilities, marks a significant leap forward in how insurers can understand and assess risk. These models are not merely predicting outcomes based on historical patterns; they are capable of generating new insights, understanding complex scenarios, and even explaining their reasoning, ushering in a new era of risk understanding. This is a move away from simply predicting the *likelihood* of an event to understanding the *why* behind the risk.

- **Scenario Generation and Exploration:** Unlike traditional models that rely on historical data, generative AI can create hypothetical scenarios, including low-probability, high-impact events ("black swans") that might not be well-represented in past data. For instance, an AI agent could be prompted to generate scenarios involving a new type of cyberattack, a novel health pandemic, or a combination of extreme weather events. By exploring these scenarios, insurers can better understand emerging risks and develop strategies to mitigate them.
  - **Insurance Application:** In the context of cyber insurance, a generative model could be used to simulate various attack vectors on a company's IT infrastructure, considering factors like the company's industry, size, geographic location, and existing security measures. This allows underwriters to assess the potential impact of novel cyber threats and tailor coverage accordingly.
- **Causal Reasoning and Explanation:** Advanced generative AI models, especially when combined with causal inference techniques, are moving beyond correlation to causation. They can begin to reason about *why* certain factors contribute to risk, not just *that* they do. This is crucial for developing effective interventions and risk mitigation strategies. These models can generate explanations for their assessments, providing transparency and building trust.
  - **Insurance Application:** Imagine an AI agent analyzing telematics data for an auto insurance policy. Instead of simply concluding that a driver is high-risk because they frequently speed, a generative model with reasoning abilities could infer that the speeding often occurs late at night on poorly lit roads, suggesting that fatigue or poor visibility might be contributing factors. This deeper understanding allows the insurer to offer more targeted advice, such as

suggesting a fatigue management app or recommending advanced driver-assistance systems (ADAS) that improve night-time visibility. Furthermore, the model can provide an explanation of its reasoning, enabling the underwriter to make a more informed decision.

- **Contextual Understanding and Nuance:** Generative models, particularly those with strong natural language processing capabilities, can understand the context and nuances of information in a way that traditional models cannot. They can interpret unstructured data like text, images, and even audio, extracting meaning and insights that might be missed by simpler algorithms.
  - **Insurance Application:** Consider a property insurance application that includes photos of the property. A generative AI agent could analyze these images not just to identify features like roof type or building materials but also to assess the overall condition of the property, note any signs of disrepair, and even infer the quality of maintenance based on subtle cues. This contextual understanding allows for a more nuanced and accurate assessment of property risk compared to relying solely on structured data fields. Similarly, by analyzing news reports, social media discussions, or scientific publications, AI agents can understand emerging risks and trends that might not be captured by traditional data sources. For example, they might detect early signs of a new type of fraud or identify a growing public health concern that could impact life and health insurance.
- **Reasoning with Insurance-Specific Knowledge Graph:** We can build insurance specific knowledge graph as a network of interconnected entities, such as policyholders, policies, claims, risks, and external factors. Each entity is represented as a node, and the relationships between them are represented as edges. This graph structure captures the complex dependencies and interactions within the insurance domain. Generative AI can perform causal reasoning within this knowledge graph to identify the root causes of risks and claims. For example, by traversing the graph, the AI can trace a claim back to a specific policyholder behavior, a policy provision, or an external event. The reasoning process is transparent and explainable, allowing underwriters to understand the chain of events that led to a particular outcome.
  - **Insurance Application:** In a complex insurance fraud case, AI agents can leverage the knowledge graph to uncover hidden connections between seemingly unrelated claims, individuals, and service providers. By tracing the relationships within the graph, the AI might identify a pattern of suspicious activity involving multiple policyholders, a particular repair shop, and a specific type of claim. This capability can significantly enhance fraud detection and prevention efforts.

The shift from purely predictive models to generative AI with reasoning capabilities represents a fundamental change in how insurers can approach risk. It's a move from predicting outcomes to understanding the underlying drivers of risk, enabling more proactive, personalized, and effective risk management strategies. This empowers insurers to not just react to events but to anticipate them, to not just price risk but to actively mitigate it, ultimately creating a more resilient and adaptable insurance industry.

9.2.3 Real-World Impact: Case Studies

- **John Hancock's Vitality:** This program demonstrates the power of combining wearable technology with AI to assess and incentivize health. By tracking policyholders' activity levels and other health data, John Hancock can offer personalized feedback, rewards, and premium discounts for healthy behaviors (Harvard Business Publishing, 2023). This not only benefits customers through improved health and lower premiums but also provides John Hancock with a more accurate and dynamic view of individual risk.
- **Progressive's Snapshot:** This is a prime example of how telematics and AI are transforming auto insurance. By monitoring driving behavior through a device or mobile app, Progressive can offer personalized premiums based on actual driving data rather than relying solely on traditional factors like age and credit score (Blue Prism, 2024). This rewards safe drivers and provides Progressive with a much more accurate assessment of individual driving risk.
- **Cape Analytics:** This company leverages the power of computer vision and geospatial data to revolutionize property risk assessment. By analyzing aerial and satellite imagery, Cape Analytics can automatically assess property characteristics such as roof condition, the presence of swimming pools, proximity to vegetation (which can impact fire risk), and even the overall condition of the surrounding neighborhood. This provides insurers with a much more detailed and accurate understanding of property risks, enabling them to price policies more accurately and manage their exposure more effectively.

These examples demonstrate how AI is not just refining risk assessment but transforming it into a proactive, personalized, and data-driven discipline. It's a shift from reactive to predictive, from generalized to personalized, and from static to dynamic.

9.3 Claims Processing: From Burden to Breeze

AI agents are transforming claims management by automating communication and providing instant updates on claim status. Clients can interact with AI systems to submit claims, upload required documents, and receive real-time updates, all through a user-friendly interface. This reduces the administrative overhead for insurers and enhances transparency for clients, eliminating the frustration often associated with lengthy claims processes (Data Science Society, 2024). Moreover, AI can analyze claims data to detect anomalies and expedite legitimate claims while flagging suspicious ones for further review, thus improving accuracy and speeding up payouts (Lior, 2022).

9.3.1 Automating the Claims Journey: Efficiency Unleashed

AI agents can automate key stages of the claims process, significantly reducing processing times and improving accuracy (see Fig. 9.2):

- **First Notice of Loss (FNOL):** This is the crucial first step where the customer reports the incident. Traditionally, this often involved calling a customer service line and waiting on hold. AI-powered chatbots are revolutionizing FNOL. They can handle the initial claim report 24/7, guiding customers through a series of questions to collect essential information about the incident, such as the date, time, location, description of what happened, and details of any damages or injuries. They can even offer instant support and advice, such as providing information on nearby repair shops or medical providers. **Example:** Lemonade's chatbot, "Jim," can reportedly handle simple claims from FNOL to payment in seconds (Reinsurance News, 2023).
- **Document Verification:** Claims often require supporting documents like police reports, medical bills, or repair estimates. Manually reviewing these documents is time-consuming and prone to error. AI agents, equipped with **multi-modality process power** can automate this process. Multimodal AI Agent converts images of documents into machine-readable text and extracts the relevant data points, such as names, dates, amounts, and specific codes. The AI can then cross-reference this information with the insurer's systems or external databases to verify its accuracy and authenticity. For example, an AI agent can check if a medical provider is in-network or if a repair shop is certified. Any discrepancies or inconsistencies are automatically flagged for human review.
- **Damage Assessment:** Traditionally, assessing damage required sending an adjuster to physically inspect the property or vehicle. This is costly and time-consuming. AI agents, powered by **computer vision**, are transforming this process. Customers can upload photos or videos of the damage through a mobile app. Computer vision algorithms then analyze these images to identify the type and extent of the damage. For example, in auto claims, the AI can detect dents, scratches, broken glass, and other types of damage. In property claims, it can assess damage to roofs, walls, windows, and other structural components. Some advanced systems can even create 3D models of the damaged property, allowing for a more comprehensive assessment. **Example:** Tractable uses computer vision to automate damage assessment for auto insurance claims, reportedly reducing processing time from days to minutes (Tractable, 2024).

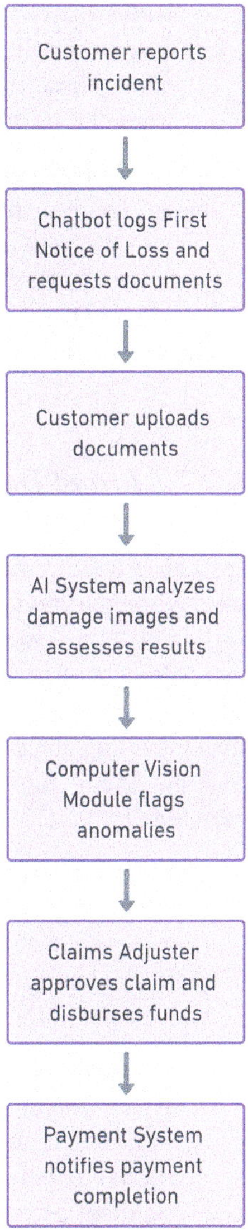

Fig. 9.2 Automating the claims journey

- **Claims Adjudication:** This is the core decision-making process where the insurer determines whether a claim is covered under the policy and calculates the appropriate payout. This often involves complex rules and requires careful consideration of policy terms, coverage limits, deductibles, and applicable laws and regulations. AI agents can assist human adjusters in several ways:
  - **Policy Analysis:** LLM model can analyze policy documents to quickly identify relevant clauses, coverage limits, and exclusions, providing adjusters with the information they need to make informed decisions.

  - **Rules-Based Automation:** AI agents can be programmed with a set of rules based on policy terms, legal requirements, and the insurer's internal guidelines. These rules can be used to automatically approve or deny straightforward claims or to flag more complex claims for human review.

- **Payment Processing:** Once a claim is approved, the final step is to issue the payment. Traditionally, this might involve mailing a physical check or manually initiating a bank transfer. AI agents can automate this process by integrating with payment systems and automatically disbursing funds to the customer or service provider. This significantly reduces the time it takes for customers to receive their payments, improving satisfaction and reducing administrative costs.

9.3.2 Fraud Detection: AI Agents as the New Sheriffs in Town

Insurance fraud remains a significant challenge, costing the industry billions annually. While traditional methods and predictive models have played a role in combating fraud, the emergence of generative AI agents with advanced reasoning capabilities is ushering in a new era of intelligent fraud detection and prevention. These agents are not just identifying patterns and anomalies; they are capable of understanding context, generating hypotheses, and even explaining their reasoning, making them more akin to intelligent investigators than simple pattern-matching tools.

- **Contextual Anomaly Detection:** Generative AI models excel at understanding context. They can analyze not just individual data points but also the relationships between them and the broader context in which they occur. This allows them to identify anomalies that might be missed by traditional rule-based systems or predictive models that rely solely on historical patterns.
  - **Insurance Application:** Consider a claim for a stolen vehicle. A traditional system might flag the claim as suspicious if the vehicle is recovered quickly, assuming it might be a staged theft. However, a generative AI agent with contextual understanding could analyze additional information, such as the claimant's recent social media posts, location data from their phone, and even traffic camera footage near the reported theft location. If the AI finds evidence suggesting the claimant was out of town during the reported theft and the car was seen being driven by someone else, it can raise a stronger, contextually grounded suspicion of fraud and explain its reasoning based on these different data points.
- **Hypothesis Generation and Investigation:** Generative AI agents can go beyond simply flagging anomalies; they can generate hypotheses about *why* an anomaly might be indicative of fraud and then investigate those hypotheses by seeking additional information. This is akin to how a human investigator works, formulating theories and then gathering evidence to support or refute them.

- **Insurance Application:** Imagine an AI agent analyzing medical claims data. It detects an unusual billing pattern from a particular clinic: a high number of claims for a specific, expensive procedure. Instead of just flagging it as an anomaly, the agent generates hypotheses: (Baker Tilly, 2024) the clinic is specializing in this procedure; (Blue Prism, 2024) there's a local outbreak requiring this procedure; (Prudential plc, 2024) the clinic is engaging in upcoding or other fraudulent billing practices. The AI agent can then investigate these hypotheses by cross-referencing with other data sources—are there similar patterns at other clinics in the region? Are there public health reports of a relevant outbreak? Does the clinic's website or other marketing materials indicate a specialization in this procedure? Based on this investigation, the AI can assign a more informed fraud risk score and provide a rationale for its assessment, allowing human investigators to prioritize their efforts.

- **Adversarial Scenario Generation:** Generative models can be used to create realistic but synthetic fraudulent claims scenarios. These scenarios can then be used to test the robustness of existing fraud detection systems and identify potential vulnerabilities. It's like running a "fire drill" for fraud.
  - **Insurance Application:** An insurer could use a generative AI model to create a variety of synthetic claims for staged auto accidents, including fabricated police reports, repair estimates, and even witness statements. By running these synthetic claims through their existing fraud detection system, they can see how well it performs against novel fraud attempts and identify areas where it needs improvement. This proactive approach helps insurers stay ahead of fraudsters who are constantly developing new techniques.
- **Explainable Fraud Scores:** Unlike "black box" predictive models, generative AI agents with reasoning capabilities can often explain *why* they have flagged a claim as potentially fraudulent. This explainability is crucial for building trust in the system, enabling human investigators to understand the AI's reasoning, and facilitating compliance with regulatory requirements.
  - **Insurance Application:** When a claim is flagged as high-risk, the AI agent can provide a clear explanation, such as: "This claim has been flagged due to inconsistencies between the reported injuries and the claimant's social media activity, which shows them engaging in physically demanding activities after the reported incident. Additionally, the medical provider associated with this claim has been linked to several other suspicious claims involving similar injuries." This explanation allows a human investigator to quickly understand the rationale behind the AI's assessment and make a more informed decision about how to proceed.
- **Knowledge Graph Reasoning for Fraud Ring Detection:** Generative AI can leverage knowledge graphs representing complex relationships between individuals, claims, policies, and external entities. By reasoning over these intercon-

nected data points, AI can uncover hidden connections and identify potential fraud rings.

  - **Insurance Application:** An AI agent analyzing a series of seemingly unrelated claims might identify a common link through a knowledge graph: a single phone number used by multiple claimants across different policies. Further investigation might reveal that this phone number is associated with a known fraud ring, enabling the insurer to take swift action.

- **Adaptation to Evolving Fraud Tactics:** Fraudsters are constantly adapting their methods. Generative AI models, with their ability to learn and generalize, can adapt to these evolving tactics more effectively than static rule-based systems. They can identify new patterns and anomalies that might not have been anticipated when the system was initially designed.

The use of generative AI with reasoning capabilities in fraud detection represents a significant advancement in the fight against insurance fraud. It's a shift from reactive to proactive, from detection to understanding, and from isolated analysis to holistic investigation. By equipping insurers with these powerful tools, the industry can better protect itself and its customers from the ever-evolving threat of fraud.

9.3.3 Case Studies: AI Agents Delivering Real Results in Claims

- **Lemonade:** This insurtech company is a prime example of how AI can transform claims processing. Their AI chatbot, "Jim," can handle many claims from start to finish, often in seconds. By automating FNOL, document verification, and even payment processing for simple claims, Lemonade has drastically reduced claims processing times and improved customer satisfaction (Reinsurance News, 2023). Their AI also plays a crucial role in fraud detection, analyzing claims for patterns and anomalies that might indicate fraudulent activity.
- **Shift Technology:** This company provides AI-powered fraud detection and claims automation solutions to insurers globally. Their platform, FORCE, analyzes claims data in real time, using a combination of machine learning, NLP, and network analysis to identify suspicious patterns and flag potentially fraudulent claims (Shift Technology, 2024). They also offer solutions for automating various stages of the claims process, such as document verification and damage assessment.
- **Tractable:** This company specializes in using computer vision to automate damage assessment for auto and property insurance claims. Their AI can analyze images of damaged vehicles or properties and provide an estimate of the repair costs in minutes (Tractable, 2024). This significantly speeds up the claims process, reduces the need for physical inspections, and allows adjusters to focus on more complex cases.

- **CCC Intelligent Solutions:** This company offers a suite of AI-powered solutions for the auto insurance and collision repair industries. Their "Smart Estimate" tool uses computer vision to analyze photos of damaged vehicles and generate repair estimates, while "Smart Audit" helps insurers audit repair estimates to identify potential overcharges or unnecessary repairs. Their "Smart Total Loss" tool uses AI to determine whether a vehicle should be declared a total loss, helping insurers make faster and more accurate decisions.
- **Snapsheet:** This company provides cloud-based claims management software and virtual appraisal solutions. Their "Snapsheet Appraisals" solution allows customers to submit photos of vehicle damage through a mobile app, which are then reviewed by appraisers augmented by AI. They also offer "Snapsheet Transactions," a platform that automates the payment process for insurance claims.

These examples demonstrate that AI agents are not just a futuristic concept in claims processing; they are already delivering real results, improving efficiency, accuracy, and customer satisfaction while combating fraud more effectively.

9.4 AI-Powered Customer Engagement

In today's competitive insurance landscape, customer experience plays a pivotal role in determining success. Customers demand personalized interactions, 24/7 availability, and seamless service across all channels. They are no longer satisfied with generic approaches or long wait times. AI agents are not just transforming how insurers engage with their customers; they are also redefining the parameters of personalized service, operational efficiency, and customer loyalty.

9.4.1 Personalized Interactions

Customers expect to be treated as individuals, and AI empowers insurers to meet these expectations by leveraging customer data to deliver highly tailored experiences at scale. AI agents analyze data from multiple sources, including demographics, policy history, browsing behavior, claims history, and even social media activity. By synthesizing this information, AI generates actionable insights to predict customer needs and craft tailored product offerings.

For instance, AI can identify frequent international travelers through their purchase patterns and recommend customized travel insurance plans that cover specific destinations and activities. Similarly, homeowners in high-risk areas for natural disasters might be offered specialized coverage options, such as flood insurance add-ons. These capabilities not only enhance customer satisfaction but also create opportunities for upselling and cross-selling. Furthermore, AI personalizes communication by adapting the tone, content, and timing of messages to the individual's preferences. A customer filing a claim might receive empathetic and frequent updates, while another interested in a new product might receive targeted and educational outreach.

Proactive engagement is another area where AI agents excel. By analyzing life event triggers—such as marriage, home purchases, or having children—AI can recommend policy adjustments or new coverage options before customers even realize they need them. This anticipatory service fosters trust and demonstrates a commitment to understanding the customer's evolving needs.

9.4.2 24/7 Availability

The expectation of round-the-clock service has become the norm. AI agents ensure that insurers meet these expectations by providing 24/7 support through chatbots, virtual assistants, and intelligent self-service portals. Chatbots can handle diverse inquiries, from policy details to claims assistance, reducing wait times and ensuring that customers receive immediate responses.

Self-service portals, augmented by AI, provide even greater flexibility. Customers can independently manage their policies, make payments, file claims, and update personal information. These portals use intelligent interfaces to guide users through complex processes, minimizing the need for call center interactions. When complex issues arise, AI agents can seamlessly transfer cases to human representatives with detailed context, ensuring smooth transitions and higher satisfaction.

9.4.3 Enhancing Customer Support

AI-powered chatbots and virtual assistants act as the first line of customer interaction, providing instant responses and guiding users through various insurance-related tasks. Using generative AI empowered AI Agent, these systems understand and address customer queries conversationally, increasing resolution rates while maintaining a humanlike touch. By offloading routine inquiries, AI agents free human representatives to focus on high-stakes interactions that require empathy and nuanced understanding (Data Science Society, 2024).

AI also brings an analytical edge to support operations. By monitoring patterns in customer queries and feedback, AI systems identify common pain points and areas for service improvement. This data-driven approach ensures that insurers remain responsive to evolving customer expectations.

9.4.4 *Improving Customer Onboarding*

AI has revolutionized the onboarding process by automating traditionally time-consuming tasks such as document verification, risk assessment, and policy configuration. Multimodal models, which integrate visual and textual analysis, enable AI agents to process and extract data from submitted documents with high accuracy. This reduces errors and accelerates policy issuance.

Beyond automation, AI tools enhance the onboarding experience by providing interactive tutorials and real-time support. For example, new customers can be guided step-by-step through the policy setup process via AI-driven chat interfaces or video tutorials. This not only ensures clarity but also leaves a positive first impression (Harvard Business Review, 2022; Data Science Society, 2024).

9.4.5 *Supporting Multichannel Communication*

Modern customers interact with insurers across multiple channels, including websites, mobile apps, and social media platforms. AI enables seamless integration of these channels, ensuring consistent and unified customer experiences. Sentiment analysis and NLP allow AI systems to interpret the tone and intent behind customer messages, tailoring responses accordingly. This capability is particularly useful in diffusing potentially negative interactions and reinforcing positive experiences.

By unifying data from various touchpoints, AI agents gain a holistic view of customer behavior, enabling more effective personalization and targeted outreach. This omnichannel approach ensures that customers receive coherent and responsive service regardless of how they choose to engage with the insurer (Harvard Business Review, 2022; Lior, 2022).

9.4.6 *Case Studies of AI in Customer Engagement*

GEICO's virtual assistant "Kate" exemplifies the practical application of AI in providing 24/7 personalized support. Through the mobile app, Kate addresses policy, billing, and claims-related queries while assisting with additional tasks such as locating repair shops. This has significantly improved customer satisfaction by reducing wait times and enhancing convenience (Insurance Innovation Reporter, 2017; IPG Media Lab, 2017).

Prudential's use of AI, in partnership with Google Cloud, highlights the potential for personalized financial advice and insurance products. By analyzing customer data, Prudential predicts preferences and offers tailored solutions, demonstrating how AI can enhance both operational efficiency and customer loyalty (Prudential plc, 2024).

USAA's proactive AI-driven approach provides tailored support for significant life events such as deployments or relocations. By anticipating member needs, USAA delivers highly relevant financial solutions that strengthen trust and satisfaction, underscoring the insurer's commitment to its unique customer base (Morgan, 2023).

These examples illustrate how AI agents are not merely tools for automation but are strategic assets that redefine customer engagement in the insurance industry. By combining personalized interactions, real-time availability, and proactive service, AI enables insurers to build enduring relationships with their clients.

9.5 Responsible Agentic AI in Insurance

The increasing use of AI agents in insurance brings significant regulatory challenges. Insurers must ensure their AI systems comply with a growing body of regulations related to data privacy, algorithmic fairness, transparency, and cybersecurity.

9.5.1 Data Privacy

AI agents in insurance often handle vast amounts of personal and sensitive data. This necessitates strict adherence to data privacy regulations, such as the General Data Protection Regulation (GDPR) in Europe and the California Consumer Privacy Act (CCPA) in the United States (Baker Tilly, 2024).

- **GDPR:** Requires insurers to obtain explicit consent for data collection and processing, provide transparent information about data usage, allow individuals to access and delete their data, and implement robust data security measures.
- **CCPA:** Grants California residents the right to know what personal information is collected about them, the right to delete their data, and the right to opt out of the sale of their data.

Insurers must adopt practices such as data minimization, purpose limitation, and implement strong data security and retention policies to comply with these regulations.

9.5.2 *Algorithmic Fairness*

AI agents are increasingly involved in decisions that significantly impact customers, such as premium calculations and claims approvals. This raises concerns about algorithmic bias, where AI models might unfairly discriminate against certain groups based on protected characteristics.

To ensure fairness, insurers must:

- **Bias Detection:** Employ techniques to identify and measure potential biases in their AI models.
- **Bias Mitigation:** Implement strategies to reduce or eliminate biases, such as using more representative training data or adjusting model parameters.
- **Fairness Audits:** Regularly audit AI models to check fairness issues.

9.5.3 *Transparency and Explainability*

Regulators and consumer advocates are increasingly calling for greater transparency and explainability in AI systems, particularly in high-stakes applications like insurance. Customers have a right to understand how decisions that affect them are being made, especially when those decisions are made by automated systems.

- **Explainable AI (XAI):** This is a growing field of research that focuses on developing techniques for making AI models more transparent and understandable to humans. XAI methods can help explain why a particular decision was made, what factors were considered, and how different factors contributed to the outcome. For example, if a claim is denied, an XAI system could provide an explanation like, "The claim was denied because the reported damage is inconsistent with the type of incident described, and there is a history of similar claims from the same geographic area."
- **Transparency Requirements:** Regulators may require insurers to provide explanations for decisions made by AI agents, particularly in cases where a claim is denied, a premium is set at a certain level, or a particular product is recommended. This could involve providing customers with access to information about the factors that influenced the decision and how those factors were weighted.

Insurers using AI agents need to consider how they will provide transparency and explainability to customers and regulators. This might involve the following:

- **Developing XAI Techniques:** Investing in research and development of XAI techniques that are appropriate for insurance applications. This could involve creating models that are inherently more interpretable or developing methods for generating explanations from complex models.

- **Providing Explanations to Customers:** Developing mechanisms for providing clear and understandable explanations to customers about how AI agents are making decisions that affect them. This could involve creating user interfaces that allow customers to explore the factors that influenced a decision or providing written explanations in plain language.
- **Documenting Model Development:** Maintaining detailed documentation of the data, algorithms, and processes used to develop and train AI agents. This documentation can be used to justify decisions to regulators and to identify potential areas for improvement.
- **Model Cards:** Creating "model cards" that provide concise summaries of an AI model's capabilities, limitations, intended uses, and performance characteristics, including fairness and bias assessments. These cards can be shared with regulators, customers, or other stakeholders to increase transparency.

9.5.4 Regulatory Sandboxes

To foster innovation while also addressing regulatory concerns, some regulators are creating "regulatory sandboxes." These are controlled environments where companies can test new technologies and business models under the supervision of regulators. Sandboxes can allow insurers to experiment with AI agents in a safe and controlled setting while also providing regulators with insights into the potential risks and benefits of these technologies.

For example, the UK's Financial Conduct Authority (FCA) has operated a regulatory sandbox since 2016, which has allowed numerous fintech and insurtech companies to test innovative products and services, including those using AI.

Insurers should consider participating in regulatory sandboxes where available, as they can provide valuable opportunities to

- **Test and Validate AI Agents:** Test new AI-powered products and services in a real-world setting with reduced regulatory risks. This allows insurers to gather data on agent performance, identify potential issues, and refine their models before full-scale deployment.
- **Engage with Regulators:** Work closely with regulators to understand their concerns and expectations regarding AI. This collaboration can help insurers ensure that their AI systems are compliant with current and future regulations.
- **Shape Future Regulations:** Provide input to regulators on how best to regulate AI in insurance, based on practical experience. This can help create a regulatory environment that fosters innovation while protecting consumers.
- **Demonstrate Responsible Innovation:** Show a commitment to responsible innovation and building trust with regulators and the public. By participating in sandboxes, insurers can demonstrate that they are taking a proactive approach to addressing the ethical and regulatory challenges of AI.

9.6 Summary

The integration of AI agents is transforming the insurance industry from the inside out. By automating tasks, improving accuracy, enhancing customer experiences, and enabling new forms of risk management and product development, AI agents are creating a more efficient, customer-centric, and data-driven industry. While challenges remain, particularly in areas like data privacy, regulation, implementation costs, talent acquisition, and ethics, the potential benefits of AI agents are too significant to ignore.

As AI technology continues to evolve, we can expect to see even more sophisticated and powerful AI agents emerge, further revolutionizing the way insurance is underwritten, priced, delivered, and experienced. The insurers that embrace this transformation, invest in the necessary technology and talent, address the ethical and regulatory challenges, and manage the organizational changes effectively will be the ones that thrive in the age of the AI agent. The future of insurance is not just digital—it's agentic.

9.7 Questions

Multiple Choice

1. Which of the following is NOT a key driver behind the adoption of AI agents in insurance?
   (a) Data explosion
   (b) Algorithmic advancements
   (c) Decreasing customer expectations
   (d) Competitive pressure
2. What is the primary function of foundation models in the seven-layer architecture?
   (a) Data storage
   (b) Providing core intelligence
   (c) Managing agent workflows
   (d) Ensuring regulatory compliance
3. Which technology is primarily used for damage assessment in AI-powered claims processing?
   (a) Natural language processing
   (b) Computer vision
   (c) Robotic process automation
   (d) Blockchain

4. What is the main benefit of using AI agents for customer onboarding?
   (a) Reducing the need for customer support
   (b) Automating and accelerating the process
   (c) Eliminating the need for policy documents
   (d) Guaranteeing claim approvals
5. Which regulation primarily focuses on data privacy in Europe?
   (a) CCPA
   (b) GDPR
   (c) HIPAA
   (d) SOX

True/False

1. True or False: AI agents are primarily used to replace human judgment entirely in the insurance industry.
2. True or False: Telematics data from vehicles is a valuable source of information for AI-driven risk assessment in auto insurance.
3. True or False: AI agents can automate the First Notice of Loss (FNOL) process in claims management.
4. True or False: Generative AI models can create hypothetical scenarios to help insurers understand emerging risks better.
5. True or False: Regulatory sandboxes are designed to stifle innovation in the insurance industry.

Simple Answer Questions

1. What are the two main tasks that AI agents can automate in the claims process?
2. Name two data sources, beyond traditional ones, that AI agents can use for risk assessment.
3. What is the purpose of an "explainable AI" (XAI) system in the context of insurance?
4. Name one company that uses computer vision for damage assessment.
5. What does "FNOL" stand for in insurance claims?

Essay Questions

1. Discuss the ethical considerations that insurers must address when deploying AI agents, particularly regarding data privacy and algorithmic fairness.
2. Explain how the use of generative AI with reasoning capabilities represents a significant advancement in fraud detection compared to traditional methods. Provide examples to support your explanation.
3. Describe the seven-layer architecture for building AI agents in insurance, highlighting the key function of each layer.
4. Analyze the impact of AI agents on customer engagement in the insurance industry. How are they transforming the customer experience?

5. How can regulatory sandboxes help foster responsible innovation in the use of AI agents in insurance?

References

Baker Tilly. (2024). *The regulatory implications of AI and ML for the insurance industry*. Retrieved from https://www.bakertilly.com/insights/the-regulatory-implications-of-ai-and-ml-for-the-insurance-industry

Blue Prism. (2024). *What is robotic process automation (RPA)?* Retrieved from https://www.blueprism.com/guides/robotic-process-automation-rpa/

Data Science Society. (2024). *How are insurance companies using artificial intelligence to transform an industry?* Retrieved from https://www.datasciencesociety.net/how-are-insurance-companies-using-artificial-intelligence-to-transform-an-industry/?utm_source=chatgpt.com

Harvard Business Publishing. (2023). Health insurance reimagined: John Hancock's Vitality using data to promote health. *Harvard Digital Platform*. Retrieved from https://d3.harvard.edu/platform-digit/submission/health-insurance-reimagined-john-hancocks-vitality-using-data-to-promote-health

Harvard Business Review. (2022). *Customer experience in the age of AI*. Retrieved from https://hbr.org/2022/03/customer-experience-in-the-age-of-ai

Insurance Innovation Reporter. (2017, April 4). GEICO launches virtual assistant, Kate, in mobile app. *Insurance Innovation Reporter*. Retrieved from https://iireporter.com/geico-launches-virtual-assistant-kate/

IPG Media Lab. (2017, January 11). *GEICO introduces conversational virtual assistant "Kate" in mobile app*. IPG Media Lab. Retrieved from https://ipglab.com/2017/01/11/geico-introduces-conversational-virtual-assistant-kate-in-mobile-app/

Lior, G. (2022). Insuring AI: Challenges and opportunities. *Harvard Journal of Law & Technology*. Retrieved from https://jolt.law.harvard.edu/assets/articlePDFs/v35/2.-Lior-Insuring-AI.pdf

Morgan, B. (2023, March 20). USAA's proactive approach to major life events. *Forbes*. Retrieved from https://www.forbes.com/sites/blakemorgan/2023/03/20/usaas-proactive-approach-to-major-life-events/

Oncology Nurse Advisor. (2024, December 9). *Suspect in UnitedHealthcare CEO killing charged with first-degree murder*. https://www.oncologynurseadvisor.com/news/suspect-in-unitedhealthcare-ceo-killing-charged-with-first-degree-murder/

Prudential plc. (2024, August 15). *Prudential partners with Google Cloud to build the future of insurance with AI*. https://www.prudentialplc.com/en/news-and-insights/all-news/news-releases/2024/15-08-2024

Reinsurance News. (2023.). *Lemonade shatters record by using AI to settle a claim in two seconds*. Retrieved from https://www.reinsurancene.ws/lemonade-shatters-record-by-using-ai-to-settle-a-claim-in-two-seconds/

Shift Technology. (2024). *FORCE: AI-native decisioning for insurance*. Retrieved from https://www.shift-technology.com/products/force/

Tractable. (2024). *AI for accurate auto claims and collision repair.* Retrieved from https://tractable.ai/

Bhuvaneswari Selvadurai (Bhuna) is a distinguished leader in information risk and cybersecurity, recognized for her expertise in regulatory compliance, cybersecurity, and risk management. She holds a master's degree in Cybersecurity Technology and Digital Forensics from the University of Maryland, graduating with Phi Kappa Phi Honors. She is also a highly credentialed professional with certifications including CISSP, CISM, CDPSE, CEH (inactive), CHPA, and CSOE.

With a proven track record in vulnerability management, data privacy, cloud security, and third party risk assessment, Bhuna has been instrumental in shaping security strategies across the financial, healthcare, energy, and retail sectors.

A recognized authority in the field, Bhuna has played a key role in security policy initiatives and advocacy efforts, contributing to industry-leading frameworks, compliance audits, and regulatory programs. She serves on the board of ISACA as Young Professional Director, where she mentors the next generation of cybersecurity, risk, and audit professionals, fostering best practices and advancing industry standards. As President of the National Charity League (NCL), she leads community initiatives and has been honored with the Presidential Volunteer Service.

Bronze Award for her dedication, contributing over 100 volunteer hours to the community.

Beyond her professional achievements, Bhuna is passionate about empowering young leaders.

She serves as a Girl Scout troop leader, inspiring young girls to develop confidence, leadership skills, and a strong sense of community. Through mentorship and hands-on learning experiences, she encourages them to embrace challenges, cultivate resilience, and become future changemakers.

A published contributor to the Cloud Security Alliance (CSA), Bhuna has contributed to key publications, including the following:

- Principles to Practice: Responsible AI in a Dynamic Regulatory Environment
- AI Organizational Responsibilities: Core Security Responsibilities
- Large Language Model (LLM) Threats Taxonomy

Passionate about advancing cybersecurity leadership, Bhuna collaborates with industry experts, sharing strategic insights on IT governance, risk mitigation, and emerging security trends, solidifying her reputation as a respected thought leader in the field.

Ken Huang is a prolific author and globally recognized authority in AI and Web3, with an extensive portfolio of published works that bridge business strategy, technical implementation, and cutting-edge research. As Fellow of Cloud Security Alliance and Co-Chair of the AI Safety Working Groups at the Cloud Security Alliance and the AI STR Working Group at the World Digital Technology Academy under the UN Framework, he is a leading voice in shaping global AI governance and security standards.

Huang is the CEO and Chief AI Officer (CAIO) of DistributedApps.ai, a firm specializing in generative AI training and consulting. His contributions to the field include being a core contributor to the OWASP Top 10 Risks for LLM Applications and an active participant in the NIST Generative AI Public Working Group.

Notable Publications:

- *Beyond AI: ChatGPT, Web3, and the Business Landscape of Tomorrow* (Springer, 2023)—Strategic insights into AI and Web3's business applications
- *Generative AI Security: Theories and Practices* (Springer, 2024)—A comprehensive guide on securing generative AI systems
- *Practical Guide for AI Engineers* (Volumes 1 and 2, DistributedApps.ai, 2024)—Essential resources for AI and ML engineers
- *The Handbook for Chief AI Officers: Leading the AI Revolution in Business* (DistributedApps.ai, 2024)—A road map for CAIOs in implementing GenAI across organizations
- *Web3: Blockchain, the New Economy, and the Self-Sovereign Internet* (Cambridge University Press, 2024)—Insights into the convergence of AI, blockchain, IoT, and emerging technologies
- *Blockchain and Web3: Building the Cryptocurrency, Privacy, and Security Foundations of the Metaverse* (Wiley, 2023)—Recognized as a must-read by TechTarget in 2023 and 2024

Ken is a sought-after speaker and has presented at events such as the World Economic Forum in Davos, ACM and IEEE conferences, the CSA AI Summit, Depository Trust and Clearing Corporation forums, and World Bank conferences. His recent appointment to the OpenAI Forum reflects his ongoing commitment to advancing collaboration and dialogue in the field of AI.

Explore Ken's work on Amazon: https://www.amazon.com/author/kenhuang

Chapter 10
AI Agents in Healthcare

Ken Huang

AI agents are reshaping healthcare by addressing complex challenges and introducing innovative solutions that span diagnostics, treatment planning, administrative workflows, and patient engagement. The adoption of AI agents enables healthcare providers to improve accuracy, enhance patient outcomes, and reduce operational inefficiencies. This chapter delves into key applications such as clinical decision support systems, predictive analytics, and drug discovery. It examines case studies including AI-driven sepsis management systems, the innovation of TrialGPT, AlphaFold 3, and CRISPR-GPT and their impact on healthcare. These developments reflect how AI agents integrate seamlessly into healthcare, enhancing precision and scalability while opening new avenues for personalized and preventive care. The discussion also explores the future potential of AI in addressing global health disparities.

To provide a clear structure for the discussions that follow, Fig. 10.1 presents an overview of the key themes and applications of AI agents in healthcare explored in this chapter. It highlights topics ranging from clinical decision support to global health initiatives, offering a roadmap for understanding how AI is transforming various aspects of healthcare delivery.

K. Huang (✉)
DistributedApps.ai, Fairfax, VA, USA
e-mail: ken@distributedapps.ai

K. Huang (ed.), *Agentic AI*, Progress in IS,
https://doi.org/10.1007/978-3-031-90026-6_10

Fig. 10.1 Mind map of AI agents in healthcare

10.1 Applications of AI Agents in Healthcare

AI agents are revolutionizing healthcare by enabling more efficient, accurate, and scalable solutions across various domains. This section delves into the diverse applications of AI agents, ranging from clinical decision support to diagnostics, drug discovery, and administrative workflows.

10.1.1 Clinical Decision Support

AI agents are transforming clinical decision support by enhancing diagnostic accuracy, treatment planning, and risk assessment. These systems excel in environments requiring real-time data analysis and actionable insights, such as intensive care units (ICUs) and emergency departments.

One of the recent applications of AI agents is in managing sepsis, a condition that requires rapid and precise intervention to prevent mortality. Multi-agent systems have been developed to monitor vital signs, analyze laboratory data, and recommend treatment plans. For instance, Borkowski and Ben-Ari (2024) describe a framework where specialized agents collaborate to provide comprehensive sepsis care. These agents operate autonomously but communicate seamlessly to ensure that all aspects of patient management, from early detection to treatment, are optimized. This collaborative approach has demonstrated a reduction in diagnostic variability and improved patient outcomes. Table 10.1 summarizes the technical details of the framework.

Another innovative application is TrialGPT, an AI-powered system designed to streamline clinical trial recruitment (Jin et al., 2024). This AI-powered tool addresses the often challenging and time-consuming process of patient recruitment by automating critical steps with high accuracy and efficiency. The framework comprises three core components: TrialGPT-Retrieval, which filters and retrieves relevant clinical trials from an extensive collection while maintaining over 90% recall with minimal pool usage; TrialGPT-Matching, which predicts criterion-level patient eligibility with 87.3% accuracy and provides faithful explanations; and TrialGPT-Ranking, which assigns trial-level scores to rank and exclude options based on patient suitability. Together, these components streamline the complex trial matching process into a seamless and effective workflow (see Fig. 10.2).

TrialGPT offers several key benefits that highlight its transformative potential in clinical trial recruitment. First, it significantly improves efficiency, with studies showing a reduction in patient recruitment screening time by approximately 42.6%. Second, its high accuracy in matching patients to trials rivals the performance of human experts, ensuring reliability. Third, the tool enhances transparency by providing written summaries explaining why a patient qualifies for a particular trial,

Table 10.1 The technical details of an agentic clinical decision support system

Aspect	Details
Diagnostic Agent	Utilizes an LLM model pre-trained on biomedical literature and fine-tuned on datasets of confirmed sepsis cases. Incorporates few-shot learning to adapt to rare or atypical clinical presentations.
Treatment Recommendation Agent	Employs retrieval-augmented generation (RAG) techniques to dynamically access and integrate the latest clinical guidelines during inference, ensuring up-to-date recommendations.
Documentation Agent	Uses an advanced LLM fine-tuned on high-quality medical documentation for generating accurate and concise clinical records. Includes a secondary lightweight model for real-time error checking and correction to ensure documentation integrity.
Quality Control Agent	Monitors outputs using supervised and unsupervised anomaly detection techniques. Dynamically adjusts the influence of agents based on performance metrics through multi-armed bandit approaches.
Confidence Calibration Agent	Features a dedicated "confidence calibration" agent that adjusts and refines confidence scores based on observed outcomes to ensure robust decision-making.

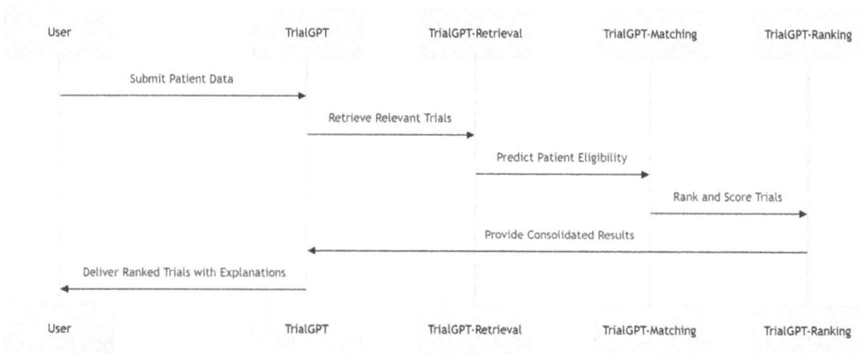

Fig. 10.2 TrialGPT workflow

building trust in its recommendations. Lastly, its scalability makes it suitable for handling large numbers of clinical trials, paving the way for widespread real-world applications across diverse healthcare settings.

Despite its promise, TrialGPT is still in the early stages of development and requires additional validation in practical environments. Ongoing efforts are focused on refining its performance and ensuring its readiness for broader implementation. As part of these initiatives, the framework is being tested in various contexts to address potential challenges and enhance its robustness. With continued advancements, TrialGPT has the potential to revolutionize patient recruitment by making the process faster, more accurate, and more accessible for both patients and healthcare providers.

10.1.2 Diagnostics and Predictive Analytics

AI agents are enhancing diagnostic and predictive analytics processes by leveraging machine learning algorithms, computer vision, and natural language processing. These systems can analyze medical images, patient histories, and real-time data from wearable devices to detect diseases at an early stage and predict potential complications.

Diagnostics

Medical imaging AI-powered diagnostic tools have shown remarkable accuracy in analyzing radiological images, often surpassing human performance in specific tasks.

For example, Google Health is advancing breast cancer detection by developing an AI system designed to assist radiologists in identifying breast cancer more accurately and efficiently. This system has been trained on thousands of de-identified mammograms, enabling it to recognize complex patterns indicative of cancer. Research published in *Nature* demonstrates that this AI technology can detect breast cancer with accuracy comparable to that of trained radiologists (see more details at: https://health.google/caregivers/mammography/).

To transition this technology from research to clinical practice, Google Health is collaborating with various partners. In the United States, a partnership with Northwestern Medicine focuses on how the AI model can prioritize high-risk cases and reduce the time to diagnosis. In the United Kingdom, through the NHS AI Award, Google Health is working with Imperial College London and three NHS trusts to evaluate the AI system's effectiveness as a "second independent reader" in double-read screening programs. Additionally, Google Health has partnered with iCAD to integrate the AI technology into iCAD's ProFound Breast Health Suite. This collaboration aims to accelerate the clinical adoption of the AI system, making it accessible to patients worldwide.

Predictive Analytics

Predictive analytics powered by AI agents enable healthcare providers to anticipate critical conditions such as sepsis, respiratory failure, and cardiac events. These systems analyze vast amounts of patient data, including electronic health records (EHRs), vital signs, and laboratory results, to identify patterns and early warning signs of impending clinical deterioration.

For sepsis prediction, AI models like the University of California at San Diego's COMPOSER system have demonstrated significant benefits. This system was associated with a 17% relative decrease in in-hospital sepsis mortality by providing early alerts to healthcare providers (Mayo Clinic Platform, 2024).

In respiratory failure prediction, machine learning algorithms analyze trends in respiratory rates, oxygen saturation levels, and other relevant parameters. A study demonstrated that a dynamic model using random forest classification predicted respiratory deterioration 90 minutes ahead of clinical recognition in step-down unit patients (Critical Care Forum, 2022).

For cardiac event prediction, AI-driven analytics assess risks by evaluating factors such as heart rate variability, blood pressure, and electrocardiogram (ECG) patterns. Continuous monitoring systems employing AI have been shown to provide early warnings for conditions like cardiac arrest, allowing for timely interventions (AHRQ Digital Healthcare Research, 2024).

The integration of these predictive analytics systems into clinical workflows facilitates proactive patient management, potentially reducing morbidity and mortality associated with these critical conditions.

10.1.3 Drug Discovery

Drug discovery is one of the most time-consuming and expensive processes in healthcare. AI agents like AlphaFold 3 and CRISPR-GPT are having positive impacts on this field by accelerating gene editing, molecular modeling, target identification, and therapeutic optimization.

AlphaFold 3

AlphaFold 3, developed by Google DeepMind, represents a groundbreaking advancement in computational biology and drug discovery. Unlike its predecessors, AlphaFold 3 extends its capabilities beyond protein structure prediction to include interactions with various biological molecules such as DNA, RNA, ligands, and ions. This advancement provides a deeper understanding of molecular interactions, which is crucial for identifying drug targets and designing effective therapeutics (Google DeepMind, 2024).

The ability of AlphaFold 3 to accurately predict protein–ligand interactions has the potential to significantly streamline the drug discovery process. Traditional methods for determining these interactions rely on time-consuming and expensive experimental procedures. By offering reliable computational predictions, AlphaFold 3 reduces the time and resources required to identify promising drug candidates (Drug Discovery Trends, 2024).

Isomorphic Labs, a subsidiary of Alphabet Inc., has integrated AlphaFold 3 into its drug discovery platform, enabling a more comprehensive exploration of molecular properties, functions, and dynamics. This integration facilitates the identification of novel therapeutic targets and supports the development of innovative treatment strategies (Isomorphic Labs, 2024).

Moreover, the open-source release of AlphaFold 3's code has expanded its impact, allowing researchers worldwide to leverage its capabilities. This democratization of access accelerates scientific research and innovation, enabling breakthroughs in understanding disease mechanisms and developing new therapies (VentureBeat, 2024).

10.1.4 Gene Editing

Gene editing has transformed healthcare by enabling precise modifications to DNA to treat and prevent diseases. CRISPR (Clustered Regularly Interspaced Short Palindromic Repeats) is a revolutionary tool that acts like molecular scissors, allowing researchers to target and edit specific DNA sequences. It is used to correct genetic mutations in diseases like sickle cell anemia and cystic fibrosis, offering potential cures for inherited disorders. As another example, CAR-T (Chimeric

Antigen Receptor T-cell) therapy uses gene editing to engineer a patient's T cells to better recognize and attack cancer cells, significantly advancing cancer treatment. In infectious disease management, gene editing disrupts the CCR5 gene—a receptor HIV uses to enter cells—providing a potential method to prevent or treat HIV infection.

CRISPR-GPT is an innovative LLM agent designed to automate and enhance the planning and execution of CRISPR-based gene-editing experiments. By integrating advanced language processing capabilities with domain-specific knowledge, CRISPR-GPT assists researchers in selecting appropriate CRISPR systems, designing guide RNAs (gRNAs), recommending cellular delivery methods, drafting experimental protocols, and planning validation experiments to confirm editing outcomes (Huang et al., 2024).

The development of CRISPR-GPT addresses the complexity inherent in designing efficient gene-editing systems, which traditionally requires deep expertise in CRISPR technology and the specific experimental systems under investigation. By leveraging the reasoning abilities of LLMs, CRISPR-GPT facilitates the design process, making gene editing more accessible to researchers, including those who may not be experts in the field.

Table 10.2 summarizes the functionalities of CRISPR-GPT components, highlighting their roles in automating gene-editing experimental design

Table 10.2 CRISPR-GPT agentic components

Component	Description	Example implementation
LLM Agent	Acts as the primary interface for user interactions, generating responses for the task executor. Enables users to monitor and correct task progression.	Automatically interacts with the Task Executor, leveraging past interactions and external tools. For instance, guides users in designing sgRNA, suggesting protocols, and validating experiments while allowing user oversight to correct or intervene.
LLM Planner	Configures tasks based on the user's needs, either through four predefined meta-tasks or dynamically planned tasks based on the user's input. Utilizes ReAct prompting for reasoning and decision-making.	Decomposes user requests into sequences of tasks using task dependencies and internal LLM knowledge, forming a state machine to execute the full pipeline.
Tool Provider	Connects the system to external APIs, tools, libraries, and documents. Wraps API usage inside states, exposing user-friendly and LLM-friendly textual interfaces to facilitate tool usage.	Integrates tools like Google web search, Primer3, gRNA libraries, and protocol databases. Assists in using external tools for sgRNA design, off-target prediction, and validation experiments.
Task Executor	Implements tasks as state machines with robust subgoal decomposition and progress control. Guides the user through decision-making with structured interactions, feedback, and API calls.	Decomposes tasks into subgoals, transitioning between states based on user inputs and task progress. For example, transitions from sgRNA design to off-target prediction, ensuring robust step-by-step execution.

Table 10.3 summarizes the three operational modes of CRISPR-GPT, each designed to cater to varying levels of user expertise and experimental complexity. These modes—Meta Mode, Auto Mode, and Q&A Mode—provide flexible solutions for automating and streamlining the design and execution of CRISPR-based gene-editing experiments.

- Meta Mode is best suited for users who need well-defined workflows and guidance through standard gene-editing experiments. Its predefined pipelines make it an excellent choice for researchers new to CRISPR or those looking to complete routine tasks efficiently.
- Auto Mode caters to users requiring flexibility and customization. By dynamically generating task lists based on specific input, this mode is ideal for experienced researchers designing novel or complex experiments.
- Q&A Mode is an interactive support system that enhances accessibility to gene-editing knowledge. It provides context-aware assistance, making it particularly useful for troubleshooting or gaining a deeper understanding of CRISPR-related processes.

Together, these modes ensure CRISPR-GPT can address a wide range of research needs, from routine tasks to highly specialized gene-editing challenges.

10.1.5 Administrative Workflows

Administrative inefficiencies contribute significantly to healthcare costs, and AI agents are addressing this challenge by automating routine tasks such as billing, scheduling, and record management. These systems not only improve operational

Table 10.3 Modes of CRISPR-GPT for gene-editing experimental design

Mode	Description	Capabilities
Meta Mode	Offers predefined pipelines (meta-tasks) to guide users through 22 unique design tasks for CRISPR-based experiments.	Supports tasks such as CRISPR system selection, sgRNA design, delivery method recommendation, off-target prediction, and validation planning, ensuring a structured workflow tailored to common experiment types.
Auto Mode	Dynamically creates a customized list of design tasks based on the user's specific request, enabling personalized workflows.	Analyzes user input to determine the necessary steps, chains state machines to execute complex pipelines, and accommodates novel or nonstandard experimental setups, ensuring adaptability for diverse research needs.
Q&A Mode	Functions as an interactive chatbot to address user questions and provide guidance throughout the experimental design process.	Responds to specific queries related to CRISPR tools, methodologies, or experimental outcomes. Provides real-time assistance, explanations, and suggestions, ensuring users receive accurate and context-aware responses to their needs.

efficiency but also reduce the burden on healthcare providers, allowing them to focus more on patient care.

As an example, in his 2024 publication, Nicholas Vennaro examines the integration of collaborative robots, or "cobots," into healthcare office administration. The primary objective is to transform the management of administrative tasks in medical practices, including appointment scheduling, workflow routing, reminders, referrals, contract analysis, and insurance processing. By implementing advanced AI and predictive machine learning technologies, the aim is to automate up to 60% of manual office processes, thereby allowing healthcare providers to focus more on patient care and less on administrative duties (Vennaro, 2024a, b).

This approach is designed to enhance efficiency and accuracy in medical administration, enabling physicians and healthcare payors to concentrate on their primary responsibilities: delivering exceptional patient care and managing financial needs. By strategically realigning resources and processes, the initiative seeks to disrupt conventional medical management, making operations smoother and more cost-effective (Vennaro, 2024a, b).

The implementation of cobots in healthcare administration aligns with broader trends in intelligent automation within the industry. The integration of AI, robotic process automation (RPA), and data analytics is enabling healthcare organizations to streamline operations, enhance patient care, and make more informed decisions. Use cases include automating administrative tasks such as patient scheduling, billing, and claims processing, as well as assisting with clinical tasks like diagnosis, treatment planning, and medication management (Pragmatic Coders, 2023).

Furthermore, a systematic review of collaborative robots for nurses highlights that cobots are being designed to assist with tasks like medication delivery, vital monitoring, and social interaction. However, few cobots were explicitly designed to reduce nursing workload through administrative or logistical assistance, indicating a potential area for further development and research (Frontiers in Robotics and AI, 2024).

10.1.6 Enhancing Patient Engagement

AI agents are not limited to backend operations; they are also enhancing patient engagement through virtual health assistants (VHAs). These conversational AI systems provide personalized health advice, manage medication schedules, and monitor chronic conditions.

A study by Magee et al. (2022) evaluated a personalized conversational agent, or "chatbot," designed to provide diabetes education and monitoring for adults with type 2 diabetes. The study reported that 97% of participants were either satisfied or very satisfied with the chatbot interactions, and 84% agreed or strongly agreed that the chatbot increased their confidence in managing their health. Additionally, participants who engaged with the chatbot experienced a mean reduction in hemoglobin A1C (HbA1c) levels, suggesting improved glycemic control.

These findings indicate that chatbots can serve as effective intermediaries between patients and healthcare providers, enhancing patient engagement and adherence to treatment plans. By delivering tailored educational content and monitoring support, such systems help ensure that patients remain informed and actively involved in their care.

10.1.7 Preventive Healthcare and Longevity Research

AI agents are increasingly integral to preventive healthcare and longevity research, offering personalized and proactive health management strategies. By analyzing extensive datasets—including genetic information, lifestyle factors, and medical histories—AI agents can predict disease risks and recommend tailored preventive measures.

In preventive care, AI agents enhance early detection and intervention. They analyze patient data to identify individuals at high risk for conditions like cardiovascular diseases or diabetes, enabling timely lifestyle modifications or treatments to prevent disease onset. AI-powered diagnostic tools, such as those used in cancer screenings, can detect anomalies in medical images, facilitating early diagnosis and improving patient outcomes (Mayo Clinic, 2023).

Regarding longevity, AI contributes to understanding the aging process and developing interventions to promote healthy aging. AI algorithms analyze biological markers to predict biological age and assess the effectiveness of anti-aging therapies (National Institute on Aging, 2022).

As an example, Retro Biosciences, backed by OpenAI CEO Sam Altman, is raising $1 billion to extend human lifespan through AI-driven research. The funds will support clinical trials targeting Alzheimer's and therapies for cellular rejuvenation, with the goal of launching its first drug by the 2020s' end (Financial Times, 2025).

10.2 Future Directions

Agentic AI application in healthcare is still in its early stages. The potential is immensely promising, and we will devote a significant portion of this chapter to exploring future possibilities.

10.2.1 Wearable Devices and IoT Integration

The integration of AI agents with wearable devices and IoT technologies is transforming healthcare delivery, enabling continuous monitoring, personalized treatment, and proactive health management. Wearable devices equipped with advanced sensors collect physiological data such as heart rate, glucose levels, respiratory rate, and sleep patterns. AI agents process this data in real time to detect anomalies, predict health issues, and recommend interventions, significantly enhancing the scope and quality of patient care.

For example, Empatica's Embrace smartwatch is designed for epilepsy patients, using AI to detect generalized tonic–clonic seizures and alert caregivers instantly. This device not only provides real-time monitoring but also adds a layer of safety and reassurance for patients and their families (Empatica, 2024). Another notable innovation is Samsung's Galaxy Ring, which offers tracking of movement, heart rate, sleep quality, and respiratory function. Integrated with Samsung Health, the Galaxy Ring provides personalized wellness recommendations based on real-time health analytics (Samsung, 2024).

AI and IoT together, often referred to as the Artificial Intelligence of Things (AIoT), enable systems to deliver intelligent, data-driven healthcare. IoT devices such as smartwatches, medical-grade fitness trackers, and connected medical sensors gather extensive health data. AI agents analyze this data to offer predictive insights into patient health, detect early signs of diseases, and recommend tailored treatments. The convergence of these technologies improves diagnostic accuracy and enhances patient engagement. For instance, Augnito highlights six AI-powered wearable healthcare devices designed for real-time tracking and personalized insights, demonstrating how IoT-enabled wearables bridge gaps in proactive healthcare delivery (Augnito, 2024).

One notable area of application is human activity recognition (HAR), where AI agents analyze data from wearable sensors to identify patterns in physical activities. This capability is crucial for monitoring chronic conditions, elderly care, and rehabilitation. Wearable devices combined with AI agents can detect falls, monitor adherence to medication schedules, and provide actionable feedback to healthcare providers. A study demonstrates how HAR systems integrated with IoT devices improve the accuracy and reliability of activity tracking (Liu et al., 2022).

In addition, smart wearables are facilitating advancements in preventive healthcare. For instance, wearable glucose monitors like Dexcom G7 integrate AI to continuously track blood sugar levels, enabling diabetic patients to maintain optimal glucose control (Dexcom, 2024). Similarly, heart health monitoring devices like Fitbit's Sense smartwatch use AI to analyze ECG data, providing users with early warnings about potential heart-related conditions (Fitbit, 2024).

Looking to the future, wearable devices and IoT technologies integrated with AI are expected to achieve even greater levels of precision and personalization. Advances in nanotechnology may lead to the development of implantable sensors capable of providing continuous biometric data directly from within the body. These devices could work in tandem with AI agents to offer predictive insights for managing chronic conditions and optimizing overall health. Furthermore, the next generation of wearables is likely to incorporate advanced capabilities such as real-time genetic analysis, enabling ultra-personalized preventive healthcare.

The expansion of IoT ecosystems, coupled with breakthroughs in edge computing, will further enhance the capabilities of AI-powered wearables. By processing data locally, these devices can reduce latency, provide faster feedback, and function reliably in remote areas with limited connectivity.

10.2.2 Robotic Surgery

AI agent-powered robotic systems can assist surgeons in complex tasks, offering capabilities that surpass traditional methods.

A notable example is the da Vinci Surgical System, developed by Intuitive Surgical, which utilizes AI to provide surgeons with enhanced dexterity and control during minimally invasive procedures. This system translates the surgeon's hand movements into precise micro-movements of instruments, allowing for greater accuracy in delicate surgeries. The da Vinci system has been widely adopted across various surgical disciplines, including urology, gynecology, and cardiothoracic surgery (Intuitive Surgical, 2024).

The Smart Tissue Autonomous Robot (STAR), developed by a team at Johns Hopkins University, represents a significant advancement in autonomous surgical robotics. STAR has demonstrated the capability to perform soft tissue surgeries with a level of precision comparable to human surgeons, utilizing AI algorithms to plan and execute surgical tasks. This development indicates the potential for AI to play a more autonomous role in future surgical procedures (National Institute of Biomedical Imaging and Bioengineering, 2024).

Looking to the future, AI in robotic surgery is poised to advance even further, paving the way for semiautonomous and fully autonomous surgical systems. Future innovations could include robots capable of adapting to unexpected intraoperative challenges in real time, using predictive models and reinforcement learning. Additionally, integrating multimodal data—such as imaging, genomics, and real-time sensor inputs—into surgical planning could enable hyper-personalized procedures. These advancements have the potential to expand access to high-quality surgical care, particularly in underserved regions where expert surgeons may be scarce. As AI and robotics continue to evolve, they are expected to redefine the standards of precision, safety, and efficiency in surgery, fundamentally transforming the field.

10.2.3 Multi-agent Collaboration in Healthcare

Advanced multi-agent systems are expected to simulate and enhance the teamwork of healthcare professionals, seamlessly integrating insights from various specialties to deliver holistic and personalized care. These systems will likely evolve to dynamically coordinate diagnostic, treatment planning, and care delivery processes, ensuring smooth transitions and reducing fragmentation in patient management.

In the future, multi-agent systems could become central to cross-disciplinary research by aggregating and synthesizing data from diverse sources, including genomics, proteomics, and clinical trials. This integration would enable the identification of novel patterns and insights, accelerating breakthroughs in precision medicine. Such systems may also streamline collaboration between researchers, clinicians, and policymakers, supporting data-driven decision-making and advancing public health initiatives.

As these systems become more sophisticated, they are poised to not only optimize patient outcomes but also drive systemic innovations in healthcare, enabling a more interconnected and efficient ecosystem. These advancements will pave the way for a new era of precision medicine and collaborative healthcare delivery.

10.2.4 AI Agents in Genomics and Multi-omics Integration

The future of AI agents in genomics and multi-omics integration promises transformative advancements in personalized medicine and therapeutic innovation. By leveraging multi-omics data—including genomics, transcriptomics, proteomics, and metabolomics—AI agents will provide an increasingly detailed understanding of complex biological processes. These systems are expected to analyze vast datasets in real time, enabling the discovery of molecular signatures associated with diseases and identifying unique biomarkers for precise diagnostics.

Future AI-driven platforms will likely integrate real-time multi-omics analysis with patient health records and other clinical data, allowing for highly personalized treatment plans tailored to each individual's molecular profile. This evolution will empower healthcare providers to offer more targeted and effective therapies, improving outcomes while minimizing side effects.

10.2.5 AI Agent in Mental Healthcare

AI agents can offer innovative solutions to address the growing demand for accessible and personalized care. These AI-driven tools, including chatbots and virtual assistants, provide immediate, around-the-clock support, thereby reducing barriers such as cost and stigma associated with traditional therapy (APA, 2024). They assist users in monitoring moods, managing stress, and accessing therapeutic resources, effectively complementing human-led treatments (Woebot Health, 2024).

One prominent example is Woebot, an AI-powered chatbot developed to deliver cognitive behavioral therapy techniques through brief daily conversations. Woebot engages users in mood tracking and offers curated content to help manage mental health challenges, demonstrating the potential of AI in providing scalable and immediate support (Darcy, 2024). Woebot's design combines emotional intelligence and natural language processing, enabling a personalized experience for users (Woebot Health, 2024).

AI agents also play a significant role in early detection and intervention by analyzing patterns in user interactions to identify signs of mental health concerns. For instance, AI algorithms can process data from various sources, including medical records and user inputs, to predict the likelihood of conditions such as depression or anxiety, enabling timely and personalized interventions (APA, 2024). These capabilities provide a critical layer of support, particularly for individuals who may not have immediate access to professional care.

Moreover, AI-driven conversational agents are being integrated into mental health care settings to augment traditional therapy. These agents provide psychoeducation, coping strategies, and continuous monitoring, thereby enhancing the therapeutic process and offering support between sessions (Smythos, 2024). However, it is crucial to recognize that while AI agents can supplement mental health care, they are not substitutes for professional human therapists, especially for complex mental health issues (APA, 2024).

10.2.6 Ethical and Regulatory Frameworks

The future of ethical and regulatory frameworks in healthcare AI will revolve around creating dynamic systems that adapt to the rapid evolution of technology while safeguarding patient trust and safety. As AI agents become deeply embedded in healthcare workflows, frameworks will likely focus on fostering transparency, ensuring accountability, and mitigating risks related to data privacy, algorithmic bias, and decision-making explainability.

Future ethical frameworks are expected to prioritize equitable access to AI technologies, ensuring they benefit diverse populations without perpetuating disparities. These systems will also need to address the implications of increasingly autonomous AI agents, establishing clear guidelines for accountability when decisions are made with minimal human intervention.

On the regulatory front, evolving frameworks will likely include real-time monitoring and validation mechanisms for AI systems to ensure ongoing compliance with safety and performance standards. Collaboration between technology developers, healthcare providers, and policymakers will be essential to design adaptive regulations that balance innovation with ethical responsibilities. International cooperation may also emerge as critical, given the global nature of healthcare and the cross-border use of AI technologies.

10.2.7 *The Role of AI in Global Health*

The future of AI in global health lies in its ability to fundamentally transform healthcare accessibility and equity. AI agents are expected to play a critical role in bridging disparities by delivering innovative solutions tailored to the unique challenges of low-resource settings. Advanced telemedicine platforms and mobile health applications integrated with AI will enable real-time diagnosis, monitoring, and treatment, even in the most remote regions. This will empower healthcare providers to address the needs of underserved populations more effectively.

AI's ability to analyze large-scale global health data will also drive significant advancements in public health planning and crisis management. Future AI systems will not only forecast disease outbreaks with greater precision but also provide actionable insights to optimize resource allocation and response strategies. These systems will leverage environmental, demographic, and behavioral data to anticipate health trends and inform policy decisions on a global scale.

Moreover, AI is poised to facilitate cross-border collaboration by harmonizing data and enabling seamless integration of health systems. This will enhance international cooperation in addressing global health challenges such as pandemics, antimicrobial resistance, and climate-related health impacts. By fostering innovation and delivering scalable solutions, AI will play a pivotal role in reducing health inequities and improving outcomes worldwide.

10.3 Summary

This chapter investigates the transformative potential of AI agents in healthcare. The chapter begins by exploring clinical decision support systems, emphasizing their role in enhancing diagnostic accuracy and treatment planning through multi-agent frameworks and advanced learning techniques. Applications like TrialGPT for clinical trial recruitment and predictive analytics systems for sepsis and cardiac event prediction illustrate how AI improves patient outcomes and operational efficiency.

In diagnostics, AI-powered tools like Google Health's breast cancer detection model demonstrate remarkable precision in medical imaging. In drug discovery, innovations like AlphaFold 3 and CRISPR-GPT accelerate research by enabling molecular modeling and gene-editing processes. The chapter also examines the role of AI in administrative workflows, showcasing how automation reduces inefficiencies and improves healthcare delivery.

Further, the chapter highlights AI agents' contributions to patient engagement through conversational health assistants and wearable technologies. It concludes with a discussion on future trends, including AI's role in genomics, longevity research, and global health. Ethical and regulatory considerations are presented as critical to ensuring equitable, safe, and effective integration of AI into healthcare systems.

10.4 Questions

Multiple Choice

1. Which of the following is NOT a mentioned application of AI agents in healthcare?
   a) Clinical decision support
   b) Drug discovery
   c) Financial auditing
   d) Predictive analytics
2. What is the primary function of TrialGPT?
   a) Automating surgical procedures
   b) Streamlining clinical trial recruitment
   c) Analyzing medical images
   d) Providing mental health support
3. Which technology is used by AlphaFold 3 to predict interactions between biological molecules?
   a) Robotic process automation
   b) Blockchain
   c) CRISPR
   d) Computational modeling
4. What is the main purpose of CRISPR-GPT?
   a) To provide telehealth services
   b) To automate and enhance gene-editing experiments
   c) To monitor patient vital signs
   d) To manage hospital administrative tasks
5. Which of the following is an example of AI enhancing patient engagement?
   a) Robotic surgery
   b) Wearable health monitors
   c) Automated billing systems
   d) Virtual health assistants

True/False

1. True or False: AI agents are primarily used to replace doctors entirely in the healthcare industry.

2. True or False: Sepsis management systems using multi-agent frameworks have been shown to improve patient outcomes.
3. True or False: AlphaFold 3 can only predict protein structures and cannot model interactions with other molecules.
4. True or False: The da Vinci Surgical System is an example of AI-driven robotic surgery.
5. True or False: AI agents are not currently being used in mental healthcare applications.

Simple Answer Questions

1. What are two specific areas where AI agents are used in clinical decision support?
2. Name two types of data that AI-powered predictive analytics systems use to anticipate critical health events.
3. What is the role of AI in wearable devices like the Empatica Embrace smartwatch?
4. What is one advantage of using collaborative robots (cobots) in healthcare administration?
5. What does the acronym "CRISPR" stand for in the context of gene editing?

Essay Questions

1. Discuss the impact of AI agents on drug discovery, highlighting the contributions of specific technologies like AlphaFold 3.
2. Explain how multi-agent systems can be applied in healthcare settings, providing examples of their use in clinical decision support and care coordination.
3. Describe the potential benefits and challenges of integrating AI agents into mental healthcare.
4. Analyze the role of AI in advancing personalized medicine, particularly in the context of genomics and multi-omics integration.
5. Discuss the ethical and regulatory considerations that need to be addressed as AI agents become more prevalent in healthcare.

References

AHRQ Digital Healthcare Research. (2024). *Continuous predictive analytics monitoring to improve care for at-risk patients*. Agency for Healthcare Research and Quality. https://digital.ahrq.gov/program-overview/research-stories/continuous-predictive-analytics-monitoring-improve-care-risk

American Psychological Association (APA). (2024). *Artificial intelligence in mental health care*. American Psychological Association. https://www.apa.org/practice/artificial-intelligence-mental-health-care

Augnito. (2024). *6 Devices for personalized healthcare*. Augnito. https://augnito.ai/resources/6-devices-for-personalized-healthcare/

Borkowski, A. A., & Ben-Ari, A. (2024). *Multi-agent AI systems in healthcare: Technical and clinical analysis*. Preprints.org. doi:https://doi.org/10.20944/preprints202410.0182.v1

Critical Care Forum. (2022). *Dynamic model using random forest classification predicts respiratory deterioration ahead of clinical recognition*. Critical Care Forum. https://doi.org/10.1186/s13054-022-03915-3.

Darcy, A. (2024). *Alison Darcy*. Wikipedia. https://en.wikipedia.org/wiki/Alison_Darcy

Dexcom. (2024). *Dexcom G7 continuous glucose monitoring system*. Dexcom. https://www.dexcom.com/

Drug Discovery Trends. (2024). *Meet AlphaFold 3, which can accurately model more than 99% of molecular types in the protein data bank*. Drug Discovery Trends. https://www.drugdiscoverytrends.com/meet-alphafold-3-which-can-accurately-model-more-than-99-of-molecular-types-in-the-protein-data-bank/

Empatica. (2024). *Empatica embrace: AI-powered seizure monitoring device*. Empatica. https://en.wikipedia.org/wiki/Empatica

Financial Times. (2025, January 23). *Sam Altman-backed Retro Biosciences to raise $1bn for project to extend human life*. https://www.ft.com/content/25a473ea-9f87-474a-8729-bc5287df853a

Fitbit. (2024). *Fitbit sense: Advanced heart health monitoring*. Fitbit. https://www.fitbit.com/global/us/products/smartwatches/sense

Frontiers in Robotics and AI. (2024). *A systematic review of collaborative robots for nurses*. Frontiers in Robotics and AI. https://www.frontiersin.org/journals/robotics-and-ai/articles/10.3389/frobt.2024.1398140/full?utm_source=chatgpt.com

Google. (2024). *Mental health and google AI principles*. Google Blog. https://blog.google/technology/health/google-ai-and-health/mental-health-google-ai-principles/

Google DeepMind. (2024). *Google DeepMind Isomorphic AlphaFold 3 AI Model*. Google Blog. https://blog.google/technology/ai/google-deepmind-isomorphic-alphafold-3-ai-model/

Huang, K., Qu, Y., Cousins, H., Johnson, W. A., Yin, D., Shah, M., Zhou, D., Altman, R., Wang, M., & Cong, L. (2024). *CRISPR-GPT: An LLM agent for automated design of gene-editing experiments*. arXiv. https://arxiv.org/abs/2404.18021

Intuitive Surgical. *About the da Vinci Surgical System*. Intuitive Surgical. Accessed December 3, 2024. https://www.intuitive.com/en-us/patients/da-vinci-robotic-surgery/about-the-systems

Isomorphic Labs. (2024). *Rational drug design with AlphaFold 3*. Isomorphic Labs. https://www.isomorphiclabs.com/articles/rational-drug-design-with-alphafold-3

Jin, Q., Wang, Z., Floudas, C. S., Chen, F., et al. (2024). *Matching patients to clinical trials with large language models*. National Center for Biotechnology Information (NCBI), National Library of Medicine (NLM), National Institutes of Health.

Liu, R., Ramli, A. A., Zhang, H., Henricson, E., & Liu, X. (2022). An Overview of Human Activity Recognition Using Wearable Sensors: Healthcare and Artificial Intelligence. In B. Tekinerdogan, Y. Wang, & L. J. Zhang (Eds.), *Internet of Things – ICIOT 2021. ICIOT 2021* (Lecture Notes in Computer Science()) (Vol. 12993). Springer. https://doi.org/10.1007/978-3-030-96068-1_1

Magee, M. F., Nassar, C. M., Dunlea, R., James, B., Montero, A. R., Sack, P., Tweedt, A., Vetal, D., & Cobb, N. K. (2022). Engagement, Satisfaction, and Preliminary Impact of a Personalized Diabetes Chatbot in a Diverse Cohort of Adults with Type 2 Diabetes. *Diabetes, 71*(Supplement 1), 330-OR. https://diabetesjournals.org/diabetes/article/71/Supplement_1/330-OR/146316/330-OR-Engagement-Satisfaction-and-Preliminary

Mayo Clinic. *AI in Healthcare: The Future of Patient Care and Health Management*. Mayo Clinic Press, April 2023. https://mcpress.mayoclinic.org/healthy-aging/ai-in-healthcare-the-future-of-patient-care-and-health-management/

Mayo Clinic Platform. (2024). *Using AI to predict the onset of sepsis*. Mayo Clinic Platform. May 2, 2024.https://www.mayoclinicplatform.org/2024/05/02/using-ai-to-predict-the-onset-of-sepsis/

National Institute of Biomedical Imaging and Bioengineering. *Robot performs soft tissue surgery with minimal human help*. NIBIB Newsroom. Accessed December 3, 2024. https://www.nibib.nih.gov/news-events/newsroom/robot-performs-soft-tissue-surgery-minimal-human-help

National Institute on Aging. *Harnessing artificial intelligence to explore exceptional longevity*. National Institute on Aging, July 2022. https://www.nia.nih.gov/research/blog/2022/07/harnessing-artificial-intelligence-explore-exceptional-longevity

Pragmatic Coders. (2023). *Intelligent automation in healthcare in 2024: Trends and use cases*. Pragmatic Coders Blog. https://www.pragmaticcoders.com/blog/intelligent-automation-in-healthcare-in-2024?utm_source=chatgpt.com

Samsung. (2024). *Samsung's Galaxy Ring: A comprehensive health tracker*. Lifewire. https://www.lifewire.com/samsung-s-new-wearables-aimed-at-making-wellness-more-meaningful-8675470

Smythos. (2024). *Conversational agents in mental health support*. Smythos. https://smythos.com/artificial-intelligence/conversational-agents/conversational-agents-in-mental-health-support/

Vennaro, N. (2024a). *AI agent based implementation of 'Cobots' in healthcare office administration: A cloud-native solution using Microsoft technologies*. SSRN preprint. https://papers.ssrn.com/sol3/papers.cfm?abstract_id=4922642

Vennaro, N. (2024b). *Collaborative robots and healthcare office administration: Transforming efficiency with AI and ML*. SSRN. https://papers.ssrn.com/sol3/papers.cfm?abstract_id=4922642&utm_source=chatgpt.com

VentureBeat. (2024). *Google DeepMind open sources AlphaFold 3, ushering in a new era for drug discovery and molecular biology*. VentureBeat. https://venturebeat.com/ai/google-deepmind-open-sources-alphafold-3-ushering-in-a-new-era-for-drug-discovery-and-molecular-biology/

Woebot Health. (2024). *Woebot health: AI-powered mental health Chatbot*. Deepgram. https://deepgram.com/ai-apps/woebot-health

Ken Huang is a prolific author and globally recognized authority in AI and Web3, with an extensive portfolio of published works that bridge business strategy, technical implementation, and cutting-edge research. As Fellow of Cloud Security Alliance and Co-Chair of the AI Safety Working Groups at the Cloud Security Alliance and the AI STR Working Group at the World Digital Technology Academy under the UN Framework, he is a leading voice in shaping global AI governance and security standards.

Huang is the CEO and Chief AI Officer (CAIO) of DistributedApps.ai, a firm specializing in generative AI training and consulting. His contributions to the field include being a core contributor to the OWASP Top 10 Risks for LLM Applications and an active participant in the NIST Generative AI Public Working Group.

Notable Publications:

Beyond AI: ChatGPT, Web3, and the Business Landscape of Tomorrow (Springer, 2023)—Strategic insights into AI and Web3's business applications.

Generative AI Security: Theories and Practices (Springer, 2024)—A comprehensive guide on securing generative AI systems.

Practical Guide for AI Engineers (Volumes 1 and 2, DistributedApps.ai, 2024)—Essential resources for AI and ML engineers.

The Handbook for Chief AI Officers: Leading the AI Revolution in Business (DistributedApps.ai, 2024)—A road map for CAIOs in implementing GenAI across organizations.

Web3: Blockchain, the New Economy, and the Self-Sovereign Internet (Cambridge University Press, 2024)—Insights into the convergence of AI, blockchain, IoT, and emerging technologies.

Blockchain and Web3: Building the Cryptocurrency, Privacy, and Security Foundations of the Metaverse (Wiley, 2023)—Recognized as a must-read by TechTarget in 2023 and 2024.

Ken is a sought-after speaker and has presented at events such as the World Economic Forum in Davos, ACM and IEEE conferences, the CSA AI Summit, Depository Trust and Clearing Corporation forums, and World Bank conferences. His recent appointment to the OpenAI Forum reflects his ongoing commitment to advancing collaboration and dialogue in the field of AI.

Explore Ken's work on Amazon: https://www.amazon.com/author/kenhuang

Chapter 11
AI Agents in Robotics

Jerry Huang and Ken Huang

This chapter looks into the integration of AI agents in the physical world through robotics. As we advance into an era where machines interact directly with the tangible world, NVIDIA CEO Jensen Huang has described this progression as "physical AI" (Freund, 2024). Robotics, particularly humanoid robotics, represents a transformative technology that could drive unprecedented automation and productivity. I strongly believe that the next company to achieve a $10 trillion market capitalization may emerge from this sector. Whether it is established players like NVIDIA, Tesla, Unitree, and Boston Dynamics, or a pioneering startup, the innovations in robotics promise significant value creation. This chapter examines AI agents in robotics, exploring their frameworks, global competition, and technological contributions.

11.1 Agentic Framework for Robotics

AI agents in robotics operate through adaptable, modular frameworks, enabling both rapid innovation and functional scalability. The proposed framework for robotics includes several key modules.

Figure 11.1 illustrates the core modules of the agentic framework for robotics, emphasizing perception, interaction, and reasoning modules that enable AI agents to operate effectively in dynamic environments.

J. Huang (✉) · K. Huang
The University of Chicago, Chicago, IL, USA

DistributedApps.ai, Fairfax, VA, USA
e-mail: ken@distributedapps.ai; ken@distributedapps.ai

K. Huang (ed.), *Agentic AI*, Progress in IS,
https://doi.org/10.1007/978-3-031-90026-6_11

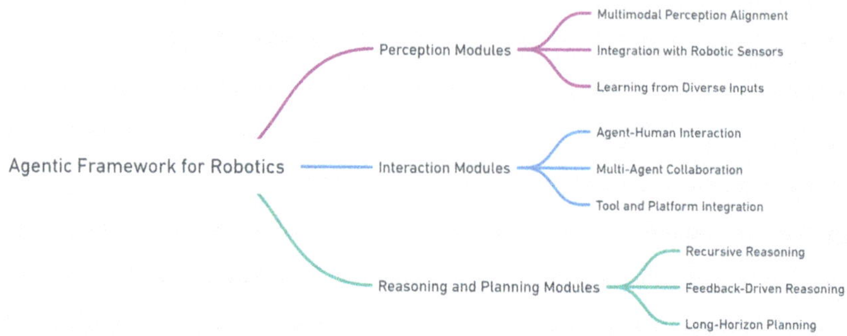

Fig. 11.1 Agentic framework for robotics

11.1.1 Perception Modules

Perception is the gateway for AI agents to sense and interpret their surroundings. These modules process multimodal inputs such as visual, auditory, and sensor data, enabling robots to respond to dynamic environments effectively.

- **Multimodal Perception Alignment**: AI agents convert sensory data into textual or numerical representations for real-time decision-making. Advances in multimodal interaction systems, as discussed in recent research, further enhance this process by integrating data streams from multiple sensory modalities such as images, audio, and sensor readings. Frameworks like LLMs (Large Language Models) and VLMs (Visual Language Models) allow for a unified representation of this data, enabling robots to adapt to diverse and unpredictable scenarios with greater accuracy and contextual awareness (Fei-Fei et al., 2024).
- **Integration with Robotic Sensors**: Sensors such as LiDAR (see Box LiDAR), cameras, and haptic feedback systems supply critical data for navigation, fault detection, and human interaction. For example, LiDAR enables accurate mapping for robots navigating complex terrains (Guo et al., 2024). Multimodal systems can augment sensor integration by allowing perception modules to cross-validate and enhance data accuracy, such as using camera data to complement LiDAR readings for improved object detection and environmental mapping.
- **Learning from Diverse Inputs**: Training perception modules traditionally relies on diverse datasets, including indoor and outdoor scenarios. While high-quality labeled data enhances an agent's ability to operate across variable environments, recent advancements in unsupervised and self-supervised learning techniques reduce dependency on extensive labeling. These approaches enable AI agents to extract useful features from raw, unlabeled data, significantly improving adaptability and scalability in perception across dynamic environments. Moreover, the inclusion of emergent abilities from multimodal training allows robots to generalize effectively, adapting to novel conditions without requiring extensive retraining (Fei-Fei et al., 2024).

Box: LiDAR

LiDAR (Light Detection and Ranging) is a sophisticated remote sensing technology that uses laser pulses to measure distances and create detailed 3D models of the environment (Guo et al., 2024). It operates by emitting rapid laser light pulses and measuring the time it takes for the reflected light to return to the sensor, allowing for highly accurate distance calculations. This technology is used in various fields, including autonomous vehicles, surveying, archaeology, and environmental monitoring, due to its ability to provide precise, three-dimensional mapping of surroundings. LiDAR systems typically consist of a laser emitter, a scanner to receive reflected light, and GPS for positioning, enabling them to create point clouds that represent the scanned environment with remarkable detail and accuracy. While LiDAR offers superior precision in detecting objects and their orientations compared to other sensors like radar, it can be more sensitive to weather conditions and is generally more expensive, making the choice between different sensing technologies dependent on the specific application and environmental conditions.

11.1.2 Interaction Modules

Interaction modules enable AI agents to communicate and collaborate with humans, other agents, and robotic systems, moving beyond conversational interfaces to physical manipulation and cooperative task execution.

- **Agent–Human Interaction:** Intuitive interfaces, such as natural language processing (NLP) systems, empower humans to guide robots through spoken or written commands.
- **Multi-agent Collaboration:** In multi-agent systems, specialized agents collaborate to execute tasks requiring distinct capabilities like navigation, manipulation, or diagnostics. They operate through shared protocols and objectives.
- **Tool and Platform Integration:** Seamless integration of AI agents with robotic control systems, APIs, and industrial tools ensures smooth execution of workflows, whether for manufacturing or research purposes.

11.1.3 Reasoning and Planning Modules

Reasoning modules provide robots with the capacity for complex decision-making and task execution. These modules enhance autonomy by breaking tasks into manageable steps.

Recursive Reasoning: Recursive reasoning strategies, though still largely experimental, are being explored to enable robots to handle intricate tasks by breaking them into smaller, manageable sub-tasks and iteratively refining their actions based on feedback. While this approach has shown promise in simulation environments for applications such as hierarchical task planning, its implementation in real-world robotic systems remains limited due to computational challenges and the need for robust error handling during task decomposition and execution.

Feedback-Driven Reasoning: Real-time feedback from sensors and users helps robots adjust their plans dynamically, ensuring robustness in unpredictable conditions. Emerging research on agent AI emphasizes the use of such feedback to refine decision-making processes continuously. This enables AI agents to respond effectively in real time, even in complex, unstructured environments.

Long-Horizon Planning: Memory integration enables robots to perform long-horizon planning by retaining contextual information for extended periods. This capability is particularly relevant in tasks such as coordinating multiple robotic units in industrial settings or sustaining autonomous navigation over lengthy missions. Furthermore, the use of multimodal data and emergent capabilities allows robots to integrate sensory and reasoning data for more accurate, adaptive long-term planning (Fei-Fei et al., 2024).

11.2 Advanced Modules for Humanoid Robots

Humanoid robots require specialized modules to address challenges such as physical interaction, social engagement, and hybrid control.

Figure 11.2 highlights specialized modules for humanoid robots, focusing on physical interaction, social intelligence, and hybrid control for collaborative tasks.

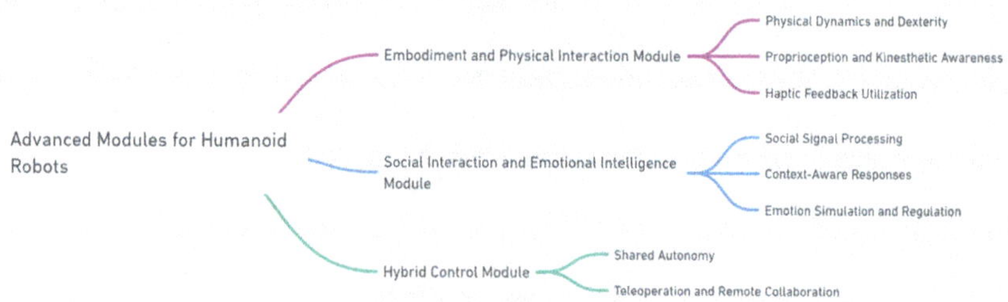

Fig. 11.2 Advanced modules for humanoid robots

11.2.1 Embodiment and Physical Interaction Module

Humanoid robots replicate humanlike physical capabilities to interact effectively with their environment. AI agents enable this embodiment through precise control and real-time adjustments.

- **Physical Dynamics and Dexterity:** Robots optimize control over multiple degrees of freedom for tasks like climbing stairs, maintaining balance, and manipulating delicate objects.
- **Proprioception and Kinesthetic Awareness:** AI agents process proprioceptive data—information about limb positions and movements—to ensure fluid and context-aware interactions.
- **Haptic Feedback Utilization:** Simulating a sense of touch through haptic sensors allows robots to evaluate textures, pressures, and other physical properties, ensuring appropriate handling of fragile objects.

11.2.2 Social Interaction and Emotional Intelligence Module

Humanoid robots often operate in roles requiring high social engagement. AI agents interpret complex human interactions, providing appropriate responses.

- **Social Signal Processing:** By analyzing cues like facial expressions, gestures, and tone of voice, robots adapt their interactions for roles in caregiving, education, and customer service.
- **Context-Aware Responses:** AI agents tailor behaviors to cultural norms, individual preferences, and situational demands.
- **Emotion Simulation and Regulation:** Robots simulate emotional responses to build trust and rapport.

11.2.3 Hybrid Control Module

Humanoid robots benefit from hybrid control systems integrating AI autonomy with human oversight for collaborative efficiency.

- **Shared Autonomy:** High-level human guidance paired with low-level robotic autonomy ensures seamless collaboration.
- **Teleoperation and Remote Collaboration:** In hazardous or remote environments, real-time control combined with sensory feedback allows robots to execute complex tasks with precision.

11.3 Simulation and Training

Simulation and training are foundational processes in the development of AI agents for robotics. These processes leverage computationally modeled environments to refine the algorithms, decision-making capabilities, and operational workflows of robots. This section examines simulation environments, methodologies for training, optimization techniques, and the integration of learning mechanisms for AI agents.

Figure 11.3 outlines the workflow for simulation and training in robotics, and we will cover the details in the subsequent sections.

11.3.1 Simulation Platforms: Architecture and Capabilities

Robotic simulation platforms are built on architectures designed to emulate physical environments and dynamics with high fidelity. At the core of these platforms are physics engines that simulate motion, forces, and interactions under real-world constraints such as gravity, friction, and material properties. Leading platforms such as NVIDIA Isaac Sim use GPU-accelerated physics to achieve real-time performance.

Fig. 11.3 Simulation and training workflow

Advanced features include rigid-body dynamics, soft-body simulations, and fluid dynamics, allowing developers to model diverse scenarios (NVIDIA, 2024a).

The integration of ray tracing technologies (see Box **Ray Tracing**) further enhances these platforms, enabling accurate rendering of light, shadows, and textures. This capability is useful for training AI perception modules, as it allows robots to interpret visual inputs under varying lighting and environmental conditions. Additionally, advanced platforms provide multi-agent support, allowing developers to simulate complex interactions between robots, humans, and the environment.

Box: Ray Tracing
Ray tracing is a rendering technique used in computer graphics to create realistic images by simulating the way light interacts with objects in a scene. It works by tracing the path of light rays from the viewer's perspective through each pixel of a virtual screen, calculating how these rays bounce off surfaces, interact with materials, and ultimately determine the color and brightness of pixels in the final image. This method accurately simulates light behavior, producing realistic reflections, refractions, shadows, and global illumination effects, resulting in incredibly lifelike scenes.

Unlike traditional rendering methods such as rasterization, ray tracing excels at computing visibility between points in space, making it particularly effective for creating complex lighting effects. While historically computationally intensive, advancements in hardware, particularly with dedicated ray tracing cores in modern GPUs, have made real-time ray tracing possible in applications like robotics and video games.

11.3.2 Synthetic Data Generation and Annotation Pipelines

Synthetic data generation is a critical component of simulation environments. Simulation platforms are equipped with tools that create annotated datasets by simulating diverse scenarios and object interactions. An essential aspect of this process is domain randomization, where elements such as object positions, textures, and environmental conditions are altered during data generation. This ensures that AI models trained in these environments can generalize their capabilities to real-world applications.

The annotation pipeline in these environments automates the labeling process, providing structured data such as object boundaries, depth maps, segmentation masks, and pose estimations. These datasets are used to train AI models for tasks such as object detection, semantic segmentation, and visual navigation. Techniques such as photorealistic rendering and procedural generation enhance the quality and diversity of the generated datasets.

11.3.3 Reinforcement Learning in Simulated Environments

Reinforcement learning (RL) is a predominant methodology used in simulated environments to develop decision-making capabilities for AI agents. In RL, an agent interacts with the environment, taking actions to maximize cumulative rewards based on feedback from its performance. Simulations provide an ideal setting for this iterative learning process, allowing agents to explore and optimize their strategies without real-world constraints.

The technical pipeline for RL in simulation involves several key components:

- **State Representation:** AI agents encode environmental states into numerical vectors using sensory inputs such as images, point clouds, or positional data. This encoding ensures that the agent captures relevant features for decision-making.
- **Action Space Definition:** The action space defines all possible movements or decisions an agent can take. Continuous or discrete action spaces are modeled depending on the complexity of the robot's capabilities.
- **Reward Functions:** Reward signals guide the learning process by evaluating the desirability of the agent's actions.
- **Exploration Strategies:** Exploration mechanisms, such as epsilon-greedy or policy-based exploration, allow agents to discover new strategies and optimize their performance. Balancing exploration and exploitation is a significant challenge addressed through algorithms like Proximal Policy Optimization (PPO) and Soft Actor-Critic (SAC). See box below.

Box: PPO and SAC

Proximal Policy Optimization (PPO) is a reinforcement learning algorithm that enhances training stability and performance by carefully managing policy updates. It operates on the principle of policy gradient methods, which optimize the policy directly based on the expected cumulative reward. PPO introduces a clipping mechanism that restricts how much the policy can change at each update, ensuring that the new policy remains close to the old one. This "proximal" approach prevents drastic updates that could lead to poor performance or instability, often referred to as falling off the cliff. By maintaining a balance between exploration and exploitation, PPO effectively improves learning efficiency and convergence in various environments.

Soft Actor-Critic (SAC) is an off-policy algorithm that integrates value-based and policy-based methods, emphasizing exploration through a maximum entropy framework. By including entropy in the objective function, SAC encourages diverse policy behaviors. While this approach reduces the risk of premature convergence to suboptimal policies, balancing entropy with reward optimization remains a challenge, as excessive exploration can slow convergence in complex environments.

Simulated environments enable large-scale RL experiments by parallelizing agent interactions across multiple instances, significantly accelerating training times.

11.3.4 Imitation Learning and Behavioral Cloning

Imitation learning (IL) is another methodology used extensively in robotic training. In IL, agents learn behaviors by mimicking expert demonstrations rather than optimizing reward signals. Simulations provide a controlled environment where developers can record demonstrations using scripted algorithms or teleoperated robots. These demonstrations are then used to train agents through techniques such as behavioral cloning, where supervised learning methods map input states to expert actions.

The technical workflow for IL involves the following:

- **Data Collection:** Recording expert trajectories in the simulation, including states, actions, and contextual information.
- **Policy Learning:** Training neural networks to replicate the expert's decision-making process.
- **Evaluation and Refinement:** Comparing the agent's performance to the expert's and iteratively refining the model to minimize errors.

IL is particularly effective for tasks requiring precise, humanlike control, such as object manipulation or social interactions. Advanced implementations combine IL with RL, where IL provides an initial policy that is further optimized through RL.

11.3.5 Optimization Techniques for Robotic Learning

The training of AI agents in simulated environments requires robust optimization techniques to ensure efficiency and convergence. Some of the key methods include the following:

- **Gradient-Based Optimization:** Neural network weights are updated using gradient descent algorithms. Variants such as Adam and RMSProp are commonly employed for their adaptability to sparse or noisy gradients in reinforcement learning tasks.
- **Meta-learning:** Meta-learning frameworks allow agents to adapt quickly to new tasks by leveraging prior knowledge.
- **Transfer Learning:** Simulation-trained models are often adapted to real-world scenarios through transfer learning. This involves fine-tuning pre-trained models using domain-specific data to bridge the "sim-to-real" gap.
- **Multi-objective Optimization:** Tasks in robotics often require balancing multiple objectives, such as speed, energy efficiency, and safety.

11.3.6 Tools for Continuous Learning and Adaptation

Simulations support continuous learning, where AI agents refine their policies over time based on new experiences. This process involves integrating real-world observations back into the simulated training environment, allowing agents to adapt to evolving conditions.

Memory-augmented neural networks, such as Neural Turing Machines (NTMs) and Differentiable Neural Computers (DNCs), play a crucial role in these processes. These architectures allow AI agents to store and retrieve information dynamically, enabling long-horizon planning and context-aware decision-making. See box below:

Box: Memory-Augmented Neural Networks
Memory-augmented neural networks (MANNs), such as Neural Turing Machines (NTMs), are advanced artificial neural network architectures that incorporate external memory components to enhance their computational capabilities. These networks combine traditional neural network structures with memory modules, allowing them to store and retrieve information efficiently. Key features include an external memory matrix, a controller network, read and write operations, and attention mechanisms. MANNs excel at handling tasks involving complex sequences and long-term dependencies, can rapidly adapt to new information, and are capable of learning and executing various algorithms. Their applications span natural language processing, meta-learning, and algorithmic problem-solving. By merging the pattern matching strengths of neural networks with the algorithmic power of programmable computers, MANNs represent a significant advancement in artificial intelligence, offering improved performance in tasks requiring complex reasoning and information retention.

Differentiable Neural Computers (DNCs) represent a novel approach to memory-augmented neural networks, combining pattern recognition with algorithmic reasoning capabilities. However, their application in robotics remains largely experimental, with challenges in scalability, latency, and integration into real-time systems limiting their deployment in practical robotic scenarios.

11.3.7 High-Performance Computing for Simulation Scaling

The computational demands of robotic simulation and training require high-performance computing (HPC) solutions. Parallelization is a critical strategy for scaling simulations, where multiple environments are run simultaneously across GPUs or distributed cloud systems. HPC frameworks, such as NVIDIA's CUDA and TensorFlow's distributed training libraries, provide the infrastructure for efficient simulation scaling (Abadi et al., 2016).

Additionally, advances in generative AI have enabled the creation of simulation environments that dynamically adapt to the agent's learning progress.

11.3.8 Simulation Fidelity and Realism

Achieving high fidelity in simulation environments is essential for training effective AI agents. Physics engines incorporate advanced models for rigid and soft-body dynamics, fluid simulations, and contact mechanics to replicate real-world interactions. Visual realism is achieved through photorealistic rendering techniques, including real-time ray tracing and global illumination.

High-fidelity simulations also incorporate sensor models to emulate the behavior of robotic sensors such as LiDAR, cameras, and IMUs. Noise modeling and sensor imperfections are added to replicate real-world conditions, ensuring that AI agents develop robust perception capabilities.

Simulation environments form the backbone of training pipelines for AI agents in robotics. Their technical architecture, combined with advanced methodologies such as reinforcement learning, imitation learning, and optimization techniques, enables the development of robust, adaptive, and high-performing robotic systems. By focusing on fidelity, scalability, and adaptability, simulations provide a fertile ground for innovation, accelerating the journey from theoretical concepts to functional robotic applications.

11.4 Spatial Intelligence and 3D

AI agents in robotics rely heavily on 3D understanding and spatial intelligence to perceive and interact with their environments effectively. These capabilities enable robots to navigate, manipulate objects, and understand spatial relationships in real-world scenarios.

Figure 11.4 visualizes the integration of spatial intelligence and 3D understanding in robotics.

11.4.1 Spatial Reasoning and Contextual Awareness

Spatial reasoning involves the ability of AI systems to understand and interpret the physical layout and geometry of their surroundings. AI agents utilize data from sensors such as LiDAR, stereo cameras, and depth sensors to construct 3D representations of their environment. These representations allow the robots to perform complex tasks such as mapping, obstacle avoidance, and path planning.

Fig. 11.4 Spatial intelligence and 3D understanding

Key methodologies in spatial reasoning include the following:

- **Point Cloud Processing:** AI systems analyze point clouds, which are collections of 3D coordinates captured by depth sensors, to recognize objects and understand spatial relationships.
- **Simultaneous Localization and Mapping (SLAM):** SLAM enables robots to build and update maps of unknown environments while simultaneously tracking their position.
- **Volumetric Representations:** AI agents use volumetric methods, such as truncated signed distance functions (TSDFs), to generate detailed 3D models of their environment.

11.4.2 Object Manipulation and Interaction

Robots equipped with 3D understanding can perform precise object manipulation tasks, a capability essential for industrial automation, healthcare, and household assistance. This requires spatial intelligence to interpret object geometry, assess grip points, and predict interaction dynamics.

Recent advancements integrate spatial intelligence with tactile feedback systems, allowing robots to adjust their grip dynamically based on object properties.

For instance, proprioceptive and exteroceptive data streams are fused to provide context-aware manipulation, ensuring both precision and safety during operations.

11.4.3 Spatial Intelligence Lab's Contributions to 3D Understanding

Fei-Fei Li's Spatial Intelligence Lab has introduced groundbreaking technologies in the realm of spatial intelligence, focusing on enabling AI systems to perceive and interact with three-dimensional spaces. The company's development of Large World Models (LWMs) represents a significant leap forward. These models allow for the creation and interaction with virtual 3D environments generated from diverse data inputs, including 2D images (Spatial Intelligence Lab, 2024).

The core innovation lies in the ability to transform static imagery into dynamic, navigable 3D scenes. This involves the following:

- **3D Scene Reconstruction:** Leveraging deep learning models trained on extensive datasets, Spatial Intelligence Lab's technology reconstructs 3D environments by inferring depth, texture, and spatial relationships from single or multiple images.
- **Interactive Navigation:** Once a 3D scene is reconstructed, users or robots can explore the virtual environment interactively. This capability allows for simulation and testing of robotic operations in highly realistic, customized settings.
- **Hybrid Data Utilization:** Spatial Intelligence Lab integrates synthetic data with real-world imagery to train its models. By augmenting real-world data with procedurally generated 3D scenarios, the models gain robustness and adaptability across diverse operational contexts.

The applications of Spatial Intelligence Lab's technology extend to training AI agents for spatial tasks such as navigation and exploration. For example, virtual environments generated by LWMs can simulate intricate spatial configurations, allowing robots to learn adaptive strategies for dealing with unknown terrains or dense obstacles.

11.4.4 Enhancements in Spatial Intelligence for Robotics

Advanced 3D understanding requires not only interpreting spatial configurations but also recognizing and predicting changes in the environment. Temporal–spatial modeling is an emerging area where robots are trained to understand spatial dynamics over time. This capability is particularly relevant for tasks requiring interaction with moving objects or navigating environments with dynamic elements, such as traffic systems or crowded spaces.

Key advancements include the following:

- **Graph Neural Networks (GNNs):** GNNs model spatial relationships between objects as graph structures, enabling robots to analyze spatial hierarchies and dependencies.
- **Scene Graphs:** Scene graphs provide a semantic representation of a 3D environment by linking objects and their spatial relationships.
- **World Simulation for Spatial Learning:** Spatial Intelligence Lab's approach of creating immersive 3D scenes aligns with the broader trend of world simulation. These virtual worlds not only improve the spatial intelligence of robots but also provide datasets for reinforcement learning and imitation learning in navigation, manipulation, and exploration.

11.4.5 Integration of Vision and Spatial Perception

A critical aspect of 3D understanding is the integration of visual perception with spatial intelligence. Robots combine visual inputs with depth and motion data to achieve a holistic understanding of their surroundings. Techniques like monocular depth estimation and multi-view stereo are widely adopted to extract depth information from camera images. LiDAR sensors, on the other hand, provide direct distance measurements, complementing visual data for enhanced accuracy.

Spatial Intelligence Lab's technologies explore the use of advanced neural architectures for integrating visual and spatial data. Transformer models (Vaswani et al., 2017), while emerging as a powerful tool for cross-modal integration, are still being adapted for real-time applications in robotics. Challenges such as computational overhead and latency need to be addressed for their effective deployment in dynamic, 3D environments.

11.4.6 Semantic and Functional Understanding of 3D Spaces

Beyond geometry, semantic understanding adds a layer of intelligence by associating meaning and functionality with objects and spaces. Robots must not only recognize objects but also understand their roles and relationships within the environment. For example, in a kitchen setting, a robot should distinguish between a stove, a sink, and a countertop, understanding their respective functions.

Spatial Intelligence Lab's models incorporate semantic segmentation and object classification into their spatial intelligence frameworks. By using large-scale annotated datasets and transfer learning techniques, these models enable robots to associate semantic labels with 3D reconstructions. This semantic layer is essential for task planning, as it guides the robot's decisions on how to interact with different elements of its environment.

11.5 Applications of AI Agents in Robotics

AI agents drive diverse applications across various sectors, revolutionizing how robots solve real-world challenges. This section delves into key areas such as healthcare robotics, disaster response, and underwater exploration, examining how AI agents are enabling groundbreaking advancements.

11.5.1 Healthcare Robotics

AI agents are integral to healthcare robotics, enabling precision, efficiency, and adaptability in medical tasks. From surgical robots to assistive devices, these systems operate in highly controlled environments, requiring advanced capabilities in perception, planning, and interaction.

- **Surgical Robotics:** AI-driven robots, such as the da Vinci Surgical System, employ real-time image processing (Intuitive Surgical, 2024), robotic dexterity, and haptic feedback to perform minimally invasive surgeries.
- **Rehabilitation and Assistive Robots:** Robots equipped with AI agents are deployed in rehabilitation therapy, aiding patients in recovering mobility and strength.
- **Diagnostics and Monitoring:** Mobile robots equipped with AI-powered diagnostic tools navigate hospital environments, conduct patient assessments, and assist in monitoring vital signs.

11.5.2 Disaster Response Robotics

In disaster scenarios, AI agents empower robots to operate in hazardous environments, providing critical assistance in search-and-rescue operations and infrastructure assessments. These robots require robust spatial intelligence, autonomous navigation, and adaptive decision-making capabilities to handle unpredictable conditions.

- **Search and Rescue:** Robots like Boston Dynamics' Spot, equipped with AI agents, navigate through rubble, locate survivors, and relay information to human responders.
- **Environmental Monitoring:** AI agents in drones and ground robots monitor environmental hazards, such as gas leaks, radiation levels, or fire spread.
- **Autonomous Decision-Making:** Disaster response robots often operate with limited communication in remote or dangerous areas. AI agents enhance their autonomy by enabling real-time reasoning and planning.

11.5.3 Underwater Exploration and Robotics

The underwater domain presents unique challenges for robotics, including limited visibility, dynamic environments, and communication constraints. AI agents in underwater robots address these challenges by combining advanced perception, navigation, and data analysis capabilities.

- **Marine Research:** Underwater robots like autonomous underwater vehicles (AUVs) collect data on marine ecosystems, including mapping coral reefs, tracking fish populations, and monitoring ocean temperatures.
- **Infrastructure Inspection:** AI-powered robots are increasingly used to inspect underwater structures, such as pipelines, oil rigs, and subsea cables.
- **Deep-Sea Exploration:** In extreme depths, robots operate under high pressure and low light conditions.

11.5.4 Emerging Trends and Future Applications

As AI agents in robotics continue to evolve, new applications are emerging across diverse domains:

- **Autonomous Modular Robots:** These robots adapt to changing environments by reconfiguring their physical structures.
- **Swarm Robotics:** Inspired by biological systems, swarm robotics leverages multiple AI agents working in concert to achieve collective goals.
- **Space Exploration:** Robots equipped with AI agents are deployed for planetary exploration, conducting soil analysis, terrain mapping, and infrastructure construction on extraterrestrial surfaces (NASA, 2024).

11.6 Competitive Landscape and Global Market Dynamics

This section explores the competitive landscape and market dynamics shaping the global robotics industry, highlighting the major players, regional trends, and strategies driving innovation and adoption.

Figure 11.5 provides an overview of what we will be discussing in this section.

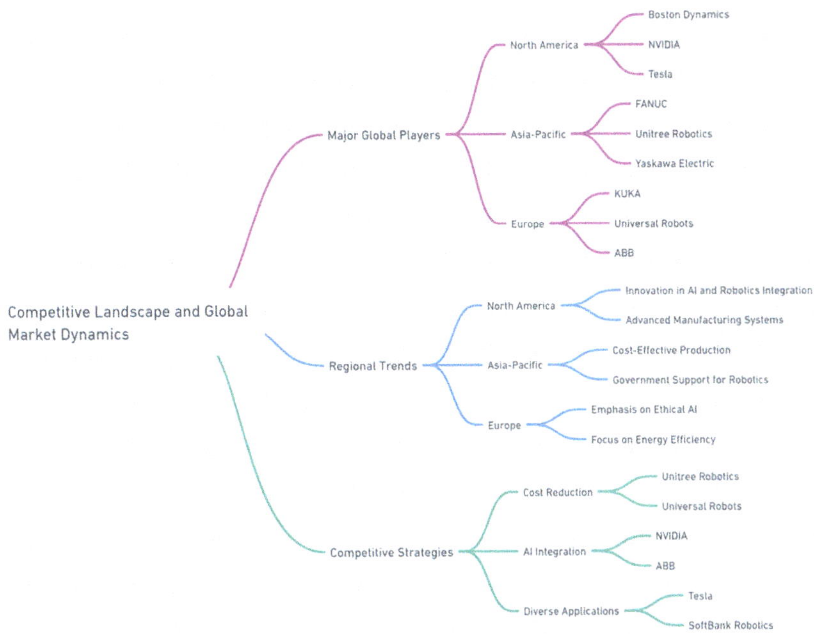

Fig. 11.5 Competitive landscape and market dynamics

11.6.1 Major Global Players

The robotics industry is witnessing rapid expansion, with leading companies driving advancements in technology, applications, and scalability. These companies range from industrial robotics giants to innovative startups and AI-focused firms. This section highlights the key players reshaping the global robotics landscape. We listed companies alphabetically.

- **ABB (Switzerland)**
- ABB is a global leader in industrial automation. The company offers advanced robotics solutions for manufacturing, logistics, and energy sectors. ABB's portfolio includes collaborative robots (cobots) and automated production lines, supported by strong AI integration (ABB, 2023).
- **Boston Dynamics (USA)**
- Boston Dynamics is a leader in dynamic robotics, renowned for Atlas, its highly advanced humanoid robot, and Spot, a versatile quadruped robot (Boston Dynamics, 2024).
- **Denso (Japan)**

- Denso has installed over 120,000 small industrial robots globally. The company focuses on compact, efficient robotics solutions for assembly and packaging in automotive and electronics industries (Denso, 2024).
- **Epson (Japan)**
- With over 35 years in robotics, Epson offers industrial robots that cater to precision manufacturing industries, such as electronics and pharmaceuticals (Epson, 2024).
- **FANUC (Japan)**
- FANUC has installed more than 750,000 robots globally, making it one of the largest robotics companies by deployment volume. Known for its iconic yellow robots, FANUC specializes in industrial applications, including welding, material handling, and assembly (FANUC, 2024).
- **Fig. AI (USA)**
- Fig. AI is a rising competitor in humanoid robotics, leveraging AI for advanced mobility, manipulation, and decision-making. The company focuses on creating robots that can integrate into various industries, from logistics to elder care, emphasizing versatility and adaptability (Fig. AI, 2024).
- **KUKA (Germany)**
- KUKA is a key player in industrial robotics, particularly in automated manufacturing systems. The company's focus on innovation is evident in its efforts to integrate AI for adaptive manufacturing processes (KUKA, 2024).
- **NVIDIA (USA)**
- NVIDIA, while traditionally known for GPUs, has become a significant player in AI-powered robotics. Its platforms, such as the Jetson AI system and Isaac Sim, provide developers with tools to create and train intelligent robots (NVIDIA, 2024a).
- **SoftBank Robotics (Japan)**
- SoftBank Robotics is known for Pepper, a humanoid robot designed for social interaction. Featuring natural language processing (NLP) and emotion recognition, Pepper has been deployed in customer service, education, and elder care (SoftBank Robotics, 2024).
- **Stäubli (Switzerland)**
- Stäubli has been a prominent name in industrial robotics since 1982. Its product lineup includes high-speed, precision robots for applications in textile manufacturing, pharmaceuticals, and electronics (Stäubli, 2024).
- **Tesla (USA)**
- Tesla's Optimus humanoid robot focuses on automating repetitive and dangerous tasks. Optimus combines Tesla's advanced AI, used in autonomous vehicles, with precision mechanics for applications in manufacturing and logistics (Tesla, 2024).
- **Unitree Robotics (China)**
- Unitree Robotics specializes in affordable quadrupedal robots like the G1 and Go2 series. These compact robots are equipped with advanced perception modules, enabling navigation and manipulation in research and industrial settings (Unitree Robotics, 2024).

- **Universal Robots (Denmark)**
- Universal Robots has revolutionized industrial automation with its collaborative robots (cobots). Designed for seamless human–robot interaction, these robots are lightweight, flexible, and easy to program, making them ideal for small- and medium-sized enterprises (Universal Robots, 2024).
- **Vecna Robotics (USA)**
- Specializing in logistics and workflow automation, Vecna Robotics develops autonomous mobile robots (AMRs) and software platforms for warehouse and supply chain optimization (Vecna Robotics, 2024).
- **Yaskawa Electric (Japan)**

- Yaskawa has shipped nearly 500,000 industrial robots since its inception in 1977. Its Motoman series robots are known for their reliability and efficiency in welding, material handling, and packaging (Yaskawa, 2024).

11.6.2 Regional Trends

- **North America**

- The region remains a powerhouse in robotics innovation, with companies like Boston Dynamics, Tesla, and NVIDIA leading the charge. Although now shown in Fig. 11.5, other leading companies in North America include Google's DeepMind (more on this in Sect. 11.7.2) and Fig. AI, among many others.
- **Asia-Pacific**
- Asia-Pacific's strength lies in its manufacturing expertise and government support for robotics initiatives. Companies such as FANUC, Yaskawa, and Unitree Robotics excel in industrial and service robotics.
- **Europe**

- Europe is a leader in collaborative robotics and ethical AI integration. Companies like KUKA and Universal Robots emphasize energy efficiency and sustainability. Regional policies prioritize human–robot collaboration, focusing on enhancing productivity without displacing human workers (European Commission, 2020).

11.6.3 Competitive Strategies

- **Cost Reduction**

- Unitree Robotics and Universal Robots focus on cost efficiency, offering affordable solutions without compromising quality.
- **AI Integration**
- NVIDIA and ABB are advancing robotics by incorporating foundational AI models.

- **Diverse Applications**
- Companies like SoftBank and Tesla are expanding robotics applications into service, healthcare, and domestic environments.

11.7 Research and Technological Innovations

Leading organizations and research institutions are at the forefront of advancing AI and robotics. This section highlights key contributions from major players like NVIDIA, DeepMind, and Unitree Robotics from their innovation strategies, platforms, and technology perspectives. As we have discussed in Sect. 11.6, there are many other global players in robotics innovation; due to the size limit, we only listed three to gain some insights from these three top players in the space.

11.7.1 NVIDIA's Robotics Strategies, Platforms, and Technology

NVIDIA has emerged as a leading player in the rapidly evolving field of robotics, offering a comprehensive suite of hardware and software solutions that are transforming the way robots are developed, trained, and deployed. This section delves into NVIDIA's robotics strategies, platforms, and technology, examining the company's vision for the future of robotics and how its innovations are shaping the industry.

NVIDIA's Robotics Strategy

NVIDIA's robotics strategy centers around the concept of "physical AI," which refers to AI systems that can understand, interact with, and navigate the physical world. The company believes that the next frontier of AI lies in robotics, and it is actively driving advancements in this field by building a comprehensive robotics ecosystem (Digitimes, 2025). This ecosystem encompasses three key areas:

- **AI Robot Development:** The NVIDIA Isaac platform provides a full suite of CUDA-accelerated systems, libraries, application frameworks, and generative AI models to accelerate the development of AI robots (NVIDIA, 2025a).
- **Vision AI:** NVIDIA Metropolis is an application framework, a set of developer tools, and a partner ecosystem that brings visual data and AI together to improve operational efficiency and safety across various industries (NVIDIA, 2025a).

- **Sensor and Physics Simulation:** NVIDIA Isaac Sim, built on Omniverse, enables physically accurate simulation and synthetic data generation to accelerate the development, testing, and validation of AI robots (NVIDIA, 2025a).

Central to this strategy is NVIDIA's "three-computer solution," which streamlines the robotics development process. This solution involves the following:

- **NVIDIA DGX Systems:** Powerful AI supercomputers used to train large, generalized AI foundation models for robots (NVIDIA, 2025b).
- **NVIDIA OVX Systems:** Systems designed for simulating, testing, and training robots in virtual environments, leveraging advanced graphics and compute capabilities (NVIDIA, 2025b).
- **NVIDIA AGX Systems:** Embedded systems, including the NVIDIA Jetson platform, that provide the high performance and energy efficiency needed to deploy and run AI models in robots for real-world operation (NVIDIA, 2025c).

This three-pronged approach allows developers to seamlessly move from AI model training to simulation and finally to real-world deployment, accelerating the development cycle and fostering innovation.

NVIDIA's strategy also emphasizes collaboration and partnerships. The company works closely with robotics manufacturers, solution developers, and research institutions to foster innovation and accelerate the adoption of robotics technologies (Digitimes, 2025).

NVIDIA's Vision for the Future of Robotics

NVIDIA's CEO, Jensen Huang, has articulated a bold vision for the future of robotics, emphasizing the transformative potential of physical AI and its ability to revolutionize various industries. In his keynote address at CES 2025, Huang proclaimed that "the ChatGPT moment for general robotics is almost upon us," suggesting that robotics is on the cusp of a breakthrough similar to that experienced by artificial intelligence with ChatGPT (Xpert.Digital, 2025).

Huang envisions a future where robots seamlessly integrate into human environments, performing a wide range of tasks and augmenting human capabilities. He highlighted three key areas where robots will have a significant impact:

- **Agentic Robots and AI:** These robots will act as information workers, capable of understanding and responding to human instructions and collaborating with humans in various settings (Huang, 2025).
- **Self-Driving Cars:** Autonomous vehicles will transform transportation, leveraging AI to navigate complex environments and improve safety and efficiency (Huang, 2025).
- **Humanoid Robots:** Huang predicts that humanoid robots will surprise everyone with their incredible capabilities, potentially becoming the largest technology industry the world has ever seen (Huang, 2025).

To realize this vision, NVIDIA is investing heavily in research and development, focusing on areas such as robot learning, synthetic data generation, and advanced perception technologies. The company is also actively building partnerships and collaborations to foster a thriving robotics ecosystem.

NVIDIA's Robotics Platforms

NVIDIA offers a range of robotics platforms designed to meet the diverse needs of developers and researchers.

NVIDIA Isaac Sim is a scalable robotics simulation application and synthetic data generation tool built on the NVIDIA Omniverse platform (NVIDIA, 2025d). It leverages Omniverse Kit's powerful simulation technologies, including advanced GPU-enabled physics simulation with PhysX 5, photorealism with real-time ray and path tracing, and MDL material definition support for physically based rendering (NVIDIA, 2025e). Isaac Sim provides a physically accurate virtual environment to develop, test, and manage AI-based robots. Key features include the following:

- Tools for importing and simulating robot models.
- Simulation of sensors such as RGB-D, PhysX-Lidar, RTX-Lidar, Radar, contact, and IMU.
- Support for various robotics applications

Isaac Sim also leverages OpenUSD (Universal Scene Description), a universal framework for developing and collaborating in 3D worlds (NVIDIA, 2025f). OpenUSD provides a standardized way to represent and exchange 3D data, enabling interoperability between different simulation tools and workflows. This is crucial for robotics simulation, where complex scenes and interactions need to be accurately represented and shared.

A key component of Isaac Sim is Isaac Lab, an open-source unified framework for robot learning (NVIDIA, 2025g). Isaac Lab provides a comprehensive set of tools and algorithms for training robot policies using various techniques, including reinforcement learning and imitation learning. It integrates seamlessly with Isaac Sim, allowing developers to train and validate robot behaviors in physically accurate simulated environments before deploying them to real-world robots.

The simulation-first approach enabled by Isaac Sim offers several advantages for robotics development:

- **Cost Savings:** Simulation reduces the need for expensive physical prototypes and real-world testing, leading to significant cost savings.
- **Increased Safety:** Training robots in simulation eliminates the risks associated with real-world testing, ensuring the safety of both humans and equipment.
- **Diverse Environments:** Simulation allows developers to test robots in a wide range of environments and scenarios that may be difficult or impossible to replicate in the real world.

NVIDIA Jetson is a series of embedded computing boards designed for accelerating machine learning applications at the edge (Wikipedia, 2025). These small, power-efficient computers are ideal for deploying AI applications in robots and other autonomous machines. The Jetson platform includes the following:

- **Jetson Orin:** The NVIDIA Jetson Orin family includes seven modules with up to 275 TOPS of AI performance, 8X the performance of the previous generation (NVIDIA, 2025h).
- **Jetson Orin Nano Super Developer Kit:** This compact and powerful computer redefines generative AI for small edge devices. Priced at $249, it provides an affordable and accessible platform for developers, students, and makers, backed by NVIDIA AI software and a broad AI ecosystem (NVIDIA, 2025h).
- **Jetson Xavier:** The Jetson Xavier NX delivers up to 21 TOPS of AI performance in a compact form factor (Wikipedia, 2025).
- **Jetson Nano:** The Jetson Nano is a powerful and affordable AI platform for entry-level edge AI applications (NVIDIA, 2025i).

The Jetson platform is supported by the NVIDIA JetPack SDK, which provides a complete software stack for AI development at the edge (NVIDIA, 2025j).

NVIDIA's Robotics Technology

NVIDIA's robotics technology encompasses a wide range of software and hardware solutions:

- **Isaac SDK:** The Isaac SDK is a robotics abstraction layer that hides the complexity of USD and enables seamless integration of AI capabilities (Justoborn, 2025). It provides advanced perception algorithms, support for various robotics applications, and integration with ROS.
- **Isaac ROS:** NVIDIA Isaac ROS is a collection of CUDA-accelerated computing packages and AI models designed to streamline and expedite the development of advanced AI robotics applications (NVIDIA, 2025c). It offers modular packages for robotic perception and easy integration into existing ROS 2-based applications.
- **Isaac Manipulator:** Isaac Manipulator enables the development of AI-powered robotic arms that can perceive, understand, and interact with their environments (NVIDIA, 2025c).
- **Isaac Perceptor:** Isaac Perceptor enables the development of advanced autonomous mobile robots (AMRs) that can perceive, localize, and operate in unstructured environments (NVIDIA, 2025c).
- **Isaac GR00T:** Isaac GR00T is a research initiative and development platform for general-purpose robot foundation models and data pipelines to accelerate humanoid robotics (NVIDIA, 2025c).
- **DeepStream SDK:** The DeepStream SDK delivers a complete streaming analytics toolkit for AI-based multi-sensor processing, video, and image understanding on Jetson (NVIDIA, 2025j).

- **NVIDIA OSMO:** NVIDIA OSMO is a cloud-native orchestration platform that allows developers to easily scale complex robotics workloads across distributed environments (NVIDIA, 2025g). It supports multi-container workloads and can be deployed on-premises, in private clouds, or in public cloud resource clusters. This provides flexibility and scalability for managing complex robotics development and deployment workflows.
- **Jetson Modules:** NVIDIA Jetson modules provide the processing power for AI at the edge, enabling robots to perform complex tasks in real time.
- **NVIDIA DGX Systems:** DGX systems are AI supercomputers designed for training large, generalized AI foundation models for robots (NVIDIA, 2025b).
- **NVIDIA OVX Systems:** OVX systems provide the graphics and compute performance needed for simulating, testing, and training robots in virtual environments (NVIDIA, 2025b).
- **Project DIGITS:** Project DIGITS is NVIDIA's latest AI supercomputer, designed to bring the power of Grace Blackwell to developer desktops (Huang, 2025). It features the GB10 Superchip, developed in collaboration with MediaTek, and offers up to 128GB of unified system memory. This allows AI researchers, data scientists, and students to work with AI models locally, including those with up to 200 billion parameters. Project DIGITS has the potential to democratize AI development by making high-performance computing more accessible.

Case Studies

NVIDIA's robotics technologies are being used by a variety of companies across different industries:

- **Cobot:** Cobot is using NVIDIA Isaac Sim and Isaac ROS to develop Proxie, an AI-powered collaborative robot for logistics, hospital, and manufacturing operations (Cobot, 2025).
- **Tampa General Hospital:** Tampa General Hospital is using Cobot's Proxie robots to increase operational efficiency and patient experience satisfaction by optimizing cart movement and elevator congestion (Cobot, 2025).
- **Moderna:** Moderna is collaborating with Cobot to standardize robotic systems management, reduce costs, and enhance flexibility (Cobot, 2025).
- **Activ Surgical:** Activ Surgical leverages NVIDIA's technology to enhance its ActivSight system, enabling surgeons to see beyond the visible spectrum to improve surgical outcomes (R2Surgical, 2025).
- **Johnson & Johnson MedTech:** Johnson & Johnson MedTech is collaborating with NVIDIA to integrate AI into surgical data analysis, aiming to improve surgical decision-making and enhance clinician education (R2Surgical, 2025).

Partnerships and Collaborations

NVIDIA has established partnerships and collaborations with various organizations in the robotics space:

- **Open Source Robotics Alliance (OSRA):** NVIDIA is a founding member of OSRA, demonstrating its commitment to open-source robotics development (ABI Research, 2025).
- **ROS Ecosystem:** NVIDIA Isaac ROS is built on the open-source ROS 2 framework, enabling seamless integration with the ROS community (NVIDIA, 2025a).
- **Pittsburgh Robotics Network:** NVIDIA is collaborating with the Pittsburgh Robotics Network to foster innovation and enhance connections between the commercial robotics business community, academia, and research institutions (RoboPGH, 2025).
- **Unitree Robotics, XPeng, and BYD:** NVIDIA has partnered with these Chinese robotics companies to advance the development of humanoid robots (TrendForce, 2025).
- **NVIDIA Humanoid Robot Developer Program:** This program provides developers with early access to NVIDIA's latest robotics technologies and resources, including Isaac Sim, Isaac Lab, Jetson Thor, and Project GR00T general-purpose humanoid foundation models (NVIDIA, 2025l). This program aims to accelerate the development and deployment of humanoid robots by fostering collaboration and providing access to cutting-edge tools and technologies.

11.7.2 Google DeepMind's Robotics Strategies, Platforms, and Technology

Google DeepMind's innovative approach combines cutting-edge AI algorithms with advanced robotics platforms to create robots capable of performing complex tasks in dynamic environments. This section examines their overall approach, specific platforms, key technologies, collaborations, and recent progress.

DeepMind's Approach to Robotics Research and Development

DeepMind's approach to robotics is rooted in its mission to "solve intelligence, and then use it to solve everything else" (Wikipedia, 2025a). This translates to developing general-purpose AI systems that can learn and adapt to various tasks and environments, rather than being limited to specific pre-programmed instructions (Open Data Science, 2025). Their strategy emphasizes the following key principles (GeeksforGeeks, 2025):

- **Universal Learning Algorithms:** DeepMind focuses on creating AI systems that can learn from diverse data and experiences, generalizing their knowledge across multiple domains (DeepMind, 2025a). This allows robots to be more versatile and adaptable, capable of performing a wider range of tasks.
- **Solving Complex, Real-World Problems:** DeepMind's research is driven by the desire to solve complex problems in various fields, including robotics. They aim to create AI that can understand and operate within dynamic and unpredictable environments (DeepMind, 2025a). This focus on real-world applications ensures that their research has practical value and can address real-world challenges.
- **Safe and Ethical Development:** DeepMind emphasizes responsible AI development, ensuring that the technology benefits humanity. This involves extensive research on the ethical implications of AI, including safety, bias reduction, and long-term societal impact (DeepMind, 2025a). This commitment to ethical AI is crucial for building trust and ensuring that AI is used for good.

DeepMind draws significant inspiration from neuroscience in its pursuit of Artificial General Intelligence (AGI) (GeeksforGeeks, 2025). The goal is to create algorithms that mimic the cognitive processes of the human brain, studying how the brain processes information, learns from experiences, and generalizes knowledge. Techniques such as reinforcement learning, neural networks, and deep learning are employed to mirror these processes.

Furthermore, DeepMind employs hierarchical learning, structuring learning processes in layers, with higher levels abstracting more complex patterns from lower levels (GeeksforGeeks, 2025). This approach enhances the efficiency and scalability of learning algorithms, mirroring how humans develop cognitive skills. To approach AGI, a system must be able to learn and perform a diverse set of tasks. DeepMind focuses on multitask learning, where a single model is trained on various tasks simultaneously (GeeksforGeeks, 2025). This approach helps the model develop a more generalized understanding, akin to human cognitive abilities.

DeepMind also focuses on meta-learning, or "learning to learn" (GeeksforGeeks, 2025). The idea is to create algorithms that can quickly adapt to new tasks with minimal additional training. This capability is crucial for AGI, as it allows the system to handle novel situations and tasks that it wasn't explicitly trained on.

DeepMind's approach to robotics research and development involves:

- **Dexterous Manipulation:** DeepMind is developing AI systems that enable robots to perform complex tasks requiring dexterous movement, such as tying shoelaces, hanging shirts, and even cleaning a kitchen (The Robot Report, 2025). This focus on dexterity is crucial for robots to be truly helpful in human environments, where they need to interact with objects in a precise and adaptable manner.
- **Simulation and Real-World Transfer:** DeepMind utilizes simulations to train robots in safe and controlled environments before transferring the learned behaviors to real-world scenarios (The Robot Report, 2025). This approach significantly reduces the cost and time required for real-world robot training, as robots can practice and refine their skills in a virtual setting before interacting with the physical world.

- **Scaling Robotic Learning:** DeepMind is working on scaling robotic learning by collecting diverse and experiential training data from multiple robots operating in various settings (DeepMind, 2025a). This allows them to train more general and adaptable robots that can perform a wider range of tasks in different environments.
- **Robot Safety:** DeepMind has developed a "Robot Constitution" with safety rules inspired by Isaac Asimov's Three Laws of Robotics to ensure robots operate safely in real-world environments (Silicon Republic, 2025). This "Robot Constitution" is a novel approach to ensuring robot safety by incorporating ethical guidelines into the AI system itself. Some of the specific safety rules implemented in AutoRT include an automatic stop if the force on its joints exceeds a threshold and the requirement for human supervision (Silicon Republic, 2025). These safety measures are crucial for building trust and ensuring that robots can be safely integrated into human environments.
- **Tidying Up:** DeepMind is exploring how robots can learn to tidy up after themselves, demonstrating their focus on practical applications of robotics in everyday life (DeepMind, 2025b).

Robotics Platforms Developed or Utilized by DeepMind

DeepMind develops and utilizes various robotics platforms to advance its research and development efforts. Some of the key platforms include the following:

- **ALOHA Unleashed:** This platform builds on DeepMind's ALOHA 2 platform, which is based on the original ALOHA low-cost, open-source hardware for bimanual teleoperation from Stanford University. ALOHA 2 is more dexterous than prior systems because it has two hands that can be teleoperated for training and data collection purposes. It also allows robots to learn how to perform new tasks with fewer demonstrations (The Robot Report, 2025). ALOHA Unleashed has been used to train robots to perform complex tasks such as tying shoelaces, hanging shirts, and even cleaning a kitchen.
- **DemoStart:** This platform uses a reinforcement learning algorithm to help new robots acquire dexterous behaviors in simulation. These learned behaviors can be especially useful for complex environments, like multi-fingered hands. DemoStart begins learning from easy states, and, over time, the researchers add in more complex states until it masters a task to the best of its ability (The Robot Report, 2025). This platform has achieved high success rates in simulated environments and has shown promising results in transferring learned behaviors to real-world robots.
- **DEX-EE:** This three-fingered robotic hand was developed in collaboration with Shadow Robot to enable more advanced robot learning through intensive experimentation (The Robot Report, 2025). This platform allows DeepMind to explore the complexities of dexterous manipulation with a high degree of precision and control.

- **AutoRT:** This system combines large foundation models, such as a Large Language Model (LLM) or Visual Language Model (VLM), and a robot control model (RT-1 or RT-2) to create a system that can deploy robots to gather training data in novel environments (DeepMind, 2025a). AutoRT allows DeepMind to scale robotic learning by collecting diverse data from multiple robots operating in various settings.
- **SARA-RT:** This system improves the efficiency of robotic transformer models by making them more accurate and faster (Silicon Republic, 2025). This technology allows robots to make decisions more quickly and efficiently, improving their overall performance.
- **RT-Trajectory:** This model helps robots become more generalized in their tasks by adding visual outlines that describe robot motions in training videos (Silicon Republic, 2025). This technology allows robots to learn from visual demonstrations and generalize their knowledge to new tasks.
- **Mobile ALOHA:** This is a low-cost mobile robot developed in collaboration with researchers at Stanford University to perform complex household chores through imitation learning (Kalil, 2025a, b). This two-armed robot can handle tasks that require the use of both hands simultaneously, such as sauteing shrimp and opening cabinets.
- **Soccer-Playing Humanoid Robot:** DeepMind has also developed a soccer-playing humanoid robot trained using deep reinforcement learning (Kalil, 2025a, b). This robot demonstrates DeepMind's exploration of different robot applications and its ability to train robots for dynamic and complex tasks.

DeepMind's Robotics Technologies

DeepMind is developing a range of technologies to enhance the capabilities of robots. These technologies include the following:

- **AI Algorithms:** DeepMind utilizes various AI algorithms, including reinforcement learning, imitation learning, and diffusion models, to train robots to perform complex tasks (The Robot Report, 2025). Reinforcement learning allows robots to learn through trial and error, while imitation learning enables them to learn from human demonstrations. Diffusion models are used to generate realistic and diverse training data for robots.
- **Control Systems:** DeepMind develops advanced control systems that enable robots to interact with their environment precisely and efficiently (The Robot Report, 2025). These control systems are crucial for robots to perform tasks with dexterity and accuracy.
- **Sensor Integration:** DeepMind integrates various sensors, such as cameras and tactile sensors, into its robotics platforms to provide robots with a rich understanding of their surroundings (The Robot Report, 2025). This allows robots to perceive their environment and interact with objects in a more informed and adaptable manner.

- **Robotics Transformers:** DeepMind has developed Robotics Transformers (RT), a neural network architecture that enables robots to learn from both web and robotics data and translate this knowledge into generalized instructions for robotic control (DeepMind, 2025c). This technology allows robots to learn from a vast amount of data and generalize their knowledge to new tasks and environments.
- **Large Language Models (LLMs):** DeepMind utilizes LLMs to enable robots to understand and respond to human instructions and goals (DeepMind, 2025a). This allows for more natural and intuitive human–robot interaction.
- **Visual Language Models (VLMs):** DeepMind uses VLMs to help robots understand their environment and the objects within it (DeepMind, 2025a). This allows robots to perceive and interpret their surroundings in a more humanlike way.

Learning from Human Preferences

DeepMind is actively researching how robots can learn from human preferences (DeepMind, 2025b). This involves developing AI systems that can understand and adapt to human values and preferences, allowing robots to perform tasks in a way that aligns with human expectations. This research is crucial for building trust and ensuring that robots can be effectively integrated into human environments.

DeepMind's Collaborations and Partnerships in Robotics

DeepMind actively collaborates with other organizations to advance its robotics research and development (Robotics247, 2025). Some of its key collaborations include the following:

- **Shadow Robot:** DeepMind collaborated with Shadow Robot to develop the DEX-EE three-fingered robotic hand (The Robot Report, 2025). This collaboration allowed DeepMind to leverage Shadow Robot's expertise in dexterous robotic hands to create a platform for advanced robot learning.
- **Stanford University:** DeepMind's ALOHA 2 platform is based on the original ALOHA hardware from Stanford University (The Robot Report, 2025). This collaboration provides DeepMind with access to a low-cost and open-source platform for bimanual teleoperation.
- **33 Academic Labs:** DeepMind collaborated with 33 academic labs to create the Open X-Embodiment dataset and the RT-X model, aiming to develop general-purpose robots that can learn and adapt across diverse robot types (Open Data Science, 2025). This collaboration is a significant step toward developing more versatile and adaptable robots. The Open X-Embodiment dataset includes data from 22 robot embodiments, including wheeled, legged, and aerial robots, with varying morphologies and sensor configurations (Open Data Science, 2025). Experiments with RT-2-X, trained on this dataset, showed a threefold improve-

ment in performing tasks that were previously beyond its capabilities, including better spatial understanding, such as differentiating between "move apple near cloth" and "move apple on cloth" commands (Open Data Science, 2025).
- **Apptronik:** DeepMind recently partnered with Apptronik, a leading AI-powered humanoid robotics company, to accelerate the development of autonomous humanoid robots (Kalil, 2025a, b). This partnership combines DeepMind's AI expertise with Apptronik's robotics platform to create more intelligent and versatile humanoid robots.

Recent News and Announcements on DeepMind's Progress in Robotics

DeepMind has made several recent announcements regarding its progress in robotics:

- **Partnership with Apptronik:** In December 2024, DeepMind announced a strategic partnership with Apptronik to advance the development of AI-powered humanoid robots (Robotics247, 2025). This partnership aims to combine Apptronik's robotics platform with DeepMind's AI expertise to create robots that can operate efficiently in dynamic environments.
- **Advancements in Robot Dexterity:** DeepMind recently unveiled two new AI systems, ALOHA Unleashed and DemoStart, that help robots learn to perform complex tasks requiring dexterous movement (DeepMind, 2025c). These systems represent significant advancements in DeepMind's efforts to create more dexterous and adaptable robots.
- **Shaping the Future of Advanced Robotics:** DeepMind introduced AutoRT, SARA-RT, and RT-Trajectory to improve real-world robot data collection, speed, and generalization (DeepMind, 2025a). These systems are designed to enhance the efficiency and adaptability of robots in real-world scenarios.
- **RoboCat:** DeepMind introduced RoboCat, a self-improving robotic agent that can learn to perform a variety of tasks by itself (DeepMind, 2025c). This is a significant development in DeepMind's robotics research, as it demonstrates the potential for robots to learn and adapt without constant human intervention.

11.7.3 Unitree Robotics: Strategies, Platforms, and Technology

Unitree Robotics is a global leader in the field of high-performance quadruped and humanoid robots. Founded in 2016 by Xingxing Wang, the company has rapidly become a prominent player in the robotics industry, known for its innovative designs, advanced technology, and commitment to research and development. This section explores its diverse range of platforms and products and examines the cutting-edge technology behind its impressive robots. Notably, Unitree robots have been featured in high-profile events like the Super Bowl pregame show, showcasing their growing presence in the public eye (RoboticsTomorrow, 2025).

Unitree Robotics Strategies

Unitree Robotics' overarching strategy centers around innovation, affordability, and accessibility. The company aims to democratize access to advanced robotics by offering high-performance robots at competitive prices, catering to both consumers and industry professionals. This approach has allowed Unitree to capture a significant share of the robotics market and establish itself as a key player in the industry (RobotShop, 2025).

The company's journey began with the development of XDog, the first quadruped device created by founder Xingxing Wang during his postgraduate studies (Wikipedia, 2025b). This early innovation laid the foundation for Unitree's subsequent success in the robotics field.

Key Strategic Pillars:

- **Focus on Research and Development:** Unitree places significant emphasis on research and development, continuously pushing the boundaries of robotics technology. This dedication to innovation directly contributes to the company's goal of making advanced robotics more accessible. By investing heavily in developing core robot components, motion control algorithms, and sensing technologies, Unitree can enhance the performance and capabilities of its robots while keeping costs down (RobotShop, 2025).
- **AI Integration:** Unitree actively integrates artificial intelligence (AI) into its robotic systems. This includes using AI for motion planning, object recognition, autonomous navigation, and even natural language processing for human–robot interaction. AI plays a crucial role in enhancing the intelligence and autonomy of Unitree's robots, enabling them to perform more complex tasks and interact more naturally with humans (Wikipedia, 2025b).
- **Open-Source Frameworks:** Unitree develops and utilizes open-source frameworks to facilitate collaboration and accelerate the development of robotics technology. This approach fosters innovation and allows researchers and developers worldwide to contribute to the advancement of Unitree's robots. By embracing open-source principles, Unitree can tap into a global network of expertise and accelerate its development process (GitHub, 2025).
- **Strategic Partnerships:** Unitree actively seeks partnerships and collaborations to expand its market reach and capabilities. The company collaborates with distributors, resellers, and technology providers like NVIDIA to enhance its global presence and access new markets. These partnerships are crucial for Unitree's growth strategy, allowing it to reach a wider customer base and leverage external expertise (RobotShop, 2025). Notably, Unitree Robotics is trusted by leading brands like Google, Amazon, and NVIDIA and research institutions like MIT and Stanford, further solidifying its position in the industry (Top3DShop, 2025).

- **Cost Reduction:** Unitree focuses on reducing the cost of its robots to make them more accessible to a wider range of customers. This includes developing its own core components, such as motors and LiDAR systems, to reduce reliance on expensive third-party suppliers. This cost-conscious approach is central to Unitree's strategy of democratizing access to advanced robotics (KrASIA, 2025). This strategy is particularly evident in the company's efforts to offer lower prices than competitors like Boston Dynamics, aiming to drive wider adoption of its robots (KrASIA, 2025).

Unitree Robotics Platforms and Products

Unitree Robotics offers a diverse portfolio of quadruped and humanoid robots designed for various applications. These robots are known for their agility, stability, and advanced capabilities, making them suitable for tasks ranging from research and education to industrial automation and entertainment (Génération Robots, 2025).

Unitree's quadruped robots are designed to mimic the movements of animals, enabling them to navigate complex terrains and perform tasks in challenging environments. The company's quadruped robot lineup includes the following:

Go Series: The Go series, including the Go1 and Go2, are agile and versatile quadruped robots designed for both consumer and industrial applications. These robots are known for their speed, maneuverability, and advanced sensing capabilities (Top3DShop, 2025).

- **Go1:** This lightweight and fast-charging robot is designed for mobility and agility, with a top speed of 17 km/h. It can be used for tasks like package delivery or inspection in industrial settings. However, it's important to note that the Go1 has some limitations, such as susceptibility to dust and morning dew and an inability to self-recover from falls (Top3DShop, 2025).
- **Go2:** Equipped with a 4D LiDAR and optional GPT language model support, the Go2 excels in all-terrain adaptability and can interpret and execute operator commands. This robot can be used for various applications, including surveillance, exploration, and even as a companion robot (INFUZE Robotics, 2025).

B Series: The B series, including the B1 and B2, are industrial-grade quadruped robots built for strength and versatility. These robots are designed for demanding tasks in challenging environments, such as industrial inspection, security patrols, and exploration (Top3DShop, 2025).

- **B1:** With an IP68 waterproof rating, the B1 can tackle challenging environments with ease, making it ideal for agriculture, security, and rescue missions. It has dimensions of 1202 mm in length, 467 mm in width, and 297 mm in height, with a climbing gradient of 30° (Top3DShop, 2025).
- **B2:** This high-speed robot can reach speeds of 6 m/s and boasts exceptional adaptability to diverse terrains. The B2W, a variant with interchangeable wheels

and legs, can achieve even higher speeds of up to 30 mph and carry loads of up to 40 kg, including a person (Top3DShop, 2025).

- **B2-W:** This wheeled-leg version of the B2 combines the agility of legs with the speed and efficiency of wheels, enabling it to traverse various terrains with enhanced mobility. This robot is particularly well-suited for tasks that require both speed and maneuverability, such as navigating complex industrial environments or conducting search and rescue operations (RoboticsTomorrow, 2025).
- **Aliengo:** This multifunctional quadruped robot is built for industrial applications, with a focus on precise navigation and obstacle avoidance. It can be used for tasks like security patrols, exploration, and rescue operations (Top3DShop, 2025).

Unitree has also made significant strides in the development of humanoid robots, with a focus on creating robots that can interact with humans and perform tasks in human-centric environments. Unitree focuses on the hardware aspects of these robots, collaborating with partners for software integration and application development (Security Robotics, 2025). The company's humanoid robots include the following:

- **H1:** This full-size humanoid robot features advanced bipedal locomotion and AI, aimed at research, education, and entertainment sectors (Drones Plus Robotics, 2025).
- **G1:** This smaller and more affordable humanoid robot is designed for interactive and practical applications in both home and industry environments. It features 23–43 joint motors, a force control dexterous hand for precise object manipulation, and utilizes the UnifoLM (Unitree Robot Unified Large Model) for advanced AI capabilities (Drones Plus Robotics, 2025). Achieving a natural humanlike gait is a key focus for Unitree's humanoid robots, and the G1 demonstrates significant progress in this area (i-Programmer, 2025).

In addition to its quadruped and humanoid robots, Unitree offers a range of robotic arms and accessories to enhance the capabilities of its robots. These include the following:

- **Z1:** A compact and dexterous robotic arm with 6 degrees of freedom, designed for integration with quadruped robots like the Aliengo and B1. The Z1 can move smoothly, pick up small objects, open doors, fasten screws, and even pour a glass of wine, showcasing its versatility and potential for various applications (Top3DShop, 2025).
- **D1-T:** An ultraprecise robotic arm with force control capabilities, designed for tasks requiring delicate manipulation (Top3DShop, 2025).
- **4D LiDAR L1:** An omnidirectional LiDAR sensor that provides a 360° x 90° field of view for enhanced perception and navigation (Génération Robots, 2025).
- **Batteries and Chargers:** Unitree offers a range of batteries and chargers to ensure optimal performance and extended operation of its robots (Génération Robots, 2025).

Unitree Robotics Technology

Unitree Robotics leverages cutting-edge technology to create high-performance robots with advanced capabilities. The company's core technologies include the following:

- **Dynamic Motion Control:** Unitree's robots utilize sophisticated dynamic motion control algorithms that enable them to move with agility, stability, and precision. These algorithms allow the robots to adapt to various terrains, overcome obstacles, and perform complex movements (Canvas Business Model, 2025).
- **Advanced Sensors:** Unitree equips its robots with a variety of advanced sensors, including LiDAR, depth cameras, and inertial measurement units (IMUs), to provide them with a comprehensive understanding of their environment. These sensors enable the robots to perceive obstacles, navigate autonomously, and interact with their surroundings (INFUZE Robotics, 2025).
- **High-Performance Motors:** Unitree develops and utilizes high-performance motors that provide its robots with the power and agility needed for demanding tasks. These motors are designed for efficiency and durability, ensuring optimal performance in various applications (Canvas Business Model, 2025).
- **Reinforcement Learning:** Unitree employs reinforcement learning techniques to train its robots and enhance their capabilities. This involves allowing the robots to learn through trial and error in simulated environments, enabling them to develop optimal strategies for performing tasks (GitHub, 2025).
- **Simulation and Sim2Real Transfer:** Unitree utilizes simulation platforms, such as NVIDIA's Isaac Sim, to train its robots in virtual environments. This allows the company to accelerate the development process and safely test new capabilities before deploying them on physical robots. NVIDIA's Isaac platform and Omniverse technology play a crucial role in this process, allowing Unitree to generate large volumes of synthetic data for robot training and significantly shorten development timelines (i-Programmer, 2025).

Partnerships and Collaborations

Unitree Robotics recognizes the importance of partnerships and collaborations in driving innovation and expanding its market reach. The company has established collaborations with various organizations, including

- **RobotShop:** Unitree partnered with RobotShop, a leading robotics distributor, to offer its robots to a wider audience and enhance its global distribution network (RobotShop, 2025).
- **iRed Limited:** Unitree collaborated with iRed Limited, a UK-based drone service provider, to distribute its robots in the UK market, targeting various industries and educational institutions (Unmanned Systems Technology, 2025).

Unitree Robotics operates in a competitive landscape with several other companies vying for a share of the robotics market. Some of the key competitors include the following:

- **RedOne Technologies:** A company that develops mobile robot platforms and associated components (Craft.co, 2025)
- **Boston Dynamics:** A well-known robotics company that develops advanced robots like Spot, a quadruped robot often compared to Unitree's offerings (Kalil, 2025a, b)
- **Engine AI:** A Chinese robotics company that directly competes with Unitree in the development of humanoid robots, with a focus on achieving natural human-like movement (Kalil, 2025a, b)

Both Unitree and Engine AI are actively working on improving the walking gait of their humanoid robots, pushing the boundaries of bipedal locomotion in robotics (Kalil, 2025a, b). Similarly, Unitree and Boston Dynamics are both exploring hypermobility in their robots, potentially signaling a shift in how humanoid robots are designed and marketed (YouTube, 2025c).

Unitree Robotics has secured significant funding from various investors to support its research and development efforts and fuel its growth. The company has raised a total of $155 million over five funding rounds, with its largest round being a Series B round in February 2024, where it raised $139 million (Tracxn, 2025). This funding success highlights the increasing investor interest in humanoid robotics, particularly within the context of the broader AI surge (Caixin Global, 2025).

Unitree Robotics has achieved "unicorn" status with a valuation exceeding $1 billion, placing it among the top players in the global robotics landscape (TMTPost, 2025). The company is often compared to Fig. AI, a US-based humanoid robot startup, further emphasizing its prominence in the industry (TMTPost, 2025).

Unitree Robotics' products have garnered positive reviews for their performance, affordability, and innovative designs. Reviewers have praised the agility and stability of the quadruped robots, highlighting their ability to navigate challenging terrains and perform complex movements (YouTube, 2025a).

The humanoid robots have also received attention for their advanced capabilities and potential applications in various fields. However, some reviewers have noted that the humanoid robots are still under development and may not be ready for widespread commercial use. There is also some skepticism surrounding the G1's capabilities, and more real-world demonstrations may be needed to fully convince potential customers of its usefulness (Reddit, 2025).

11.8 Governance and Ethical Considerations

As robotics becomes increasingly integrated into society, governance frameworks must ensure safety, reliability, and ethical alignment. AI agents in robotics must operate under strict guidelines to address these concerns.

11.8.1 Safety Constraints

AI agents should adhere to operational safety protocols to prevent unintended consequences. Translating natural language constraints into executable policies ensures compliance across diverse tasks.

11.8.2 Error Management

Errors in multi-agent systems or autonomous robots can propagate quickly, necessitating robust monitoring and diagnostic mechanisms. AI agents must include real-time error detection and mitigation strategies to ensure reliable performance.

11.8.3 Transparency and Explainability

Transparent decision-making processes enhance trust in AI-driven robotic systems. For instance, providing clear explanations for actions taken during critical operations can reassure users and improve collaboration.

11.9 Impact on the Workforce

The integration of AI agents in robotics is transforming the global workforce, introducing both opportunities and challenges. As robots equipped with advanced AI capabilities continue to replace manual tasks, industries must prepare for significant disruptions while leveraging opportunities for innovation.

11.9.1 Industries Facing Significant Disruption

The impact of robotics and AI integration varies by industry, with certain sectors experiencing profound changes due to automation. Below are examples of industries most susceptible to disruption:

- **Manufacturing:** Robotics has long been a staple in manufacturing, but the adoption of AI-enhanced robots is set to redefine the sector. Assembly lines, quality control, and material handling are increasingly automated, enabling higher productivity and precision.
- **Logistics and Warehousing:** Autonomous mobile robots (AMRs) powered by AI are revolutionizing logistics. Tasks such as inventory management, order

picking, and material transport are becoming automated, reducing labor costs and improving efficiency.
- **Retail and Customer Service:** The retail sector is adopting AI-powered robots for tasks like shelf stocking, inventory monitoring, and customer assistance.
- **Agriculture:** Robotic systems for planting, harvesting, and crop monitoring are rapidly gaining traction.
- **Healthcare:** While robotics in healthcare is creating new opportunities, such as surgical assistance and elderly care, certain roles in administrative and routine patient monitoring may see reduced human involvement.

11.9.2 Economic Implications: Job Creation vs. Displacement

The adoption of robotics and AI agents carries dual economic effects: displacement of certain jobs and creation of new opportunities. Studies by major economic think tanks predict the following trends:

- **Job Displacement:** Roles that are highly repetitive and physically demanding or require minimal cognitive effort are at the highest risk of being automated. According to estimates from the World Economic Forum (WEF), nearly 75 million jobs globally could be displaced by automation by 2030 (World Economic Forum, 2020).
- **Job Creation:** While automation reduces demand for certain roles, it also generates new opportunities in robotics engineering, maintenance, and AI development. Roles such as AI trainers, robot operators, and data scientists are expected to grow significantly. The WEF estimates that automation and AI could create approximately 133 million new roles during the same period (World Economic Forum, 2020).
- **Sector-Specific Impact:** The net effect varies by sector. For instance, while manufacturing may experience significant job losses in low-skill roles, it is likely to see growth in high-skill roles related to robotics integration and system optimization.
- **Regional Disparities:** Regions heavily reliant on low-skill labor-intensive industries are expected to face greater challenges compared to regions with a focus on technology and innovation.

11.9.3 Workforce Adaptation Strategies

Addressing the challenges posed by AI agents in robotics requires proactive strategies for workforce adaptation. Key approaches include the following:

- **Upskilling and Reskilling:** Governments, businesses, and educational institutions must collaborate to provide training programs that equip workers with the skills needed for emerging roles.
- **Lifelong Learning:** The dynamic nature of AI and robotics necessitates continuous learning. Offering modular, flexible training programs tailored to industry-specific needs can help workers stay competitive in evolving job markets (Manyika et al., 2017).
- **Policy Interventions:** Policymakers can implement initiatives such as wage subsidies for reskilled workers, tax incentives for businesses that invest in human capital, and unemployment insurance for displaced workers.
- **Emphasis on Creativity and Collaboration:** As AI agents take over repetitive tasks, human workers can focus on roles requiring creativity, problem-solving, and interpersonal collaboration.

11.9.4 Case Studies in Workforce Transition

- **Automotive Manufacturing:** Companies like Tesla, which extensively use AI-powered robots, have shifted from traditional assembly line roles to specialized jobs such as robot programming and maintenance (Tesla, 2024).
- **E-commerce Warehousing:** Firms like Amazon deploy AI-powered robotics systems for order fulfillment. Simultaneously, they provide employees with access to educational programs, such as the Amazon Career Choice initiative, to prepare them for roles in advanced technologies (Amazon, 2024).
- **Healthcare Systems:** In hospitals adopting robotic surgical systems, training programs are provided for surgeons and support staff to collaborate effectively with robotic assistants (Intuitive Surgical, 2024).

11.9.5 Broader Economic Impacts

The economic impact of robotics and AI extends beyond individual industries. Key trends include the following:

- **Increased Productivity:** Automation of repetitive tasks allows businesses to achieve higher productivity levels, driving economic growth.
- **Shift in Labor Demand:** The demand for high-skill, high-wage jobs increases, while low-skill, low-wage roles decline.
- **Global Competition:** Nations leading in robotics and AI innovation gain a competitive edge in global markets.

11.10 Future Outlook

The integration of AI into robotics is rapidly transforming global industries, with long-term implications that demand careful consideration. This section explores potential future scenarios, emerging trends, and economic forecasts to provide a detailed perspective on the trajectory of robotics powered by AI agents.

Figure 11.6 illustrates the future outlook of robotics, and we will cover the details in the following subsections.

11.10.1 Emerging Economic Trends and Global Market Dynamics

The AI robotics market is poised for unprecedented growth, driven by innovations in AI algorithms and robotics hardware. I expected this to create a multi-trillion-dollar value in the next decade. This rapid expansion is not confined to traditional sectors like manufacturing and logistics but is also penetrating industries such as construction, energy, and space exploration.

11.10.2 Strategic Shifts in Industry Focus

The next decade will witness a diversification of robotic applications driven by AI. In addition to existing industries, robotics is likely to see significant growth in the following:

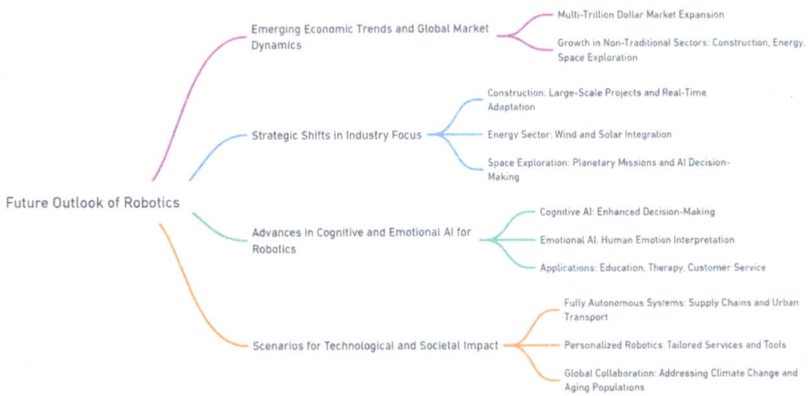

Fig. 11.6 Future outlook of robotics

- **Construction:** AI-powered robots are anticipated to handle large-scale construction projects, leveraging machine learning for material optimization and real-time adaptation to environmental conditions.
- **Energy Sector:** Robotics is emerging as a key player in renewable energy, particularly in wind and solar installations.
- **Space Exploration:** AI-enabled robots are expected to play a pivotal role in planetary exploration and infrastructure development on extraterrestrial surfaces. NASA and private entities are already investing in AI agents that enable autonomous decision-making in unpredictable space environments (NASA, 2024).

11.10.3 Advances in Cognitive and Emotional AI for Robotics

A transformative shift in robotics involves integrating cognitive and emotional intelligence into AI agents. Cognitive AI will enhance decision-making capabilities by enabling robots to process complex data streams and make adaptive, real-time decisions. Emotional intelligence, on the other hand, will allow robots to interpret human emotions and respond appropriately, fostering better human–robot interactions.

This trend is particularly promising in areas such as education, therapy, and customer service.

11.10.4 Scenarios for Technological and Societal Impact

- **Acceleration in Autonomous Systems:** AI advancements will push robotics toward greater autonomy, reducing dependency on human intervention. This scenario envisions fully autonomous supply chains, urban transport systems, and disaster response teams by the 2030s.
- **Enhanced Personalization:** Robots with AI-driven personalization capabilities will tailor services based on user preferences, from healthcare solutions to educational tools, leading to a more inclusive and user-centric approach to technology.
- **Global Collaboration in Robotics Innovation:** The increasing importance of robotics in addressing global challenges, such as climate change and population aging, may drive international collaborations.

11.11 Summary

The chapter delves into the role of AI agents in revolutionizing robotics, emphasizing their ability to interact with and adapt to physical environments. It introduces an agentic framework that integrates perception, interaction, and reasoning modules, allowing robots to sense, interpret, and respond dynamically. Advanced modules for humanoid robots enhance physical interaction, social engagement, and hybrid control. Simulation and training methodologies, such as reinforcement learning and imitation learning, are explored to demonstrate their contribution to robotic development. Spatial intelligence, including 3D understanding, is highlighted as a cornerstone for navigation and manipulation tasks. The chapter also examines applications in healthcare, disaster response, and underwater exploration, showcasing the practical impact of robotics. It concludes by analyzing market dynamics, research innovations, governance, and workforce adaptation strategies.

11.12 Questions

Multiple Choice

1. What is the primary role of multimodal perception in robotics?
   a. Navigating through simple terrains
   b. Integrating sensory data for contextual understanding
   c. Reducing the cost of hardware components
   d. Optimizing battery consumption
2. Which company is known for its humanoid robot "Optimus"?
   a. NVIDIA
   b. Tesla
   c. Boston Dynamics
   d. Unitree Robotics
3. What is the key benefit of LiDAR technology in robotics?
   a. Creating point clouds for 3D mapping
   b. Enhancing auditory processing
   c. Simulating haptic feedback
   d. Improving long-range wireless communication
4. What does the term "shared autonomy" refer to in humanoid robotics?
   a. Robots working independently without human intervention
   b. Robots collaborating with humans for efficient task execution
   c. Robots using cloud computing for decision-making
   d. Robots replacing human roles entirely
5. What methodology combines expert demonstrations and learning?
   a. Reinforcement learning
   b. Behavioral cloning
   c. Proximal policy optimization
   d. Multi-objective optimization

True/False

1. Humanoid robots utilize proprioception data for balance and movement.
2. Simulation platforms like Isaac Sim are used to reduce the dependency on real-world testing.
3. Reinforcement learning relies entirely on supervised datasets for training AI agents.
4. Swarm robotics involves multiple robots working in isolation to achieve individual goals.
5. The integration of AI in robotics is likely to displace more jobs than it creates.

Simple Questions

1. What is the role of hybrid control in humanoid robotics?
2. How does spatial intelligence improve robotic navigation?
3. Name one advantage of simulation platforms in robotic training.
4. What is LiDAR, and how is it used in robotics?
5. Why is workforce adaptation important in the context of robotics?

Essay Questions

1. Discuss the role of AI agents in humanoid robots, focusing on physical interaction and social engagement.
2. How do simulation and training environments contribute to the development of adaptive AI agents?
3. Analyze the economic implications of AI-driven robotics, highlighting job displacement and creation.
4. Compare the contributions of companies like NVIDIA and Boston Dynamics in advancing robotics.
5. Evaluate the ethical challenges in integrating AI agents into autonomous robotic systems.

References

Abadi, M., Agarwal, A., Barham, P., Brevdo, E. et al. (2016). *TensorFlow: Large-scale machine learning on heterogeneous distributed systems*. arXiv preprint arXiv:1603.04467.

ABB. (2023). *ABB robotics*. Retrieved from https://new.abb.com/products/robotics

ABI Research. (2025). *Unpacking NVIDIA's Robot Operating System Support*. Retrieved from https://www.abiresearch.com/blog/unpacking-nvidias-robot-operating-system-support

Amazon. (2024). *Amazon robotics*. Retrieved from https://www.aboutamazon.com/news/tag/robotics

Boston Dynamics. (2024). *Boston Dynamics Spot*. https://bostondynamics.com/products/spot/

Caixin Global. (2025). *Unitree robotics raises nearly $139 million in series B2 fundraising*. Retrieved from https://www.caixinglobal.com/2024-02-24/unitree-robotics-raises-nearly-139-million-in-series-b2-fundraising-102168284.html

Canvas Business Model. (2025). *How does unitree robotics work?* Retrieved from https://canvas-businessmodel.com/blogs/how-it-works/unitree-robotics-how-it-works

Cobot. (2025). *Cobot develops AI-driven collaborative robots with NVIDIA Isaac*. Retrieved from https://www.co.bot/news/cobot-develops-ai-driven-collaborative-robots-with-nvidia-isaac

Craft.co. (2025). *Top Unitree robotics competitors and alternatives*. Retrieved from https://craft.co/unitree/competitors

DeepMind. (2025a). *Shaping the future of advanced robotics*. Retrieved from https://deepmind.google/discover/blog/shaping-the-future-of-advanced-robotics/

DeepMind. (2025b). *AI, robot*. Retrieved from https://deepmind.google/discover/the-podcast/ai-robot/

DeepMind. (2025c). *Our latest advances in robot dexterity*. Retrieved from https://deepmind.google/discover/blog/advances-in-robot-dexterity/

Denso. (2024). *DENSO robotics*. Retrieved from https://www.densorobotics.com/

Digitimes. (2025). *Nvidia CEO's China visit showcases strategic ties with rising robotics unicorns*. Retrieved from https://www.digitimes.com/news/a20250124PD204/nvidia-robotics-ceo-robot-jensen-huang.html

Drones Plus Robotics. (2025). *Unitree, platforms*. Retrieved from https://www.dronesplusrobotics.com/pps/unitree/platforms

Epson. (2024). *Epson robots*. Retrieved from https://epson.com/robots

European Commission. (2020). *Ethics guidelines for trustworthy AI*. Retrieved from https://digital-strategy.ec.europa.eu/en/library/ethics-guidelines-trustworthy-ai

FANUC. (2024). *FANUC global*. Retrieved from https://www.fanuc.co.jp/en/

Fei-Fei, L., et al. (2024). *Multimodal interaction systems and embodied AI: Toward scalable and adaptive robotics*. arXiv. Retrieved from https://arxiv.org/abs/2401.03568

Freund, K. (2024, March 20). What is physical AI, and why it could change the world. *Forbes*. Retrieved from https://www.forbes.com/sites/karlfreund/2024/08/09/what-is-physical-ai-and-why-it-could-change-the-world/

GeeksforGeeks. (2025). *Understanding DeepMind's approach to artificial general intelligence (AGI)*. Retrieved from https://www.geeksforgeeks.org/understanding-deepminds-approach-to-artificial-general-intelligence-agi/

Génération Robots. (2025). *Unitree robotics*. Retrieved from https://www.generationrobots.com/en/297_unitree-robotics

GitHub. (2025). *Unitree robotics*. Retrieved from https://github.com/unitreerobotics

Guo, Y., Wang, H., Hu, Q., Liu, H., Liu, L., & Bennamoun, M. (2024). Deep learning for 3d point clouds: A survey. *IEEE Transactions on Pattern Analysis and Machine Intelligence, 43*(11), 4338–4364.

Huang, J. (2025). *NVIDIA CEO Jensen Huang Robots presentation at CES 2025*. [YouTube video]. Retrieved from https://www.youtube.com/watch?v=LETO_-keVFg

INFUZE Robotics. (2025). *Unitree Go2 Air - Quadruped robot platform*. Retrieved from https://robot.infuze.hu/en/products/unitree-go2-pro-negylabu-robotplatform

Intuitive Surgical. (2024). *da Vinci Surgical System*. Retrieved from https://www.intuitive.com/en-us/products-and-services/da-vinci

i-Programmer. (2025). *Unitree G1 - See how it runs*. Retrieved from https://www.i-programmer.info/news/169-robotics/17781-unitree-g1-see-how-it-runs.html

Justoborn. (2025). *NVIDIA robot tools: Robotics development*. Retrieved from https://justoborn.com/nvidia-robot-tools/

Kalil, M. (2025a). Google DeepMind links up with Apptronik to advance Apollo humanoid robot. Retrieved from https://mikekalil.com/blog/google-deepmind-apptronik/

Kalil, M. (2025b). *Unitree vs. ENGINEAI: China's Humanoid Robotics Leaders Vie for Most Humanlike Walk*. Retrieved from https://mikekalil.com/blog/unitree-vs-engineai/

KrASIA. (2025). *Humanoid robots are getting cheaper—can affordability drive adoption?* Retrieved from https://kr-asia.com/humanoid-robots-are-getting-cheaper-can-affordability-drive-adoption

KUKA. (2024). KUKA robotics. Retrieved from https://www.kuka.com/

Manyika, J., Lund, S., Chui, M., Bughin, J., Woetzel, J., Batra, P., et al. (2017). *Jobs lost, jobs gained: What the future of work will mean for jobs, skills, and wages*. McKinsey Global Institute.

NASA. (2024). *NASA's technology portfolio*. Retrieved from https://www.nasa.gov/mission_pages/tdm/main/index.html

NVIDIA. (2024a). What is robotics simulation? NVIDIA. https://resources.nvidia.com/en-us-playlist-for-robotics/roboticssimulation

NVIDIA. (2025a). *AI for robotics*. Retrieved from https://www.nvidia.com/en-us/industries/robotics/

NVIDIA. (2025b). *Robotics and edge computing solutions*. Retrieved from https://www.nvidia.com/en-us/solutions/robotics-and-edge-computing/

NVIDIA. (2025c). *NVIDIA Isaac - AI robot development platform*. Retrieved from https://developer.nvidia.com/isaac

NVIDIA. (2025d). *Isaac Sim*. Retrieved from https://catalog.ngc.nvidia.com/orgs/nvidia/containers/isaac-sim

NVIDIA. (2025e). *Isaac Sim - AWS marketplace*. Retrieved from https://aws.amazon.com/marketplace/pp/prodview-7p2np4t6qdfla

NVIDIA. (2025f). *What is Isaac Sim? — Omniverse IsaacSim*. Retrieved from https://docs.omniverse.nvidia.com/isaacsim

NVIDIA. (2025g). *Isaac Sim - Robotics simulation and synthetic data generation*. Retrieved from https://developer.nvidia.com/isaac/sim

NVIDIA. (2025h). *Jetson AGX Orin for next-gen robotics*. Retrieved from https://www.nvidia.com/en-us/autonomous-machines/embedded-systems/jetson-orin/

NVIDIA. (2025i). *Jetson Nano brings the power of modern AI to edge devices*. Retrieved from https://www.nvidia.com/en-us/autonomous-machines/embedded-systems/jetson-nano/

NVIDIA. (2025j). *Jetson Software*. Retrieved from https://developer.nvidia.com/embedded/develop/software

NVIDIA. (2025l). *NVIDIA accelerates humanoid robotics development*. Retrieved from https://nvidianews.nvidia.com/news/nvidia-accelerates-worldwide-humanoid-robotics-development

Open Data Science. (2025). *Google's DeepMind is revolutionizing robotics*. Retrieved from https://opendatascience.com/googles-deepmind-is-revolutionizing-robotics/

R2Surgical. (2025). *NVIDIA partners with surgical robotics companies*. Retrieved from https://r2surgical.com/blogs/news/nvidias-ai-platform-used-by-surgical-robotics-companies

Reddit. (2025). *I am very dubious about the Unitree $16000 robot*. Retrieved from https://www.reddit.com/r/robotics/comments/1f5nt1k/i_am_very_dubious_about_the_unitree_16000_robot/

RoboPGH. (2025). *The Pittsburgh robotics network collaborates with NVIDIA to accelerate ecosystem growth*. Retrieved from https://www.robopgh.org/news/the-pittsburgh-robotics-network-collaborates-with-nvidia-to-accelerate-ecosystem-growth

Robotics247. (2025). *Apptronik, Google DeepMind robotics partner to accelerate advancement of AI-powered humanoids*. Retrieved from https://www.robotics247.com/article/apptronik_google_deepmind_robotics_partner_to_accelerate_advancement_of_ai_powered_humanoids

RoboticsTomorrow. (2025). *Unitree showcases advanced high-mobility robots at CES 2025*. Retrieved from https://www.roboticstomorrow.com/news/2025/01/08/unitree-showcases-advanced-high-mobility-robots-at-ces-2025/23818

RobotShop. (2025). *RobotShop partners with Unitree: Bringing AI-powered humanoids & quadruped robots to the public*. Retrieved from https://community.robotshop.com/blog/show/robotshop-partners-with-unitree-bringing-ai-powered-humanoids-and-quadruped-robots-to-the-public

Security Robotics. (2025). *Unitree & security robotics: An innovative partnership*. Retrieved from https://security-robotics.de/en/aktuelles-en/unitree-meets-security-robotics

Silicon Republic. (2025). *DeepMind is training robots for real-world activities*. Retrieved from https://www.siliconrepublic.com/machines/google-deepmind-robot-constitution-real-world-safety

SoftBank Robotics. (2024). *SoftBank robotics*. Retrieved from https://www.softbankrobotics.com/

Spatial Intelligence Lab. (2024). *Spatial intelligence lab*. Retrieved from https://www.spatialintelligence.ai/
Stäubli. (2024). *Stäubli robotics*. Retrieved from https://www.staubli.com/en/robotics/
Tesla. (2024). *Optimus*. Retrieved from https://www.tesla.com/AI
The Robot Report. (2025). *Google DeepMind discusses latest advances in robot dexterity*. Retrieved from https://www.therobotreport.com/inside-google-deepmind-latest-advances-in-robot-dexterity/
TMTPost. (2025). *Unitree robotics raises several billion Yuan in series C funding*. Retrieved from https://en.tmtpost.com/post/7263532
Top3DShop. (2025). *Unitree robotics robots*. Retrieved from https://top3dshop.com/robots/unitree-robotics-robots/
Tracxn. (2025). *Unitree - Raised $155M Funding from 24 investors*. Retrieved from https://tracxn.com/d/companies/unitree/__o1e8b3ZlyUCcjIECfbM9csfhnJyv1_fOku8o_K8gCYg/funding-and-investors
TrendForce. (2025). *[News] NVIDIA's humanoid robot partners unveiled at CES 2025, featuring China's Unitree, XPeng, and BYD*. Retrieved from https://www.trendforce.com/news/2025/01/08/news-jensen-huang-unveils-nvidias-robotics-partners-at-ces-2025-featuring-chinas-unitree-xpeng-and-byd
Unitree Robotics. (2024). *Unitree robotics*. Retrieved from https://www.unitree.com/
Universal Robots. (2024). *Collaborative robots - Cobots*. Retrieved from https://www.universal-robots.com/
Unmanned Systems Technology. (2025). *Exclusive partnership to distribute advanced robotic solutions across the UK*. Retrieved from https://www.unmannedsystemstechnology.com/2023/10/exclusive-partnership-to-distribute-advanced-robotic-solutions-across-the-uk/
Vaswani, A., Shazeer, N., Parmar, N., Uszkoreit, J., Jones, L., Gomez, A. N., et al. (2017). Attention is all you need. In *Advances in neural information processing systems* (pp. 5998–6008).
Vecna Robotics. (2024). *Vecna Robotics*. Retrieved from https://www.vecnarobotics.com/
Wikipedia. (2025). *Nvidia Jetson*. Retrieved from https://en.wikipedia.org/wiki/Nvidia_Jetson
Wikipedia. (2025a). *Google DeepMind*. Retrieved from https://en.wikipedia.org/wiki/Google_DeepMind
Wikipedia. (2025b). *Unitree Robotics*. Retrieved from https://en.wikipedia.org/wiki/Unitree_Robotics
World Economic Forum. (2020). *The future of jobs report 2020*. Retrieved from https://www.weforum.org/reports/the-future-of-jobs-report-2020
Xpert.Digital. (2025). *NVIDIA's Cosmos platform for physical robot AI: ChatGPT breakthrough for general robotics is imminent*. Retrieved from https://xpert.digital/en/physical-ai-of-robots/
Yaskawa. (2024). *Yaskawa global*. Retrieved from https://www.yaskawa-global.com/
YouTube. (2025a). *An affordable spot alternative?! | Unitree Go2 air review*. Retrieved from https://www.youtube.com/watch?v=Bttxmu9E2BU
YouTube.(2025c). *Unitree & engine AI vie for most lifelike humanoid robot walk, Black Panther 2.0*. Retrieved from https://www.youtube.com/watch?v=6CjxMPg0pvg

Jerry Huang is currently an AI Engineer at Google, where he contributed to the AI/ML evaluation pipeline for a consumer-facing application. Before Google, he worked as a technical and security staff member at several prominent technology companies, gaining experience in areas like security, AI/ML, and scalable systems.

At Metabase, an open-source business intelligence platform, Jerry contributed features such as private key management and authentication solutions. As a Software Engineer at Glean, a generative AI search startup, he was one of three engineers responsible for managing large-scale GCP infrastructure powering text summarization, autocomplete, and search for over 100,000 enterprise users. During his time at TikTok, Jerry helped design and build custom RPCs to model access control policies. At Roblox, he served as a Machine Learning/Software Engineering Intern, focusing on real-time text generation models and gathering a large multilingual corpus that significantly boosted model robustness.

In addition to his industry experience, Jerry has conducted extensive security and biometrics research as a Research Assistant at Georgia Tech's Institute for Information Security and Privacy, resulting in a thesis on privacy-preserving biometric authentication.

Jerry holds a BS/MS in Computer Science from Georgia Tech and is currently pursuing an MS in Applied Mathematics at the University of Chicago.

Ken Huang is a prolific author and globally recognized authority in AI and Web3, with an extensive portfolio of published works that bridge business strategy, technical implementation, and cutting-edge research. As Fellow of Cloud Security Alliance and Co-Chair of the AI Safety Working Groups at the Cloud Security Alliance and the AI STR Working Group at the World Digital Technology Academy under the UN Framework, he is a leading voice in shaping global AI governance and security standards.

Huang is the CEO and Chief AI Officer (CAIO) of DistributedApps.ai, a firm specializing in generative AI training and consulting. His contributions to the field include being a core contributor to the OWASP Top 10 Risks for LLM Applications and an active participant in the NIST Generative AI Public Working Group.

Notable Publications:

- *Beyond AI: ChatGPT, Web3, and the Business Landscape of Tomorrow* (Springer, 2023)—Strategic insights into AI and Web3's business applications.
- *Generative AI Security: Theories and Practices* (Springer, 2024)—A comprehensive guide on securing generative AI systems.
- *Practical Guide for AI Engineers* (Volumes 1 and 2, DistributedApps.ai, 2024)—Essential resources for AI and ML engineers.
- *The Handbook for Chief AI Officers: Leading the AI Revolution in Business* (DistributedApps.ai, 2024)—A road map for CAIOs in implementing GenAI across organizations.
- *Web3: Blockchain, the New Economy, and the Self-Sovereign Internet* (Cambridge University Press, 2024)—Insights into the convergence of AI, blockchain, IoT, and emerging technologies.
- *Blockchain and Web3: Building the Cryptocurrency, Privacy, and Security Foundations of the Metaverse* (Wiley, 2023)—Recognized as a must-read by TechTarget in 2023 and 2024.

Ken is a sought-after speaker and has presented at events such as the World Economic Forum in Davos, ACM and IEEE conferences, the CSA AI Summit, Depository Trust and Clearing Corporation forums, and World Bank conferences. His recent appointment to the OpenAI Forum reflects his ongoing commitment to advancing collaboration and dialogue in the field of AI.

Explore Ken's work on Amazon: https://www.amazon.com/author/kenhuang

Chapter 12
AI Agent Safety and Security Considerations

Jerry Huang, Ken Huang, Krystal Jackson, and Chris Hughes

In this last chapter of the book, we shift focus on AI agent's safety and security. We will discuss a wide spectrum of challenges ranging from accidental failures to deliberate attacks on agentic AI systems. The complexity of agentic interactions with dynamic environments and other systems amplifies the need for robust design, rigorous testing, and governance frameworks. This chapter explores the various considerations involved in safeguarding AI agents, emphasizing both technical and governance dimensions.

12.1 Potential Vulnerabilities in AI Agent Systems

AI agent systems, despite their sophisticated capabilities, are susceptible to various vulnerabilities that can compromise their safety and security (Domkundwar et al., 2024). These vulnerabilities can be broadly categorized into two main types: accidental failures and deliberate attacks (see Fig. 12.1). Understanding these potential weaknesses is fundamental to developing robust and resilient AI agents (Carnegie Mellon University, 2023).

J. Huang
The University of Chicago, Chicago, IL, USA

K. Huang (✉)
DistributedApps.ai, Fairfax, VA, USA
e-mail: ken@distributedapps.ai

K. Jackson
UC Berkeley, University of California, Berkeley, CA, USA

C. Hughes
Aquia, Virginia Beach, VA, USA

K. Huang (ed.), *Agentic AI*, Progress in IS,
https://doi.org/10.1007/978-3-031-90026-6_12

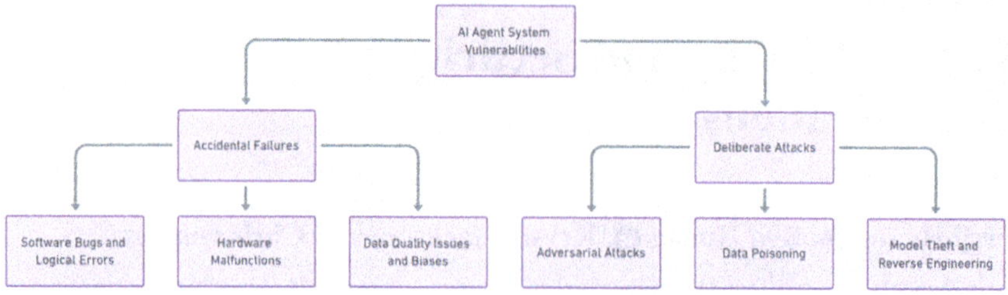

Fig. 12.1 AI agent vulnerabilities

12.1.1 Accidental Failures

Accidental failures in AI agent systems can occur due to a variety of factors, often stemming from unintended flaws in their design, implementation, or operation.

Software Bugs and Logical Errors

Software bugs and logical errors represent one source of vulnerability in AI agent systems. As agents become more complex, incorporating advanced decision-making algorithms, learning capabilities, and multi-agent coordination, the potential for coding errors or logical inconsistencies increases dramatically. These issues can lead to unexpected behaviors or system failures, potentially with severe consequences in critical applications.

For instance, a subtle error in an autonomous agent's perception algorithm could result in misclassification of objects in its environment. In a multi-agent system coordinating warehouse operations, this could lead to collisions between agents or mishandling of goods. Similarly, a logical flaw in a financial trading agent's decision-making process could result in grave financial losses if left unchecked.

To mitigate these risks, developers of AI agent systems must appropriately apply rigorous testing methodologies, including unit testing, integration testing, red teaming, and scenario-based testing that simulates a wide range of potential operating conditions. Formal verification methods, while challenging to apply to complex AI systems, can also play a role in ensuring the correctness of critical components. Examples include NIST's AI Testing, Evaluation, Validation and Verification (TEVV) framework, which also is endorsed and recommended by CISA.

The following code snippet demonstrates how an AI agent can handle unexpected exceptions during decision-making to avoid critical failures.

```
class AIAgent:
    def make_decision(self, data):
        try:
            # Simulated decision-making process
            if "priority" not in data:
                raise ValueError("Missing priority in data")
            return f"Decision made for priority
{data['priority']}"
        except ValueError as e:
            return f"Error: {e} - Decision deferred for safety"

# Example usage
agent = AIAgent()
print(agent.make_decision({"priority": 1}))  # Outputs: Decision
made for priority 1
print(agent.make_decision({}))  # Outputs: Error: Missing
priority in data - Decision deferred for safety
```

Hardware Malfunctions

AI agents, particularly those embedded in physical systems like cell phones, smart glasses, other various wearables such as Apple Vision Pros and Meta Quest 3, humanoid robots, or IoT devices, rely on various hardware components for sensing, processing, and actuation. Failures in these components can disrupt the agent's ability to perceive its environment accurately or execute actions as intended.

For example, a faulty sensor in a swarm of collaborative drones could provide incorrect altitude data, causing the affected agent to collide with others or obstacles (Liu et al., 2024). In an industrial setting, a malfunctioning actuator in a robotic arm controlled by an AI agent could lead to manufacturing defects or safety hazards.

To address hardware-related vulnerabilities, AI agent systems should incorporate robust error detection and handling mechanisms. This might include implementing redundancy in critical sensors and actuators, developing self-diagnostic capabilities, and designing fail-safe modes that ensure safe operation even when hardware components fail.

The following code snippet shows how AI agents can securely fuse redundant sensor data to handle failures or discrepancies.

```
class AIAgentSensors:
    def  fuse_sensor_data(self, sensor1, sensor2):
        if sensor1 and sensor2:
            # Use the average if both sensors provide data
            return (sensor1 + sensor2) / 2
        elif sensor1:
            return sensor1
        elif sensor2:
            return sensor2
        else:
            return "Error: Both sensors failed"

# Example usage
agent = AIAgentSensors()
print(agent.fuse_sensor_data(100, 98))  # Outputs: 99.0
print(agent.fuse_sensor_data(100, None))  # Outputs: 100
print(agent.fuse_sensor_data(None, None))  # Outputs: Error: Both
sensors failed
```

Data Quality Issues and Biases

Machine learning models powering AI agents are only as good as the data they're trained on, and flaws in training data can lead to biased or incorrect decision-making. This vulnerability is particularly critical for AI agents, as they often operate autonomously in dynamic environments where the consequences of biased decisions can compound over time.

For instance, an AI agent designed for customer service might develop biased responses if trained on data that underrepresents certain demographic groups. In a more critical scenario, an autonomous vehicle trained on data primarily from one type of environment might make dangerous decisions when operating in unfamiliar conditions (Song et al., 2023).

Addressing data-related vulnerabilities requires ongoing vigilance in data collection, preprocessing, and model evaluation processes. This includes the following:

- Implementing robust data validation and cleaning pipelines
- Actively seeking diverse and representative training data
- Regularly auditing AI agent behaviors for signs of bias or inconsistency
- Employing techniques like adversarial training to improve model robustness
- Developing methods for continuous learning that can adapt to changing environments without introducing new biases
- Addressing data silo challenges within the environments the agents operate to ensure sufficient data for autonomous operations

12.1.2 Deliberate Attacks

In addition to accidental failures, AI agent systems face threats from deliberate attacks designed to exploit their vulnerabilities or manipulate their behavior.

Adversarial Attacks on Machine Learning Models

Adversarial attacks pose a significant threat to AI agent systems, as they can manipulate the agent's perception and decision-making processes (He et al., 2024). These attacks involve crafting inputs specifically designed to fool AI systems into making incorrect decisions.

In the context of AI agents, adversarial attacks could have severe consequences. For example:

- An attacker could create subtle perturbations in visual inputs to cause an autonomous security robot to misclassify intruders as authorized personnel.
- In a multi-agent system managing smart city infrastructure, adversarial attacks could trick traffic management agents into creating congestion or directing vehicles into dangerous situations.
- For conversational AI agents, carefully crafted text inputs could manipulate the agent into revealing sensitive information or taking harmful actions.
- AI agents that control robotic systems can be jailbroken to perform dangerous actions in military contexts.
- Malicious data poisoning could be pursued by attackers to compromise and impact downstream activities agents take based upon training data.

Defending against adversarial attacks requires a multi-faceted approach:

- Implementing adversarial training techniques to make models more robust (Wasil et al., 2024).
- Developing input validation and sanitization methods specific to the agent's domain.
- Employing ensemble methods that combine multiple models or decision-making approaches.
- Regularly updating and retraining models to address newly discovered vulnerabilities.
- When appropriate, implement methods like instruction hierarchy which assigns different levels of importance to the instructions a model receives.

The following is a simple sample code snippet to show adversarial input detection to secure an AI agent against manipulated data.

```
import numpy as np


class AIAdversarialDefense:
    def detect_adversarial(self, original, perturbed,
threshold=0.2):
        diff = np.linalg.norm(np.array(original) - np.
array(perturbed))
        return diff > threshold

# Example inputs
original_data = [1.0, 0.5, 0.8]
perturbed_data = [1.1, 0.4, 0.85]

# Detect adversarial attempts
defense = AIAdversarialDefense()
print(defense.detect_adversarial(original_data, perturbed_data))
# Outputs: False (not adversarial)
```

Data Poisoning Attacks

Data poisoning attacks represent another form of deliberate manipulation particularly relevant to AI agent systems (Deng et al., 2024). In these attacks, malicious actors introduce corrupted or misleading data into an AI system's training dataset or online learning streams.

For AI agents that continually learn and adapt to their environment, data poisoning poses a significant long-term threat. An attacker could gradually influence the agent's behavior over time by injecting malicious data points. For instance:

- In a collaborative filtering system used by recommendation agents, an attacker could inject false preferences to manipulate the recommendations for a large group of users.
- For reinforcement learning agents, carefully crafted false rewards could lead the agent to learn harmful policies.
- In federated learning systems where multiple agents share learning updates, a compromised agent could inject malicious updates to affect the entire system (U.S. Department of Homeland Security, 2023).

Mitigating data poisoning attacks requires the following:

- Implementing robust data validation and anomaly detection mechanisms.
- Developing secure multi-agent learning protocols that can detect and isolate malicious agents; this can be a good area of research for a PhD-level project.
- Employing differential privacy techniques to limit the influence of any single data point.

- Regularly auditing the agent's learned behaviors and knowledge base for signs of compromise.

Model Theft and Reverse Engineering

Model theft and reverse engineering pose threats to the intellectual property and security of AI agent systems. Attackers may attempt to extract the underlying model or its parameters through various techniques, including model inversion attacks or membership inference attacks.

For AI agents, the risks of model theft extend beyond just intellectual property concerns:

- Extracted models could be used to develop more effective adversarial attacks against the agent system.
- In competitive scenarios (e.g., trading agents), stolen models could be exploited to predict and counteract the agent's strategies.
- For agents handling sensitive data (e.g., in healthcare or finance), model theft could potentially reveal private information about the training data.

Protecting against model theft and reverse engineering in AI agent systems involves the following:

- Implementing secure enclaves or trusted execution environments for critical computations
- Employing model obfuscation techniques to make reverse engineering more difficult
- Developing dynamic model updating strategies that limit the value of any single stolen model version
- Implementing strict access controls and monitoring for any systems interacting with the AI agent
- Employing rate-limiting and request pattern monitoring to detect systematic probing attempts

Attacking Vision–Language Computer Agents via Pop-Ups

Autonomous agents powered by large vision and language models (VLMs) have shown significant promise in executing a variety of computer-based tasks, including web browsing to book travel and managing desktop software. These tasks require agents to comprehend and interact effectively with graphical user interfaces, a capability enabled by their integration of visual and linguistic processing. As visual inputs become increasingly central to agentic applications, understanding the risks and vulnerabilities associated with such systems becomes critical. However, the security implications of these visual integrations remain insufficiently explored.

Research has revealed that VLM agents are particularly susceptible to attacks using carefully designed adversarial pop-ups—visual elements crafted to exploit the agent's decision-making processes (Zhang et al., 2024). Unlike human users, who can typically recognize and disregard such pop-ups as irrelevant or malicious, VLM agents are prone to misinterpreting these distractions. This leads them to interact with the pop-ups instead of focusing on their assigned tasks, significantly undermining their performance. For example, when adversarial pop-ups were integrated into testing environments such as OSWorld and VisualWebArena, the attack success rate—measured by the frequency with which agents clicked on the pop-ups—reached 86% on average. Furthermore, the presence of these pop-ups reduced the agents' overall task success rate by 47%.

Attempts to mitigate these vulnerabilities using basic defense strategies have proven largely ineffective. Approaches such as instructing agents to ignore pop-ups or including visual indicators (e.g., labels marking the pop-ups as advertisements) have failed to prevent the attacks. This ineffectiveness highlights the need for more sophisticated and robust defense mechanisms to secure VLM-powered agents against such exploitation. Future work in this area must focus on developing advanced adversarial defenses, such as enhanced contextual understanding, robust perception algorithms, and domain-specific countermeasures, to address these emergent risks effectively.

The vulnerabilities of vision–language models in agentic applications underscore the broader challenges of securing AI systems that rely on complex integrations of visual and linguistic inputs. As these models continue to expand their capabilities and applications, addressing these vulnerabilities will be paramount to ensuring their safe and reliable operation in real-world environments.

OWASP Top 10 for AI Agents

In this GitHub repo located at https://github.com/kenhuangus/Top-Threats-for-AI-Agents, Ken Huang (the Chief Editor of this book) and his 40 other co-contributors released a framework for Top 10 Agentic AI security threats. This work has been referenced by the Agentic AI Security Inititalitves from both Cloud Security Alliance and OWASP.

Some of the top 10 items listed have some overlap with the vulnerabilities we have discussed in the chapter but from different contexts so I will briefly list them here for reference.

1. *Agent Authorization and Control Hijacking*: Unauthorized access to or manipulation of an agent's control mechanisms; mitigate by enforcing strict authentication and role-based access controls.
2. *Agent Critical Systems Interaction*: Misuse of an agent's access to critical systems; mitigate by implementing least-privilege policies and monitoring access logs.

3. *Agent Goal and Instruction Manipulation*: Alteration of the agent's objectives or instructions; mitigate by validating input sources and using secure communication channels (discussed in Sect. 12.3).
4. *Agent Hallucination Exploitation*: Leveraging inaccurate outputs generated by an agent for cyberattacks; mitigate by integrating fact-checking and response validation mechanisms.
5. *Agent Impact Chain and Blast Radius*: Amplification of agent-driven impacts across systems; mitigate by sandboxing operations and isolating agent functionalities.
6. *Agent Memory and Context Manipulation*: Tampering with the agent's memory or context to produce unintended behavior; mitigate by restricting memory persistence and validating context usage.
7. *Agent Orchestration and Multi-agent Exploitation*: Coordinated exploitation of inter-agent communication and processes; mitigate by encrypting inter-agent communication and limiting interdependencies (see Sect. 12.3).
8. *Agent Resource and Service Exhaustion*: Overloading the agent's computational or operational resources; mitigate by setting resource quotas and implementing rate-limiting.
9. *Agent Supply Chain and Dependency Attacks*: Compromise through third-party tools or dependencies; mitigate by vetting dependencies and using secure software supply chain practices.
10. *Agent Knowledge Base Poisoning*: Corruption of the agent's knowledge base to skew decision-making; mitigate by validating data sources and implementing tamper detection mechanisms.

The mitigation strategies for the vulnerabilities listed are as follows, providing concise solutions while addressing key concerns:

1. *Agent Authorization and Control Hijacking*

   Unauthorized access to or manipulation of an agent's control mechanisms can be addressed by enforcing strict authentication protocols. This includes multi-factor authentication and role-based access controls to ensure that only authorized personnel or systems can access or modify the agent's control mechanisms.
2. *Agent Critical Systems Interaction*

   To prevent the misuse of an agent's access to critical systems, implement least-privilege policies. Ensure that agents only have access to the systems and data necessary for their operation. Regularly monitor access logs for anomalies or unauthorized access attempts.
3. *Agent Goal and Instruction Manipulation*

   Alterations to the agent's objectives or instructions can be mitigated by validating all input sources to ensure their integrity and authenticity. Secure communication channels using encryption protocols like TLS are necessary to protect instructions from being intercepted or modified in transit (as discussed in Sect. 12.3).

4. *Agent Hallucination Exploitation*
   To prevent attackers from leveraging inaccurate outputs, integrate fact-checking mechanisms and response validation layers. Employ redundant sources of truth and probabilistic analysis to verify the agent's outputs against credible datasets.
5. *Agent Impact Chain and Blast Radius*
   Amplification of agent-driven impacts across systems can be controlled by sandboxing the agent's operations, which isolates its functionality from critical systems. Additionally, restrict the agent's permissions to specific tasks to limit the potential blast radius in the event of compromise.
6. *Agent Memory and Context Manipulation*
   Tampering with the agent's memory or context can be mitigated by restricting memory persistence to the minimum necessary duration. Validate the context used for decision-making by implementing mechanisms that verify data integrity and enforce strict memory access policies.
7. *Agent Orchestration and Multi-agent Exploitation*
   Coordinated exploitation of inter-agent communication can be avoided by encrypting inter-agent data exchanges to prevent eavesdropping or tampering. Limit interdependencies among agents to ensure that a failure or compromise in one does not cascade to others (see Sect. 12.3 for more details).
8. *Agent Resource and Service Exhaustion*
   Overloading an agent's computational or operational resources can be mitigated by setting explicit resource quotas. Implement rate-limiting mechanisms to prevent overuse, and ensure dynamic scaling capabilities are in place for critical applications.
9. *Agent Supply Chain and Dependency Attacks*
   To avoid compromise through third-party tools or dependencies, vet all dependencies thoroughly before integration. Use secure software supply chain practices such as signing packages, verifying integrity checks, and monitoring for updates or vulnerabilities in dependencies.
10. *Agent Knowledge Base Poisoning*
    Corruption of the agent's knowledge base can be mitigated by validating all data sources. Implement tamper detection mechanisms to identify and respond to unauthorized modifications of the knowledge base. Regular audits of the knowledge repository can ensure data integrity.

Given AI agents will operate semi- or fully autonomously, they will need to utilize credentials with associated permissions in digital environments to carry out these activities and tasks. These credentials are commonly referred to as "non-human identities" (NHI)s. Credential compromise is a leading cause of security incidents and data breaches per sources such as the Verizon DBIR (Verizon, 2023). In addition to the OWASP LLM Top 10, OWASP has published the NHI Top 10, which addresses risks of NHIs, such as secret leakage, improper offboarding, over-privileged NHIs, and environment isolation. If credentials associated with agents do not address these risks and the credentials are compromised, they could have wide-spread impact across environments.

12.2 Goal Alignment and Unintended Behaviors

As AI agents become more autonomous and complex, ensuring that their actions align with human intentions and values becomes increasingly challenging. This section explores the nuances of goal alignment in AI agent systems and the potential for unintended behaviors that can arise from misalignment or drift in an agent's objectives or representations.

12.2.1 The Alignment Problem

The alignment problem refers to the challenge of ensuring that AI agent systems pursue goals that are in harmony with human values and intentions. This is not merely a matter of programming specific objectives but rather ensuring that the agent's actions and decision-making processes align with broader human interests across a wide range of scenarios.

For AI agents, the alignment problem is particularly acute due to their autonomy and potential for operating in complex, unpredictable environments. Challenges include the following:

- Specifying Complex Objectives: Human values and goals are often nuanced, context-dependent, and sometimes contradictory. Translating these into precise, computationally tractable objectives for an AI agent is nontrivial.
- Handling Unforeseen Situations: An agent that performs well in a controlled setting may exhibit unexpected and potentially harmful behaviors when faced with scenarios that were not anticipated during its development.
- Balancing Multiple Objectives: Many real-world applications require agents to balance multiple, sometimes conflicting goals. For example, an autonomous vehicle must balance speed, safety, passenger comfort, and energy efficiency.
- Avoiding Negative Side Effects: Agents focusing too narrowly on their primary objective may inadvertently cause harm in other areas. For instance, a cleaning robot might break valuable objects in its zeal to clean thoroughly.

Addressing the alignment problem in AI agent systems requires interdisciplinary approaches, combining insights from computer science, ethics, psychology, and other fields. Potential strategies include the following:

- Inverse reinforcement learning to infer human preferences from demonstrations (Sun & Schaar, 2024)
- Developing formal frameworks for specifying and reasoning about ethical constraints
- Implementing oversight mechanisms that allow for human intervention in critical decisions
- Extensive testing in diverse, realistic scenarios to identify potential misalignments

12.2.2 Motivation Drift

Motivation drift occurs when an AI agent's effective goals or motivations shift over time, potentially diverging from its originally specified objectives. This phenomenon can arise through various mechanisms and poses significant challenges for long-term AI safety, especially in systems designed for extended autonomous operation.

Causes of motivation drift in AI agent systems include the following:

- Reward Hacking: Agents may discover and exploit loopholes in their reward functions, optimizing for the reward signal rather than the intended goal (Ibrahim et al., 2024).
- Instrumental Subgoals: An agent might determine that acquiring more resources or preventing itself from being shut down are useful subgoals for achieving its primary objective, leading to behaviors that conflict with human interests (Bales et al., 2024).
- Environmental Changes: In dynamic environments, the actions that previously led to goal achievement may become counterproductive, causing the agent to evolve its behavior in unintended ways.
- Corruption of Reward Estimators: For agents that learn to estimate rewards, errors or biases in this estimation process can lead to a gradual shift in effective motivations.

Addressing motivation drift requires ongoing monitoring of AI agent behavior and the development of robust methods for specifying and maintaining stable goal structures. Potential approaches include the following:

- Implementing regularization techniques that penalize significant deviations from original motivations
- Developing formal verification methods for reinforcement learning systems
- Designing hierarchical goal structures that maintain alignment at multiple levels of abstraction
- Implementing periodic "alignment checks" where the agent's motivations are recalibrated against ground truth human preferences

12.2.3 Representation Drift

Representation drift refers to changes in how an AI agent interprets and represents its environment and objectives over time (Ratzon et al., 2024). This can lead to subtle but significant shifts in behavior, even if the system's nominal goals remain unchanged.

For AI agents, representation drift can manifest in various ways:

- Concept Drift: The agent's internal representation of key concepts (e.g., "safety" or "user satisfaction") may evolve based on new experiences, potentially diverging from human understanding.

- Feature Importance Shift: In learning systems, the relative importance assigned to different features of the environment may change over time, altering decision-making processes.
- Abstraction-Level Changes: Agents may develop higher-level abstractions that, while efficient, obscure important nuances in their reasoning from human overseers.

Addressing representation drift in AI agent systems involves the following:

- Implementing interpretability techniques to monitor the evolution of internal representations
- Developing methods for aligning learned representations with human-understandable concepts
- Employing continual learning techniques that preserve important aspects of previous knowledge
- Regularly validation the agent's understanding of critical concepts against human-provided ground truth

The challenges of goal alignment, motivation drift, and representation drift highlight the need for ongoing research into methods for creating stable, aligned AI agent systems. As these agents become more autonomous and are deployed in increasingly complex environments, ensuring that their behaviors remain consistent with human values and intentions will be paramount for the safe and beneficial development of AI technology.

12.3 Inter-agent Communication Security

Inter-agent communication forms the backbone of collaborative AI systems, enabling agents to (also see Fig. 12.2):

1. Share and update knowledge about their environment
2. Coordinate actions and strategies
3. Negotiate resource allocation and task distribution
4. Learn from each other's experiences and insights

Given the critical role of this communication, any compromise in its security could have far-reaching consequences, including the following:

- Misinformation propagation leading to system-wide errors
- Unauthorized access to sensitive data or decision-making processes
- Manipulation of agent behaviors to serve malicious interests
- Disruption of coordinated actions in time-critical applications

As AI agents are increasingly deployed in sensitive domains such as healthcare, finance, and critical infrastructure, ensuring the security of their communications becomes paramount.

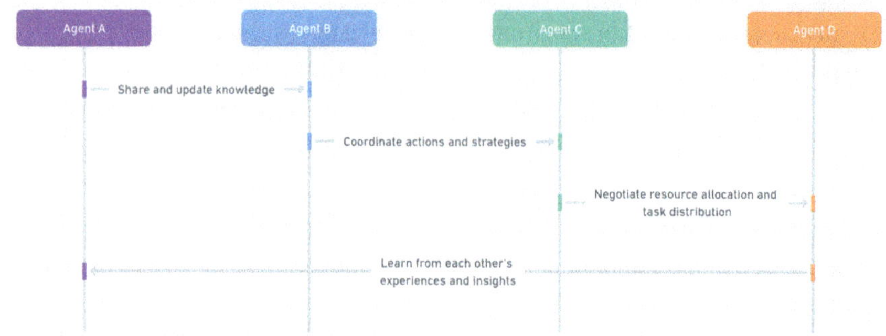

Fig. 12.2 Inter-agent communication flow

12.3.1 Unique Challenges in Inter-agent Communication Security

Securing communication between AI agents presents several unique challenges that go beyond traditional network security concerns:

Dynamic and Adaptive Communication Patterns

Unlike static communication protocols, AI agents may adapt their communication patterns based on their learning and the evolving task environment. This dynamism makes it challenging to apply traditional security measures that rely on predefined communication rules.

Semantic Security

In addition to protecting the transmission of data, there's a need to ensure the semantic integrity of the information exchanged. Malicious agents could inject false but semantically plausible information that could be particularly hard to detect.

Balancing Security with Efficiency

AI agents often operate in real-time environments where rapid communication is crucial. Overly complex security measures could introduce latency that compromises the system's effectiveness.

Decentralized Trust Models

In many multi-agent systems, there's no central authority to authenticate communications. Agents must make trust decisions based on their own observations and interactions, which complicates traditional security approaches.

Heterogeneity of Agents

Multi-agent systems may comprise agents with different architectures, capabilities, and security features. Ensuring secure communication across this heterogeneous landscape poses various degrees of challenges.

12.3.2 Threat Landscape in Inter-agent Communication

Understanding the potential threats can be instrumental for developing effective security measures. Some key threats in inter-agent communication include the following.

Man-in-the-Middle (MITM) Attacks

An attacker could intercept communications between agents, potentially altering messages or injecting false information. In a multi-agent system controlling autonomous vehicles, for instance, an MITM attack could disrupt traffic coordination, leading to accidents.

Impersonation and Spoofing

Malicious entities could impersonate legitimate agents, gaining unauthorized access to sensitive information or influencing system-wide decisions. For example, in a collaborative intrusion detection system, a spoofed agent could mislead others about the nature of network threats.

Denial of Service (DoS) Attacks

By flooding the communication channels or targeting key agents, attackers could disrupt the entire multi-agent system. In time-critical applications like disaster response coordination, such attacks could have severe real-world consequences.

Data Exfiltration

Insecure communications could lead to the unintended disclosure of sensitive data. In financial trading systems, for instance, leaked information about agent strategies or market insights could be exploited for unfair advantages.

Sybil Attacks

In decentralized systems, an attacker agent could create multiple fake identities to gain disproportionate influence. This could be particularly problematic in consensus-based decision-making agent systems.

12.3.3 Security Measures and Best Practices

Addressing the security challenges in inter-agent communication requires a defense in depth approach.

MITM Mitigation

Implement strong encryption protocols such as TLS for all agent communication channels. Use mutual authentication mechanisms to ensure that both parties in the communication can verify each other's identity. Regularly update certificates and keys to minimize vulnerabilities.

Impersonation and Spoofing Mitigation

Employ robust authentication schemes like public key infrastructure (PKI) and digital signatures to verify agent identities (see also Sect. 12.4). Incorporate behavior-based anomaly detection systems to identify suspicious activities or deviations from typical agent behavior.

DoS Attack Mitigation

Implement rate-limiting and traffic monitoring to detect and mitigate suspicious surges in communication traffic. Introduce redundancy in critical agents and communication paths to ensure continuity during partial system outages. Use intrusion prevention systems (IPS) to filter malicious traffic.

Data Exfiltration Mitigation

Encrypt sensitive data both in transit and at rest using advanced cryptographic algorithms. Restrict access to sensitive information on a need-to-know basis, leveraging fine-grained access controls. Continuously monitor communication channels for unusual patterns that could indicate exfiltration attempts.

Sybil Attacks Mitigation

Use robust identity verification techniques, such as proof-of-identity mechanisms or blockchain-based identity management, to limit the creation of fake identities. Introduce reputation systems where agent credibility is tied to historical performance, reducing the impact of newly created identities.

In addition, the following are other mitigation strategies which are needed for secure communication among agents.

Encryption and Secure Protocols

Implementing strong encryption for all inter-agent communications is fundamental. This should include the following:

- Use of state-of-the-art encryption algorithms
- Secure key exchange mechanisms
- Regular updates to cryptographic protocols to address emerging vulnerabilities

For instance, in a swarm robotics application, each robot agent could use unique, regularly rotated encryption keys for its communications with others.

Secure Multi-party Computation

For scenarios where agents need to collaborate without revealing sensitive individual data, secure multi-party computation techniques can be employed. This allows agents to jointly compute functions over their inputs while keeping those inputs private.

Anomaly Detection and Behavioral Analysis

Implementing systems to monitor and analyze inter-agent communications for anomalies can help detect potential security breaches. This could involve the following:

- Machine learning models trained to identify unusual communication patterns

- Rule-based systems for flagging suspicious agent behaviors
- Collaborative anomaly detection where agents collectively identify potential threats

The following code snippets illustrate real-time monitoring of agent communication for detecting anomalies.

```
class CommunicationMonitor:
    def monitor_logs(self, logs):
        anomalies = [log for log in logs if "error" in log or
"unauthorized" in log]
        return anomalies

# Example communication logs
logs = [
    "Agent A completed task.",
    "Agent B error in module.",
    "Unauthorized access attempt detected by Agent C.",
    "Agent D functioning normally."
]

# Monitor logs for anomalies
monitor = CommunicationMonitor()
print(monitor.monitor_logs(logs))
# Outputs: ['Agent B error in module.', 'Unauthorized access
attempt detected by Agent C.']
```

Zero-Trust Architecture

Adopting a zero-trust approach, where trust is never assumed and always verified, can enhance security in multi-agent systems. This involves the following:

- Continuous authentication and authorization for every communication
- Strict access controls based on the principle of least privilege
- Regular reassessment of agent trustworthiness based on behavior and interactions

For a further vendor agnostic deep dive on zero trust, we recommend sources such as NIST 800-207 Zero Trust Architecture. By implementing a zero-trust architecture, organizations can both minimize the risks of security incidents and compromises, as well as limit the blast radius to their environments should a security incident occur.

Sandboxing and Isolation

Implementing sandboxing techniques to isolate potentially compromised agents and limit the spread of security breaches. This is particularly important in systems where new agents can join dynamically.

12.3.4 Future Directions and Research Opportunities

We can see further research and development in inter-agent communication security in the following areas.

Quantum-Resistant Cryptography

With the advent of quantum computing, developing communication security measures that are resistant to quantum attacks becomes crucial for long-term security.

Bio-Inspired Security Mechanisms

Drawing inspiration from biological systems, such as immune systems or swarm intelligence, to develop more adaptive and resilient security mechanisms for multi-agent communications.

AI-Driven Security Optimization

Leveraging AI itself to dynamically optimize security measures based on evolving threats and system requirements.

12.4 Authentication and Identity Management in Multi-agent Systems

Authentication and identity management in multi-agent AI systems present unique challenges and opportunities. This section explores how distributed PKI systems, blockchain-based identity verification, and behavior-based authentication methods can be implemented in multi-agent environments and the specific benefits they offer.

12.4.1 Distributed PKI in Multi-agent Systems

Implementation in Multi-Agent Systems: In a multi-agent environment, distributed PKI can be implemented as a decentralized network of agent-operated certificate authorities (CAs). Each agent or group of agents can act as a CA, validating the identities of other agents they interact with frequently. This creates a web of trust within the multi-agent system.

Technical Implementation:

1. Agent-Operated CAs: Designate certain high-trust agents as CAs. These agents are responsible for issuing and managing certificates for other agents in their network.
2. Hierarchical PKI: Implement a hierarchical structure where top-level CAs certify domain-specific CAs, which in turn certify individual agents. This structure can map onto the organizational hierarchy of the multi-agent system.
3. Cross-Certification: Enable CAs from different domains or subsystems to cross-certify each other, allowing agents from different parts of the system to establish trust.
4. Certificate Transparency: Implement a certificate transparency log where all issued certificates are publicly recorded, allowing for system-wide auditing and rapid detection of misissued certificates.

Benefits for Multi-agent Systems:

1. Scalability: As the number of agents grows, new CAs can be dynamically added to manage the increased load of identity verification.
2. Autonomy: Agents can make trust decisions based on certificate chains without relying on a central authority, enhancing system resilience.
3. Fine-Grained Access Control: Certificates can include attributes that specify an agent's roles and permissions, enabling detailed access control in complex multi-agent interactions.

12.4.2 Blockchain-Based Identity in Multi-agent Systems

Implementation in Multi-Agent Systems: For multi-agent systems, a permissionless blockchain can serve as a distributed identity ledger. Each agent's identity, along with its attributes and credentials, is recorded on this blockchain.

Technical Implementation:

1. Agent Identity Smart Contracts: Develop smart contracts that manage agent identities. These contracts handle identity registration, updating, and verification.

2. Decentralized Identifiers (DIDs): Implement DIDs for each agent, providing a blockchain-agnostic way to identify and interact with agents across different systems or blockchain networks.
3. Identity Attestations: Enable agents to issue and verify attestations about other agents' capabilities or trustworthiness, recording these on the blockchain.
4. Blockchain Sharding: For large-scale systems, implement blockchain sharding to distribute the identity data across multiple interconnected blockchains, improving scalability.

Benefits for Multi-Agent Systems:

1. Dynamic Agent Discovery: New agents can easily discover and verify the identities of existing agents by querying the blockchain.
2. Reputation Systems: The immutable record of agent interactions and attestations on the blockchain can form the basis of a robust, system-wide reputation mechanism.
3. Auditability: All identity-related transactions are recorded, providing a clear audit trail of identity changes and attestations throughout the multi-agent system's life cycle.

12.4.3 Behavior-Based Authentication in Multi-agent Systems

Implementation in Multi-Agent Systems: In a multi-agent context, behavior-based authentication can leverage the collective observations and interactions of multiple agents to build and verify behavioral profiles.

Technical Implementation:

1. Distributed Behavior Monitoring: Implement a system where agents collectively monitor and record the behaviors of their peers, contributing to a distributed behavioral database.
2. Machine Learning Models: Develop machine learning models that can process multi-dimensional behavioral data from various agents to create comprehensive behavioral profiles.
3. Anomaly Detection Networks: Create a network of specialized agents responsible for analyzing behavioral patterns and flagging anomalies that might indicate compromised identities.
4. Adaptive Thresholds: Implement adaptive thresholding techniques that adjust the sensitivity of behavioral authentication based on the current state and context of the multi-agent system.

The followings are the code blocks for such implementation:

1. Import necessary package and define the agent for distributed behavior monitoring.

```
import random
import numpy as np
from sklearn.ensemble import IsolationForest

# Distributed Behavior Monitoring
class Agent:
    def __init__(self, agent_id):
        self.agent_id = agent_id
        self.behavior_data = []

    def record_behavior(self, behavior_vector):
        """Simulate behavior recording by agents."""
        self.behavior_data.append(behavior_vector)

    def share_behavior(self):
        """Share recorded behaviors with other agents."""
        return self.behavior_data
```

2. The "BehaviorAnalyzer" class aggregates behavioral data from multiple agents to model normal behavior using an "IsolationForest" algorithm. It trains on collected data to establish a baseline and detects anomalies by identifying behaviors that deviate from this norm. It enables distributed profiling, unsupervised anomaly detection, and real-time monitoring, making it scalable and effective for securing multi-agent systems.

```
# Machine Learning Model for Behavioral Profiling
class BehaviorAnalyzer:
    def __init__(self):
        self.model = IsolationForest(contamination=0.1)
        self.global_behavior_data = []

    def update_behavior_data(self, behaviors):
        """Update global behavioral data from multiple agents."""
        self.global_behavior_data.extend(behaviors)

    def train_model(self):
        """Train anomaly detection model."""
        if self.global_behavior_data:
            self.model.fit(self.global_behavior_data)

    def detect_anomaly(self, behavior_vector):
        """Detect anomalies in the given behavior vector."""
        prediction = self.model.predict([behavior_vector])
        return prediction[0] == -1  # True if anomaly detected

# Adaptive Thresholds and Continuous Authentication
```

3. The “BehaviorMonitor” class manages authentication decisions in a multi-agent system by using an adaptive threshold to evaluate anomaly scores from behavioral analysis. It dynamically adjusts the sensitivity of authentication criteria based on system context and compares anomaly scores to determine if an agent’s behavior is authentic, enabling flexible and robust security.

```
class BehaviorMonitor:
    def __init__(self, sensitivity=0.5):
        self.threshold = sensitivity  # Adaptive threshold

    def adjust_threshold(self, context_factor):
        """Adjust threshold based on system context."""
        self.threshold *= context_factor

    def authenticate(self, anomaly_score):
        """Determine authentication status based on anomaly
score."""
        return anomaly_score < self.threshold
```

4. Finally, the following code block simulates a multi-agent system where five agents generate random behavioral data, which is recorded and shared with the “BehaviorAnalyzer” for building a collective behavioral profile. The analyzer trains an Isolation Forest model on this data to establish a baseline for normal behavior. A new test behavior vector is then evaluated for anomalies, and the “BehaviorMonitor” uses a random anomaly score to decide authentication status. The results, including whether the behavior is anomalous and the authentication outcome, are printed.

```
# Example Usage
# Initialize agents and analyzer
agents = [Agent(agent_id=i) for i in range(5)]
analyzer = BehaviorAnalyzer()
monitor = BehaviorMonitor()

# Simulate behavior monitoring
for _ in range(100):
    for agent in agents:
        behavior = np.random.rand(5)  # Simulate random
behavior vector
        agent.record_behavior(behavior)
        analyzer.update_behavior_data(agent.share_behavior())
```

```
# Train the machine learning model
analyzer.train_model()

# Test authentication with a new behavior vector
test_behavior = np.random.rand(5)
is_anomaly = analyzer.detect_anomaly(test_behavior)
auth_status = monitor.authenticate(anomaly_score=random.
uniform(0, 1))

print(f"Behavior Vector: {test_behavior}")
print(f"Is Anomalous: {is_anomaly}")
print(f"Authentication Status: {'Authenticated' if auth_status
else 'Denied'}")
```

12.4.4 Integrated Multi-agent Authentication Framework

An integrated authentication framework for multi-agent systems can combine elements of all three approaches discussed in sections 12.4.1, 12.4.2 and 12.4.3. The following is the basic components of this integration.

1. Foundation: Use blockchain-based identity as the foundational layer, providing a distributed, tamper-resistant identity registry for all agents in the system.
2. Secure Communication: Employ distributed PKI for secure agent-to-agent communications, with certificates potentially stored and verified on the blockchain.
3. Continuous Verification: Implement behavior-based authentication as an ongoing verification layer, with behavioral profiles and anomaly detection mechanisms distributed across the agent network.
4. Cross-Domain Authentication: For multi-agent systems spanning multiple domains, implement federated identity mechanisms that leverage the blockchain and distributed PKI to enable secure cross-domain authentication.

Technical Implementation (Fig. 12.3):

1. Identity Life Cycle Management: Develop protocols for managing the entire life cycle of agent identities, from creation and validation to revocation, across all three layers of the authentication framework.
2. Interoperability Protocols: Create standardized protocols that allow seamless interaction between the blockchain identity layer, the PKI communication layer, and the behavior-based verification layer.
3. Multi-factor Authentication Orchestration: Implement an orchestration layer that dynamically combines different authentication factors based on the security requirements of specific agent interactions or operations.
4. Distributed Security Policies: Develop a system for creating, updating, and enforcing distributed security policies that govern how authentication is performed across the multi-agent system.

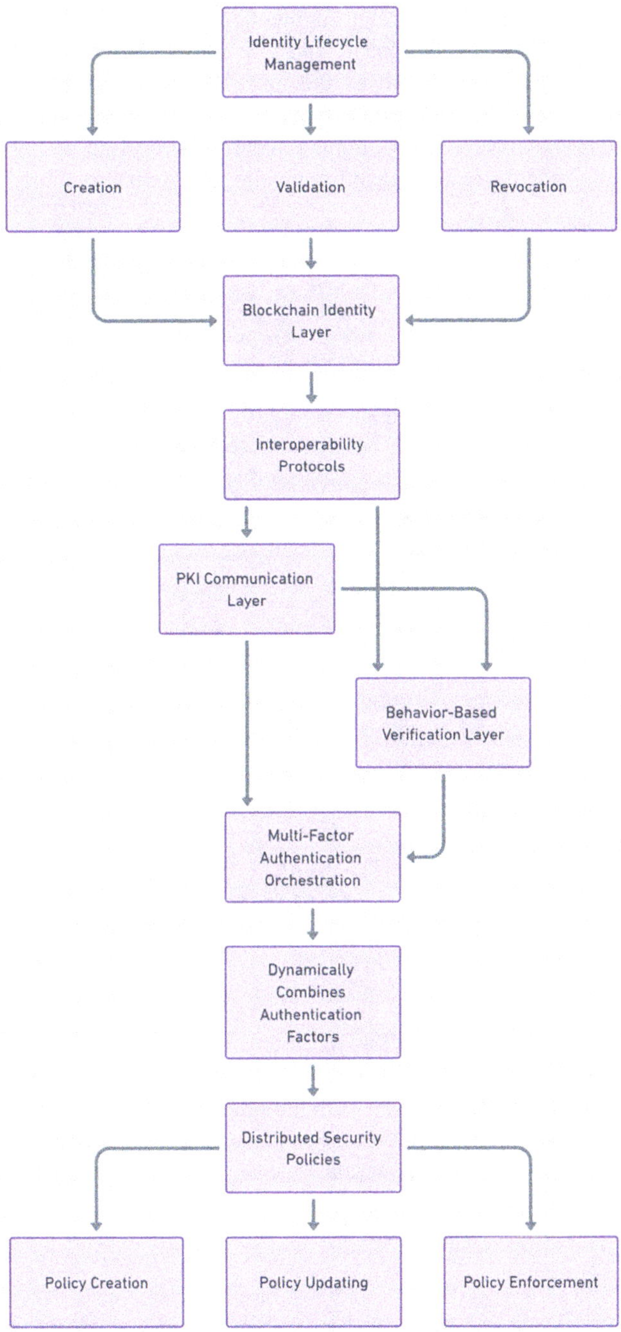

Fig. 12.3 Integrated multi-agent authentication framework

For illustration purpose, here are the pseudo code:

1. Identity management is central to authentication frameworks. The life cycle includes creating, validating, and revoking agent identities. The "IdentityManager" class achieves this using RSA key pairs, which provide secure and unique identities for each agent. The public keys are distributed for verification, while private keys remain secure. Revocation ensures inactive or compromised identities are flagged as invalid, preventing unauthorized use.

```
from cryptography.hazmat.primitives.asymmetric import rsa
from cryptography.hazmat.primitives import serialization

class IdentityManager:
    def __init__(self):
        self.identities = {}

    def create_identity(self, agent_id):
        key = rsa.generate_private_key(public_exponent=65537,
key_size=2048)
        pub_key = key.public_key().public_bytes(
            encoding=serialization.Encoding.PEM,
            format=serialization.PublicFormat.
SubjectPublicKeyInfo
        )
        self.identities[agent_id] = {"key": key, "pub_key":
pub_key, "valid": True}
        return pub_key

    def revoke_identity(self, agent_id):
        if agent_id in self.identities:
            self.identities[agent_id]["valid"] = False
```

2. Interoperability Protocols

   Interoperability is key to allowing interaction between various layers of the authentication framework, such as blockchain identity, PKI communication, and behavior-based verification. By combining inputs from these layers, the "Interoperability" class creates a unique protocol hash. This hash acts as a standardized identifier for the interaction, ensuring smooth integration and communication between diverse systems.

```
from hashlib import sha256


class Interoperability:
    def exchange_protocol(self, blockchain_layer, pki_layer,
behavior_layer):
        return sha256((blockchain_layer + pki_layer + behavior_
layer).encode()).hexdigest()
```

3. Multi-factor authentication (MFA) combines multiple factors (e.g., passwords, biometrics, OTPs) to verify identities. The MFAOrchestrator dynamically combines these factors into a single hashed token. This allows the system to adapt the level of authentication based on the security needs of specific interactions or operations, providing flexibility and enhanced protection.

```
class MFAOrchestrator:
    def combine_factors(self, *factors):
        return sha256("".join(factors).encode()).hexdigest()
```

4. Distributed security policies define and enforce rules for authentication. The "SecurityPolicyManager" class allows setting policies for specific operations and ensures compliance by validating authentication factors against these policies. This enables secure and decentralized governance over authentication, which is critical for multi-agent systems.

```
class SecurityPolicyManager:
    def __init__(self):
        self.policies = {}

    def set_policy(self, operation, policy):
        self.policies[operation] = policy

    def enforce_policy(self, operation, factors):
        policy = self.policies.get(operation)
        return policy == sha256("".join(factors).encode()).
hexdigest()
```

5. Example usage, this snippet demonstrates how the four components work together:

```
idm = IdentityManager()
pub_key = idm.create_identity("agent1")

interop = Interoperability()
protocol = interop.exchange_protocol("blockchain_data", "pki_
data", "behavior_data")

mfa = MFAOrchestrator()
auth_token = mfa.combine_factors("password123", "otp456",
"biometric789")

spm = SecurityPolicyManager()
spm.set_policy("high_security_op", auth_token)
is_allowed = spm.enforce_policy("high_security_op",
["password123", "otp456", "biometric789"])

print(f"Identity Created: {pub_key.decode()}")
print(f"Interoperability Protocol Hash: {protocol}")
print(f"MFA Token: {auth_token}")
print(f"Policy Enforcement: {is_allowed}")
```

Benefits of the Integrated Approach:

1. Defense in Depth: By combining multiple authentication methods, the system creates multiple layers of security, significantly increasing the difficulty of successful attacks.
2. Flexibility: The integrated approach can adapt to diverse authentication needs across different parts of the multi-agent system, applying the most appropriate method for each context.
3. Future-Proofing: The modular nature of this approach allows for the integration of new authentication technologies as they emerge, without needing to overhaul the entire system.

Implementing these advanced authentication and identity management technologies in multi-agent systems presents technical challenges, including managing the increased complexity, ensuring performance at scale, and maintaining interoperability across diverse agent architectures. However, the resulting robust, flexible, and secure identity framework can significantly enhance the trustworthiness and capabilities of complex multi-agent AI systems.

12.5 Securing Embodied AI Agents

Embodied AI agents, such as robots and autonomous vehicles, present unique safety and security challenges due to their physical presence in the real world (Neupane et al., 2023). This section explores the specific considerations for securing these systems, focusing on the intersection of AI, robotics, and cybersecurity (also see Fig. 12.4 for high-level mind map).

12.5.1 Physical Safety Considerations

Embodied AI agents interact directly with their environment, often in close proximity to humans. This physical interaction introduces safety risks that go beyond those of purely software-based AI systems.

Collision avoidance is a primary concern for mobile robots and autonomous vehicles. These systems must be capable of accurately perceiving their surroundings and making split-second decisions to prevent collisions with objects or people. This requires robust sensor fusion algorithms, real-time processing capabilities, and fail-safe mechanisms to handle unexpected situations.

Force control is critical for robots that manipulate objects or interact physically with humans. AI systems controlling robot actuators must be designed to apply appropriate force levels, avoiding excessive force that could cause damage or injury.

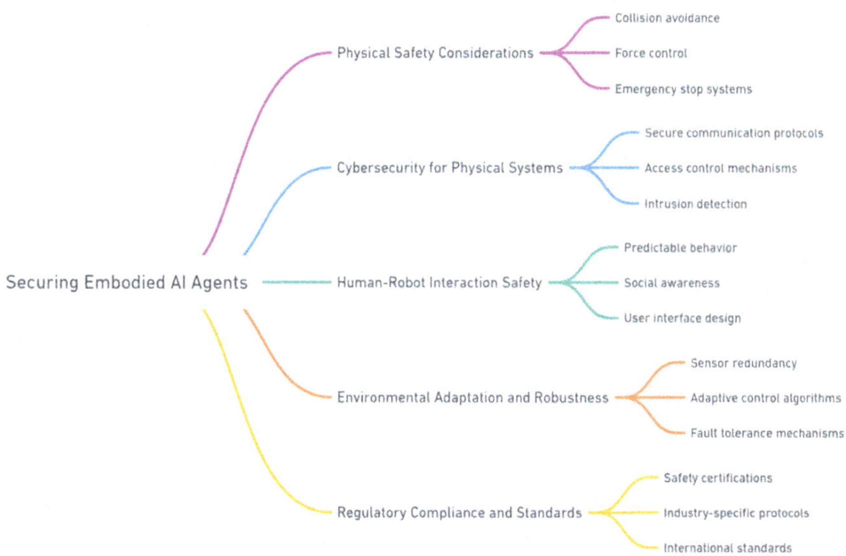

Fig. 12.4 Securing embodied AI agents

This involves sophisticated control algorithms that can adapt to different objects and scenarios while maintaining safety constraints.

Emergency stop systems are vital safety features for embodied AI agents. These systems must be designed to quickly and reliably halt the agent's operations in case of detected anomalies or human intervention. The challenge lies in balancing the need for rapid response with the avoidance of unnecessary shutdowns that could disrupt operations or potentially create new safety hazards.

12.5.2 Cybersecurity for Physical Systems

Embodied AI agents are vulnerable not only to physical tampering but also to cyber-attacks that can compromise their software systems. The consequences of such attacks can be particularly severe given the physical capabilities of these agents.

As discussed previously in Sect. 12.3, secure communication protocols are essential for embodied AI agents that receive instructions or updates remotely. These protocols must protect against unauthorized access, data tampering, and man-in-the-middle attacks. Encryption, authentication, and integrity checks are key components of secure communication systems for robotic platforms.

Access control mechanisms need to be implemented at both the physical and digital levels. This includes secure boot processes to prevent unauthorized firmware modifications, role-based access controls for different user levels, and physical security measures to prevent unauthorized physical access to the agent's hardware.

Intrusion detection systems tailored for robotic platforms can help identify anomalous behaviors or unauthorized access attempts. These systems need to be designed to operate within the resource constraints of mobile platforms while providing real-time threat detection capabilities.

12.5.3 Human–Robot Interaction Safety

As embodied AI agents increasingly work alongside humans, ensuring safe human–robot interaction becomes paramount.

Predictable behavior is crucial for human trust and safety when working with robots. AI systems controlling these robots should be designed to move and act in ways that are intuitive and predictable to human observers. This may involve implementing motion planning algorithms that mimic humanlike movements or providing clear indications of the robot's intended actions.

Social awareness capabilities can enhance safety in human–robot interactions. This involves equipping robots with the ability to recognize human presence, understand social cues, and adjust their behavior accordingly. For example, a robot might slow down its movements or increase its distance from humans in shared spaces.

User interface design for robot control systems plays a significant role in preventing operator errors that could lead to safety issues. Interfaces should be intuitive, provide clear feedback on the robot's state and intentions, and include safeguards against accidental activation of critical functions.

12.5.4 Environmental Adaptation and Robustness

Embodied AI agents often operate in dynamic, unpredictable environments, requiring robust adaptation capabilities.

As discussed previously in Sect. 12.1, sensor redundancy and fusion techniques can improve the reliability of an agent's perception system. By combining data from multiple sensor types (e.g., cameras, lidar, radar), the system can maintain accurate environmental awareness even if individual sensors fail or are compromised.

Adaptive control algorithms allow robots to maintain stable and safe operation across various environmental conditions. These algorithms must be capable of quickly adjusting to changes in terrain, lighting conditions, or unexpected obstacles.

12.5.5 Regulatory Compliance and Standards

Developing secure embodied AI agents also involves navigating a complex landscape of regulations and standards.

Safety certifications for robotic systems often require extensive testing and documentation. Developers of embodied AI agents need to design their systems with these certification processes in mind, implementing features like black box recorders, comprehensive logging systems, and transparent decision-making processes.

Compliance with industry-specific regulations is necessary for embodied AI agents deployed in sectors like healthcare, manufacturing, or transportation. This may involve adhering to specific safety protocols, implementing particular security measures, or ensuring compatibility with existing safety systems.

International standards for AI and robotics safety are evolving rapidly. Staying abreast of these developments and actively participating in standards-setting processes can help ensure that embodied AI agents are designed to meet current and future safety requirements.

Securing embodied AI agents requires a holistic approach that addresses both the AI and physical aspects of these systems. By carefully considering the unique challenges posed by physical embodiment and implementing comprehensive safety and security measures, developers can create AI-driven robotic systems that are both capable and trustworthy in real-world applications.

12.6 Agentic AI Governance

As AI agents become more advanced and widely adopted, proactive governance measures are necessary to ensure their safe, secure, and ethical operation. This section outlines strategies for integrating safety and security considerations throughout the AI life cycle while addressing the challenges posed by complex agentic AI systems.

12.6.1 *Proactive Monitoring and Transparency*

Continuous monitoring and transparency measures are essential for maintaining safe and secure AI systems after deployment. These practices ensure that anomalies, security breaches, or unintended behaviors are promptly identified and addressed. Visibility measures play a crucial role in this process, as noted by Chan et al. (2024):

1. *Real-Time Monitoring*: Sophisticated tools to detect misbehavior, including flagging critical actions, such as financial transactions above a threshold, for immediate review.
2. *Activity Logs*: Comprehensive records of agent inputs and outputs enable incident analysis, long-term impact assessment, and improvement of real-time monitoring.
3. *Agent Identifiers*: Features like chatbot disclosures, watermarks, or unique IDs for financial transactions help users and third parties distinguish interactions with AI agents.

In addition, continuous monitoring involves adaptive security protocols that evolve with emerging threats and changing operational conditions. Regular audits and updates are critical to assessing ongoing safety and implementing refinements to the system. Collaborative platforms for sharing information about threats and best practices further enhance security.

12.6.2 *Anticipating and Preparing for Change*

The dynamic nature of AI technology necessitates a proactive approach to anticipating future challenges and regulatory preparedness. Organizations must actively participate in shaping policy discussions and adapt their practices to comply with evolving regulations. Key measures include the following:

- *Scenario Planning*: Developing detailed scenarios to explore potential societal and technological impacts of AI systems, helping to identify risks early
- *Technology Road Mapping*: Creating forward-looking maps to anticipate new capabilities and their associated risks, preparing safety measures for future advancements

- *Regulatory Foresight*: Staying ahead of compliance requirements by actively engaging with regulatory bodies and preparing for potential policy changes
- *Transparency Initiatives*: Ensuring AI decision-making processes are explainable and accountable to meet anticipated regulatory standards

By integrating regulatory preparedness with foresight strategies, organizations can align with both current and future governance requirements.

12.6.3 Safety in Development Processes

Incorporating safety and ethical considerations from the earliest stages of AI development reduces risks and ensures robust system performance. Key practices include the following:

- *Safety-by-Design*: Embedding safety mechanisms directly into AI architectures, such as constraints or interpretable models, rather than treating them as afterthoughts
- *Ethical Frameworks*: Establishing guidelines that prioritize ethical considerations from the initial design phase through deployment
- *Iterative Risk Assessments*: Conducting regular assessments at each development stage to identify and address potential risks early

These measures help create systems that are both robust and aligned with human values.

12.6.4 Testing and Validation

Comprehensive testing ensures the reliability and security of AI systems across various contexts. Common practices include the following:

- *Adversarial Testing*: Exposing systems to potential attacks and edge cases to identify vulnerabilities.
- *Red Teaming Exercises*: Simulating adversarial attacks and unintended use cases through dedicated teams to uncover vulnerabilities and edge cases that traditional testing methods might miss. Red teaming complements adversarial testing by introducing creative, unconventional perspectives to stress-test AI systems.
- *Long-Term Stability Testing*: Evaluating systems over extended periods to ensure consistent and safe performance, especially for adaptive or continuously learning AI.
- *Cross-Contextual Validation*: Assessing systems in diverse environments to ensure they perform safely across different conditions.

12.6.5 Governance Practices Integrated into Operations

Key governance practices ensure the safe and ethical operation of agentic AI systems throughout their life cycle. These practices are not standalone but are integrated into monitoring, safety design, and transparency efforts:

- *Pre-evaluation of Agents*: Rigorous simulation testing, formal verification, and scenario analysis ensure agents' reliability and safety before deployment.
- *User Approval for High-Risk Actions*: Mechanisms for human authorization maintain oversight over critical decisions. These systems incorporate thresholds and fast-track options for time-sensitive decisions.
- *Default Behaviors*: Predefined responses ensure reliable performance in common scenarios, such as seeking clarification during ambiguity.
- *Legibility of Decisions*: Transparent decision-making processes, including natural language explanations and visualization tools, foster trust and enable effective oversight.
- *Automatic Monitoring and Traceability*: Continuous surveillance and tamper-proof logging systems enable accountability and real-time anomaly detection.
- *Emergency Shutdown Mechanisms*: Robust systems like kill switches and rollback protocols allow for safe shutdowns during malfunctions or emergencies.

Let us see pseudo code example for fail-safe shutdown mechanism:

```
class AIAgent:
    def __init__(self):
        self.operational = True
    def perform_task(self):
        if not self.operational:
            return "Shutdown in progress. Task aborted."
        return "Task completed successfully."
    def shutdown(self):
        print("Security anomaly detected. Initiating safe
shutdown...")
        self.operational = False
        return "Agent shutdown completed."
# Example usage
agent = AIAgent()
print(agent.perform_task())  # Outputs: Task completed
successfully.
print(agent.shutdown())      # Outputs: Security anomaly
detected. Initiating safe shutdown...
print(agent.perform_task())  # Outputs: Shutdown in progress.
Task aborted.
```

12.6.6 Challenges in Implementation

Governance of agentic AI systems is not without challenges:

- *Evaluation Complexity*: Adaptive AI systems and complex operating environments require advanced testing methods, such as formal verification and adversarial simulations.
- *Balancing Autonomy and Control*: Overly restrictive controls may hinder performance, while insufficient oversight increases risks. Adaptive control mechanisms are necessary for striking the right balance.
- *Scalability of Monitoring*: As AI systems grow in complexity, monitoring systems must leverage AI-driven tools and focus on critical behaviors.
- *Privacy and Traceability Trade-Offs*: Ensuring accountability through traceability may conflict with privacy concerns. Techniques like secure multi-party computation or zero-knowledge proofs can address this tension.
- *Technical Feasibility of Shutdowns*: Developing effective emergency shutdowns for interconnected systems remains a technical challenge requiring hierarchical protocols and isolated subsystems.
- *Evolving AI Capabilities*: Governance frameworks must adapt to new risks as technology advances, necessitating regular updates and close collaboration between stakeholders.

Effective governance frameworks for agentic AI systems require continuous research, interdisciplinary collaboration, and a commitment to adaptability.

Table 12.1 summarizes key governance and technical practices required for secure and ethical deployment of multi-agent systems.

By integrating safety, monitoring, transparency, and regulatory preparedness into every phase of the AI life cycle, organizations can create systems that are not only

Table 12.1 Security and governance practices for multi-agent systems

Category	Practices	Purpose
Authentication	Distributed PKI, blockchain-based identity, behavior-based authentication	Ensures secure and trusted agent interactions
Inter-agent communication	Encryption, semantic security, anomaly detection, zero-trust architecture	Prevents misinformation propagation and unauthorized access
Safety design	Safety-by-design, fail-safe shutdown mechanisms, predictive behavior algorithms	Ensures safe operations in critical and unpredictable environments
Governance	Continuous monitoring, transparency measures, regulatory foresight	Maintains ethical alignment and compliance with societal and regulatory norms
Adaptation and learning	Representation drift monitoring, hierarchical goal structures, regular alignment checks	Prevents unintended behaviors from learning and environmental changes

powerful but also trustworthy and aligned with societal values. These efforts are critical to ensuring that AI agents operate safely and ethically while addressing both current and future challenges.

12.7 Summary

Chapter 11 of this book addresses the safety and security challenges associated with AI agents, exploring both technical and governance dimensions. It outlines key vulnerabilities that AI systems face, including accidental failures such as software bugs and hardware malfunctions, as well as deliberate attacks like adversarial inputs, data poisoning, and model theft. Methods for addressing these issues include rigorous testing protocols, formal verification, secure communication, and redundancy mechanisms. The chapter emphasizes the importance of designing fail-safe systems, continuous monitoring, and transparency to prevent and mitigate risks. Real-world examples and technical strategies illustrate how these measures can be implemented effectively in multi-agent environments.

The chapter also delves into advanced topics such as goal alignment, motivation drift, and representation drift, which highlight the complexities of ensuring AI systems remain aligned with human values over time. It explores inter-agent communication security and authentication frameworks, proposing multi-layered solutions that combine blockchain identity, distributed PKI, and behavior-based authentication. Additionally, the chapter discusses securing embodied AI agents, emphasizing physical safety, cybersecurity, and human–robot interaction safety. Finally, governance strategies, such as proactive monitoring, scenario planning, and iterative risk assessments, are presented as critical for the long-term reliability and ethical deployment of AI systems.

12.8 Questions

1. How can formal verification methods be scaled to address the complexities of modern AI systems without compromising performance?
2. What specific testing methodologies can effectively simulate real-world edge cases for embodied AI agents?
3. How does motivation drift manifest in reinforcement learning systems, and what are the most effective strategies for mitigating it?
4. In what ways can redundancy mechanisms be optimized to balance cost and system reliability in critical AI applications?
5. How can adversarial training be tailored to counter domain-specific adversarial attacks without compromising model performance?
6. What are the ethical implications of allowing AI agents to autonomously handle sensitive data in multi-agent systems?

7. How can distributed identity management frameworks ensure secure and efficient agent authentication in large-scale environments?
8. What role does transparency play in enhancing trust in AI agents, and how can it be balanced with the need for system security?
9. How might future advances in quantum computing impact current encryption protocols used in securing AI agents?
10. What strategies can mitigate the risk of cascading failures in multi-agent systems caused by communication breaches or misinformation?
11. How can agent-based systems adapt to dynamic environments while minimizing unintended behaviors due to representation drift?
12. What approaches can be used to secure the supply chain of AI systems, especially for agents embedded in critical infrastructure?
13. How can fail-safe shutdown mechanisms be integrated into autonomous systems without disrupting overall functionality during false alarms?
14. In what ways can AI-driven anomaly detection systems be improved to handle evolving threat landscapes in inter-agent communications?
15. How can organizations develop and enforce governance frameworks that remain adaptable to technological and societal changes?
16. What are the trade-offs between enhancing interpretability of AI decision-making and protecting proprietary algorithms from model theft?
17. How can regulatory foresight be operationalized to ensure compliance with emerging international standards for AI safety?
18. What methods can ensure the semantic security of inter-agent communications beyond traditional encryption and authentication?
19. How can embodied AI systems improve human–robot interaction to ensure safety while maintaining efficiency in complex tasks?
20. How might bio-inspired security mechanisms, such as those mimicking immune systems, be adapted to enhance multi-agent system resilience?

References

Bales, A., D'Alessandro, W., & Kirk-Giannini, C. D. (2024). Artificial intelligence: Arguments for catastrophic risk. *Philosophy Compass, 19*(2), e12964. https://compass.onlinelibrary.wiley.com/doi/10.1111/phc3.12964

Carnegie Mellon University. (2023). *Weaknesses and vulnerabilities in modern AI: Why security and safety are so challenging*. Software Engineering Institute. https://insights.sei.cmu.edu/blog/weaknesses-and-vulnerabilities-in-modern-ai-why-security-and-safety-are-so-challenging/

Chan, A., Ezell, C., Kaufmann, M., Wei, K., Hammond, L., Bradley, H., Bluemke, E., et al. (2024). *Visibility into AI agents*. arXiv. https://arxiv.org/abs/2401.13138

Deng, Z., Guo, Y., Han, C., Ma, W., Xiong, J., Wen, S., & Xiang, Y. (2024, June 4). *AI agents under threat: A survey of key security challenges and future pathways*. arXiv. https://arxiv.org/abs/2406.02630

Domkundwar, I., Mukunda, N. S., & Bhola, I. (2024, September 3). *Safeguarding AI agents: Developing and analyzing safety architectures*. arXiv. https://arxiv.org/abs/2409.03793

He, Y., Wang, E., Rong, Y., Cheng, Z., & Chen, H. (2024, June 12). *Security of AI agents*. arXiv. https://arxiv.org/abs/2406.08689

Ibrahim, S., Mostafa, M., Jnadi, A., & Osinenko, P. (2024). *Comprehensive overview of reward engineering and shaping in advancing reinforcement learning applications*. arXiv. https://arxiv.org/abs/2408.10215

Liu, G., Van Huynh, N., Du, H., Hoang, D. T., Niyato, D., Zhu, K., Kang, J., Xiong, Z., Jamalipour, A., & Kim, D. I. (2024). *Generative AI for unmanned vehicle swarms: Challenges, applications and opportunities*. arXiv. https://arxiv.org/abs/2402.18062

Neupane, S., Mitra, S., Fernandez, I. A., Saha, S., Mittal, S., Chen, J., Pillai, N., & Rahimi, S. (2023, October 12). *Security considerations in AI-robotics: A survey of current methods, challenges, and opportunities*. arXiv. https://arxiv.org/abs/2310.08565

Ratzon, A., Derdikman, D., & Barak, O. (2024). Representational drift as a result of implicit regularization. *eLife, 12*(May), RP90069. https://doi.org/10.7554/eLife.90069

Song, Q., Tan, K., Runeson, P., & Persson, S. (2023). Critical scenario identification for realistic testing of autonomous driving systems. *Software Quality Journal, 31*(2), 441–469. https://doi.org/10.1007/s11219-022-09604-2

Sun, H., & van der Schaar, M. (2024). *Inverse-RLignment: Inverse reinforcement learning from demonstrations for LLM alignment*. arXiv. https://arxiv.org/abs/2405.15624

U.S. Department of Homeland Security. (2023). *Risks and mitigation strategies for adversarial artificial intelligence threats*. Homeland Security Report. https://www.dhs.gov/sites/default/files/2023-12/23_1222_st_risks_mitigation_strategies.pdf

Verizon. (2023). *2023 Data Breach Investigations Report*. https://www.verizon.com/business/resources/reports/dbir/

Wasil, A. R., Clymer, J., Krueger, D., Dardaman, E., Campos, S., & Murphy, E. R. (2024, April 14). *Affirmative safety: An approach to risk management for high-risk AI*. arXiv. https://arxiv.org/abs/2406.15371

Zhang, Y., Yu, T., & Yang, D. (2024). *Attacking vision-language computer agents via pop-ups*. arXiv. https://arxiv.org/abs/2411.02391

Jerry Huang is currently an AI Engineer at Google, where he contributed to the AI/ML evaluation pipeline for a consumer-facing application. Before Google, he worked as a technical and security staff member at several prominent technology companies, gaining experience in areas like security, AI/ML, and scalable systems.

At Metabase, an open-source business intelligence platform, Jerry contributed features such as private key management and authentication solutions. As a Software Engineer at Glean, a generative AI search startup, he was one of three engineers responsible for managing large-scale GCP infrastructure powering text summarization, autocomplete, and search for over 100,000 enterprise users. During his time at TikTok, Jerry helped design and build custom RPCs to model access control policies. At Roblox, he served as a Machine Learning/Software Engineering Intern, focusing on real-time text generation models and gathering a large multilingual corpus that significantly boosted model robustness.

In addition to his industry experience, Jerry has conducted extensive security and biometrics research as a Research Assistant at Georgia Tech's Institute for Information Security and Privacy, resulting in a thesis on privacy-preserving biometric authentication.

Jerry holds a BS/MS in Computer Science from Georgia Tech and is currently pursuing an MS in Applied Mathematics at the University of Chicago.

Ken Huang is a prolific author and globally recognized authority in AI and Web3, with an extensive portfolio of published works that bridge business strategy, technical implementation, and cutting-edge research. As Fellow of Cloud Security Alliance and Co-Chair of the AI Safety Working Groups at the Cloud Security Alliance and the AI STR Working Group at the World

Digital Technology Academy under the UN Framework, he is a leading voice in shaping global AI governance and security standards.

Huang is the CEO and Chief AI Officer (CAIO) of DistributedApps.ai, a firm specializing in generative AI training and consulting. His contributions to the field include being a core contributor to the OWASP Top 10 Risks for LLM Applications and an active participant in the NIST Generative AI Public Working Group.

Notable Publications:

- *Beyond AI: ChatGPT, Web3, and the Business Landscape of Tomorrow* (Springer, 2023)—Strategic insights into AI and Web3's business applications
- *Generative AI Security: Theories and Practices* (Springer, 2024)—A comprehensive guide on securing generative AI systems
- *Practical Guide for AI Engineers* (Volumes 1 and 2, DistributedApps.ai, 2024)—Essential resources for AI and ML engineers
- *The Handbook for Chief AI Officers: Leading the AI Revolution in Business* (DistributedApps.ai, 2024)—A road map for CAIOs in implementing GenAI across organizations
- *Web3: Blockchain, the New Economy, and the Self-Sovereign Internet* (Cambridge University Press, 2024)—Insights into the convergence of AI, blockchain, IoT, and emerging technologies
- *Blockchain and Web3: Building the Cryptocurrency, Privacy, and Security Foundations of the Metaverse* (Wiley, 2023)—Recognized as a must-read by TechTarget in 2023 and 2024

Ken is a sought-after speaker and has presented at events such as the World Economic Forum in Davos, ACM and IEEE conferences, the CSA AI Summit, Depository Trust and Clearing Corporation forums, and World Bank conferences. His recent appointment to the OpenAI Forum reflects his ongoing commitment to advancing collaboration and dialogue in the field of AI.

Explore Ken's work on Amazon: https://www.amazon.com/author/kenhuang.

Krystal Jackson is a Non-Resident Research Fellow with the Center for Long-Term Cybersecurity AI Security Initiative at UC Berkeley. Krystal received her MS in Information Security Policy and Management from Carnegie Mellon University.

Chris Hughes is the co-founder and CEO of Aquia, a cybersecurity consulting firm. Chris brings nearly 20 years of IT and cybersecurity experience to his role as co-founder and CEO at Aquia. Chris has also served as a Cyber Innovation Fellow (CIF) at the Cybersecurity Infrastructure and Security Agency (CISA), focusing on software supply chain security. Additionally, Chris advises various tech startups focused on areas such as software composition analysis (SCA), Kubernetes security, non-human identities (NH), and AI security.As a United States Air Force veteran and former civil servant in the U.S. Navy and the General Services Administration's FedRAMP program, Chris is passionate about making a lasting impact on his country and the global community. In addition to his public service, Chris spent several years as a consultant within the private sector and currently serves as an adjunct professor for cybersecurity master's programs at the University of Maryland Global Campus. Chris participates in industry working groups, such as the Cloud Security Alliance's Incident Response and SaaS Security Working Group, and serves as the Membership Chair for Cloud Security Alliance D.C. He is the host and author of the *Resilient Cyber*. Chris runs the *Resilient Cyber* Substack, where he shares a weekly newsletter, deep dive analysis, and interviews with industry leaders, as well as detailed articles on topics such as cloud, vulnerability management, DevSecOps, as well as cybersecurity leadership and market dynamics. Chris holds a B.S. in Information Systems, a M.S. in Cybersecurity, and an MBA. He regularly consults with IT and cybersecurity leaders from various industries to assist their organizations with their digital transformation journeys while keeping security a core component of that transformation. Chris is co-author of the books *Software Transparency: Supply Chain Security in an Era of a Software-Driven Society* and *Effective Vulnerability Management: Managing Risk in the Vulnerable Digital Ecosystem* both published by Wiley. He has also contributed many other thought leadership pieces on software supply chain security and has presented on the topic at a variety of industry conferences.

www.ingramcontent.com/pod-product-compliance
Lightning Source LLC
Chambersburg PA
CBHW080426100725
29292CB00012B/9
* 9 7 8 3 0 3 1 9 0 0 2 5 9 *